# A-LEVEL CHEMISTRY
## COURSE COMPANION

Bob McDuell BSc (Hons)
Deputy Headmaster, Berry Hill High School, Stoke-on-Trent

**BPP Letts Educational Ltd**

First published 1982

2nd edition 1984
Reprinted 1984
Reprinted 1985
3rd edition 1986
4th edition 1988
Revised 1991
Reprinted 1992

Illustrations: Rod Fraser

Extract on pp. 7 and 8 reprinted by
permission of *Education in Chemistry*.

**British Library Cataloguing in Publication Data**
McDuell, G. R.
A-level chemistry: course companion. —
4th ed. — (Letts study aids).
1. Chemistry — Examinations, questions, etc.
I. Title
540′.76 QD42

ISBN 1 85758 021 4

Printed and bound in Great Britain by
WM Print Ltd, Frederick Street,
Walsall, West Midlands WS2 9NE

# Preface to the fourth edition

This A-level *Course Companion* was written to provide a complete guide to students preparing for GCE A level, AS level, Scottish Higher and TEC examinations (up to level III). The previous editions have been well accepted by students and teachers.

Students entering A-level courses will now have completed GCSE Chemistry or Science courses. The gap between GCSE and A level was always a big one, but the gap between GCSE and A level is probably greater. The reduced factual content of GCSE courses obviously means students may know less, but this should be compensated by the development of other skills. This book is intended to help bridge the gap. The first seven units are vital for this purpose and should be studied thoroughly first before other units are attempted.

It is essential that the candidate is thoroughly prepared for the examination. This book provides an analysis of the requirements of each Examination Board, a concise summary of the material common to most of the syllabuses and plenty of examination questions with skeleton answers.

This new edition provides the opportunity to revise the text and increase the number and range of questions and, as a result, produce a book which is even more helpful than the previous editions.

I would like to take this opportunity to thank Tom Chamberlain, Dennis Garvie and Ron Frost for their help in the development of this project. Also the staff of Charles Letts & Co Ltd for turning the manuscript into a finished book. I thank Examination Boards for permission to reproduce past examination questions. It should be noted that the answers given are mine and not official mark schemes.

Lastly, but certainly not least, I must thank my wife Judy and my sons, Robin and Timothy, for all their help and also their patience when I become engrossed in writing.

Bob McDuell, 1991

# Contents

# Introduction and guide to using this book

Revise Chemistry has been most useful in preparing students for GCSE examinations. Students, parents and teachers have requested that similar help be made available for A-level candidates. This book is designed to provide this help. It is hoped that these students will realise the importance of long-term preparation for A-level examinations. This book is designed to provide help throughout the course and, for this reason, it is called a course companion.

The book should be useful for students preparing for A-level examinations, for AS examinations, for TEC courses up to level III and for students in Zimbabwe preparing for M-level examinations.

On pages vi-viii there is an analysis of the current syllabuses of the different Examination Boards in England, Wales and Scotland. In the table the following symbols are used.

- ● Probably studied at GCSE but required at greater depth for A level.
- ○ Completely new at A level or AS level.
- □ Option at A level or AS level.
- ○/□ Included in the core syllabus but in greater depth in an option.

At A level the syllabuses are very detailed and can provide useful information to the student. It is important to read the appropriate syllabus thoroughly during the course.

It is impossible to cover all the content of all syllabuses in this book. The choice of topics was determined by the frequency of appearance on examination papers and the difficulty likely to be encountered by the student.

Having identified the units necessary for a particular syllabus, the student should study each unit thoroughly. The units 1–7 are key units. Examination candidates frequently show they do not understand these topics, which are extensions of GCSE topics.

At the end of each unit there are examination questions with brief answers. The advice of experienced examiners should be carefully noted by the candidate.

I am grateful to the following Examination Boards for permission to reproduce past examination questions. The answers supplied are written by the author and are not the official answers:

Associated Examining Board
Welsh Joint Education Committee
Southern Universities' Joint Board for School Examinations
Oxford and Cambridge Schools Examination Board
Scottish Examinations Board
London University
Northern Ireland Schools Examinations Council.

Other Examining Boards could not agree to their questions being used in this book. I hope that they will agree to the use of these questions in future editions when they see the sympathetic use of the examination questions. Questions have been written to ensure that the questions in the book reflect the style of questions of *all* Examination Boards.

On page 13 a summary of the practical requirements of the different Examination Boards is given. Advice is given on page 9 for practical work.

* London University's new syllabus (080) has reduced the content of 'A' level. Throughout this book any section marked * does not have to be studied.

# Table of analysis of A-level examination syllabuses

| | Oxford A55 | Welsh | Oxford and Cambridge | AEB 654 | Northern Ireland |
|---|---|---|---|---|---|
| **Paper 1** | 2¼ hr 28% Physical | 3 hr 43% Short answer | A 1¼ hr Multiple choice B 1½ hr Short answer | Paper 1 2¼ hr Longer questions | 1½ hr 30% Multiple choice |
| **Paper 2** | 2½ hr 28% Inorganic | 3 hr 43% Longer questions and options | 3 hr Choice of longer questions | Paper 2 1¾ hr Short Answer | 2¼ hr 35% Compulsory structured questions |
| **Paper 3** | 2¼ hr 28% Organic | | 3 hr Choice of longer questions | Paper 3 1¼ hr Multiple choice | 3 hr 35% Free response |
| **Paper 4** | Practical 14% | — | Practical | Practical | — |
| **Units 1 – 7** | ● | ● | ● | ● | ● |
| **8** | ● | ● | ● | ● | ● |
| **9** | ○ | ○ | | ○ | ○ |
| **10** | ○ | | ○ | ○ | ○ |
| **11** | ○ | | ○ | ○ | ○ |
| **12** | ○ | | ○ | ○ | ○ |
| **13** | ● | ● | ● | ● | ● |
| **14** | ● | ● | ● | ● | ● |
| **15** | ● | ● | ● | ● | ● |
| **16** | ○ | ○ | ○ | ○ | ○ |
| **17** | ○ | ○ | ○ | ○ | |
| **18** | ○ | ○ | ○ | ○ | ○ |
| **19** | ○ | ○ | ○ | ○ | ○ |
| **20** | ● | ● | ● | ● | |
| **21** | ○ | ○ | ○ | ○ | ○ |
| **22** | ○ | ○ | ○ | ○ | ○ |
| **23** | ○ | ○ | ○ | | ○ |
| **24** | ○ | ○ | ○ | | ○ |
| **25** | ○ | ○ | ○ | ○ | ○ |
| **26** | ○ | ○ | ○ | | ○ |
| **27** | ○ | ○ | ○ | ○ | ○ |
| **28** | ○ | ○ | ○ | ○ | ○ |
| **29** | ● | ● | ● | ● | ● |
| **30** | ○ | ○ | ○ | ○ | ○ |
| **31** | ○ | ○ | ○ | ○ | ○ |
| **32** | ○ | ○ | ○ | ○ | ○ |
| **33** | ○ | ○ | ○ | ○ | ○ |
| **34** | ○ | ○ | ○ | ○ | ○ |
| **35** | ○ | ○ | ○ | ○ | ○ |
| **36** | ○ | ○ | ○ | ○ | ○ |
| **37** | ○ | ○ | ○ | ○ | ○ |
| **38** | ○ | ○ | ○ | ○ | ○ |
| **39** | | | | ○ | ○ |

| Nuffield | London | Southern Universities | Cambridge | JMB A | JMB B | Scottish Higher |
|---|---|---|---|---|---|---|
| I 2 hr 30% 7 structured questions | I 2 hr 30% 7 structured questions | I 2½ hr 15 short questions | 2½ hr 40% longer questions | 2½ hr A Multiple choice questions B Longer questions C Questions on options | 2½ hr A Multiple choice questions B Short answer | 1½ hr 43% Multiple choice |
| II 2¼ hr 15% ** III 1 hr 17% multiple choice IV ¾ hr 10%* | II 2¼ hr 30% ** III 1 hr 20% multiple choice | II 3 hr Longer questions – | 1½ hr 24% Structured questions | 2½ hr A Short answer B Longer questions C Questions on options | 2½ hr A Short answers B Longer answers | 2½ hr 57% A Short questions B Longer questions |
| Internal assessment 18% | Int. assessment OR Practical 20% | – | 1 hr 16% Multiple choice questions | Practical 3 hr | Practical 3 hr | — |
| — | Practical 20% | — | Practical 20% | — | — | — |
| ● | ● | ● | ● | ● | ● | ● |
| ○ | ○ |  | ○ | ○ | ○ | ● |
|  |  | ○ | ○ | ○ | ○ | ● |
| ○ | ○ | ○ | ○ | ○ | ○ | ● |
| ○ | ○ | ○ | ○ | O/□ |  |  |
|  |  | ○ |  |  |  |  |
| ● | ● | ● | ● | ● | ● | ● |
| ● | ● | ● | ● | ● | ● | ○ |
| ● | ● | ● | ● | ● | ● | ○ |
| ○ | ○ | ○ | ○ | ○ | ○ |  |
| ○ | ○ | ○ | ○ | ○ | ○ |  |
| ○ | ○ | ○ | ○ | □ | ○ | ○ |
| ○ | ○ | ○ | ○ | ○ | ○ |  |
|  | ● | ● |  | ● | ● | ○ |
| ○ | ○ | ○ |  | ○ | ○ |  |
| ○ | ○ | ○ | ○ | ○ | ○ | ○ |
| ○ | ○ | ○ | ○ | ○ | ○ |  |
| ○ | ○ | ○ | ○ | ○ | ○ |  |
| ○ | ○ | ○ | ○ | ○ | ○ |  |
| ○ | ○ | ○ | ○ | ○ | ○ |  |
| ○ | ○ | ○ | ○ | ○ | ○ | ○ |
| ○ | ○ | ○ | O/□ | ○ | ○ |  |
| ● | ● | ● | ● | ● | ○ | ○ |
| ○ | ○ | ○ | ○ | ○ | ○ |  |
| ○ | ○ | ○ | ○ | ○ | ○ | ○ |
| ○ | ○ | ○ | ○ | ○ | ○ | ○ |
| ○ | ○ | ○ | ○ | ○ | ○ | ○ |
| ○ | ○ | ○ | ○ | ○ | ○ | ○ |
| ○ | ○ | ○ | ○ | ○ | ○ | ○ |
| ○ | ○ | ○ | ○ | □ | ○ | ○ |
| ○ | ○ | ○ | ○ | ○ | ○ | ○ |
| ○ | ○ |  | ○ | ○ | ○ |  |
| ○ | ○ |  | ○ | ○ | ○ |  |

*Questions set on special study.    **Summary/comprehension and choice of free response/essay questions.

# Table of analysis of AS-level examination syllabuses

| | AEB | COSSEC | JMB | London | Oxford |
|---|---|---|---|---|---|
| Paper 1 | 2½ hours 80% Comprehensive structured, free response | Structured | 1½ hours multiple choice & structured | 1 hour multiple choice 25% | 1½ hours compulsory structure 40% |
| Paper 2 | Assessment of coursework 20% | Long questions | 1½ hours structured & long | 2½ hours structured & essay questions 50% | 2 hours Longer questions on 40%, two options |
| Paper 3 | — | Assessment of practical work | Assessment of practical work | Assessment of practical work 25% | Assessment of practical work 20% |
| Paper 4 | — | — | — | — | – |
| Units 1–7 | ● | ● | ● | ● | ● |
| 8 | ○ | ○ | ○ | | ○ |
| 9 | ○ | | ○ | | ○ |
| 10 | ○ | | ○ | | ○ |
| 11 | | | | | |
| 12 | | | | | |
| 13 | ○ | ○ | ○ | ○ | O/□ |
| 14 | ○ | ○ | ○ | ○ | O/□ |
| 15 | ○ | ○ | ○ | ○ | ○ |
| 16 | ○ | ○ | | ○ | ○ |
| 17 | ○ | | | | |
| 18 | | | | | ○ |
| 19 | ○ | ○ | ○ | ○ | |
| 20 | | | | | |
| 21 | ○ | ○ | ○ | ○ | ○ |
| 22 | ● | ● | ● | | ● |
| 23 | | ○ | | | |
| 24 | | ○ | | | |
| 25 | | ○ | | | |
| 26 | | ○ | | | |
| 27 | ● | ● | ● | ● | ● |
| 28 | ○ | ○ | | ○ | ○ |
| 29 | ○ | ○ | ○ | ○* | O/□ |
| 30 | ○ | ○ | ○ | ○ | ○ |
| 31 | ○ | ○ | ○ | ○ | ○ |
| 32 | ○ | ○ | ○ | | ○ |
| 33 | ○ | ○ | ○ | ○ | O/□ |
| 34 | ○ | ○ | ○ | ○ | □ |
| 35 | | ○ | ○ | ○ | □ |
| 36 | | ○ | ○ | | O/□ |
| 37 | ○ | | | | |
| 38 | | | ○ | ○ | □ |

*29.10–29.13 not required

# Starting your A-level Chemistry studies

A-level or AS-level Chemistry requires study at a much greater depth than you will have been
used to if you have completed GCSE Chemistry or Science courses. Also, there is a great deal
of subject material to be covered. It is important to try and understand principles and patterns
rather than try learning isolated facts. This book will try to emphasize the patterns that exist.

The first seven units in this book are designed to bridge the gap between GCSE and
A-level. Study them carefully before attempting the rest of the book.

We now look at different styles of questioning you might meet at A-level. These include:

    Multiple choice;          Comprehension;
    Structured questions;     Writing summaries.
    Guided essays;

TYPES OF EXAMINATION QUESTION

**Multiple choice**
You may have met multiple choice questions at GCSE in Chemistry or some other subject.
Each question consists of a *stem* together with four or five possible answers or responses
labelled either A to D or A to E. One answer (called the *key*) is correct and the others (called
the *distracters*) are incorrect. You are required to select your answer and record the answer on
a printed answer sheet. These answer sheets will be marked accurately by machine. It is
important to follow **exactly** the instructions for recording your answers on the answer sheet.

Multiple choice tests are made up from the questions tested on A level students before they
are used. These tests are designed to give a wide range of marks when used. They are not as
easy as some candidates think. You should try to do at least four multiple choice tests before
you take the examination. Doing this you will ensure that you are scoring good marks in the
examination.

It is important to remember the following points:
● Read the questions **very** carefully and use **all** of the information given. Mistakes are often
due to the candidate choosing an answer without reading them all;
● It **is** in order to guess if you cannot work out the correct answer. If you can eliminate any of
the responses you will increase your chances of guessing correctly;
● There is **no** penalty for a wrong answer. **Never** leave any question without attempting it;
● The easier questions should be at the beginning of the test. The questions towards the end
of the test are more difficult and will probably need more time to answer them. Do not spend
too long on any one question. There are different types of multiple choice questions.

**1 Simple multiple choice**
e.g. Which of the following elements is a *d*-block element?
    A    Aluminium    C    Chromium
    B    Boron        D    Sulphur
The correct answer is C.

**2 Multiple completion questions**
In this type of question you are given four answers labelled (*i*) to (*iv*). On your answer sheet
you have to answer A, B, C, D or E depending upon the combination of statements (i) to (iv)
which are correct.

For example, if you think:
(*i*), (*ii*) and (*iii*) are correct you answer A;     (*vi*) only *is* correct you answer D;
(*i*) and (*iii*) are correct you answer B;             (*i*) and (*iv*) only you answer E.
(*ii*) and (*iv*) are correct you answer C;

This can be summarized in a table:

| Combination | (*i*) (*ii*) (*iii*) | (*i*) (*iii*) | (*ii*) (*iv*) | (*vi*) | (*i*) (*iv*) |
|---|---|---|---|---|---|
| Response | A | B | C | D | E |

No other combination of responses can be recorded. If for example you think that only (*i*) is correct, you have made a mistake!
The following short multiple choice test will enable you to practice your technique:

**Multiple choice questions**

**1** The atomic number of calcium is 20. If Ar represents the electronic structure of an atom, the calcium atom may be represented by:

    A  Ar          D  Ar $3d^1$
    B  Ar $4s^1$    E  Ar $3d^2$
    C  Ar $4s^2$

**2** In which one of the following compounds is the bonding most ionic?

    A  lithium chloride    D  sodium chloride
    B  lithium bromide    E  sodium iodide
    C  lithium iodide

**3** Which of the following has the greatest polarizing power?

    A  $Li^+$    D  $Sr^{2+}$
    B  $Be^{2+}$    E  $Ba^{2+}$
    C  $Ca^{2+}$

**4** The value of $K_a$ for a weak monobasic acid at 298 K is $4 \times 10^{-5}$. Which of the following is the concentration of $H^+$ ions in a solution of the acid (0.1 mol $dm^{-3}$)?

    A  $2\times10^{-2}$ mol $dm^{-3}$    D  $2\times10^{-4}$ mol $dm^{-3}$
    B  $2\times10^{-3}$ mol $dm^{-3}$    E  $4\times10^{-5}$ mol $dm^{-3}$
    C  $4\times10^{-4}$ mol $dm^{-3}$

**5** When a group I metal atom **X** reacts to become an ion $X^+$:
    A  the diameter of the particle increases
    B  the positive charge on the nucleus increases
    C  the atomic number of **X** decreases
    D  the number of occupied electron shells decreases by 1
    E  the number of protons increases

**6** The usual method of extracting alkali metals and alkaline earth metals from their ores is:
    A  reduction of the oxide with carbon
    B  electrolysis of an aqueous solution of the chloride
    C  electrolysis of the molten chloride
    D  strongly heating the oxide in air
    E  reduction of the oxide with aluminium

**7** 1, 2-dibromo-3-chloropropane (DBCP) has been used in the control of earthworms. Which of the following would be the best method of making it?

    A  $CH_3CH_2CH_2Cl + 2Br_2 \rightarrow DBCP + 2HBr$
    B  $CH_3CHBrCH_2Br + Cl_2 \rightarrow DBCP + HCl$
    C  $CH_2{=}CHCHBr_2 + HCl \rightarrow DBCP$
    D  $CH_2{=}CHCH_2Cl + Br_2 \rightarrow DBCP$
    E  $ClCH_2CH{=}CH_2 + PBr_5 \rightarrow DBCP + PBr_3$

**8** 0.20g of a monobasic acid required 8.0cm$^3$ of 0.40 mol $dm^{-3}$ sodium hydroxide for complete reaction. What is the relative molecular mass of the acid?

    A  62.5    D  640
    B  250    E  2500
    C  625

**9** Aspirin was prepared in an industrial laboratory. Before it can be made into tablets it has to be purified by recrystallization using hot water. Which one on the following is the most important factor for choosing hot water as the solvent for recrystallization?

    A  readily available    D  large change in solubility of aspirin
    B  non-flammable           with temperature
    C  inexpensive       E  aspirin dissolves readily in hot water.

**10** What is the oxidation state of chromium in the complex ion $[Cr(H_2O)_4Cl_2]^+$?

  A  0      D  +3
  B  +1     E  +6
  C  +2

**11** Which of the following is the enthalpy change called lattice energy of sodium chloride?

  A  $Na(s) + \frac{1}{2}Cl_2(g) \rightarrow NaCl(s)$      D  $Na(s) + \frac{1}{2}Cl_2(l) \rightarrow NaCl(s)$
  B  $Na(g) + Cl(g) \rightarrow NaCl(s)$      E  $Na^+(g) + Cl^-(g) \rightarrow NaCl(s)$
  C  $Na(l) + \frac{1}{2}Cl_2(g) \rightarrow NaCl(s)$

**12** 1, 2-dibromoethane reacts with potassium iodide dissolved in methanol according to the equation:

$$C_2H_4Br_2 + 2KI \rightarrow C_2H_4 + 2KBr + I_2$$

The rate expression for this reaction is

  A  rate $= k[KI]^2[C_2H_4Br_2]$      D  rate $= k[C_2H_2Br_2]$
  B  rate $= k[KI][C_2H_4Br_2]$      E  not possible to say
  C  rate $= k[KI]^2$

Questions 13 to 16 concern the following reaction mechanisms of organic chemicals:

  A  free radical addition      D  electrophilic substitution
  B  free radical substitution      E  nucleophilic addition
  C  nucleophilic substitution

Select from A to E, the likely mechanism for each of the following reactions:

**13** $CH_3CH_2Br + OH^- \rightarrow CH_3CH_2OH + Br$

**14**

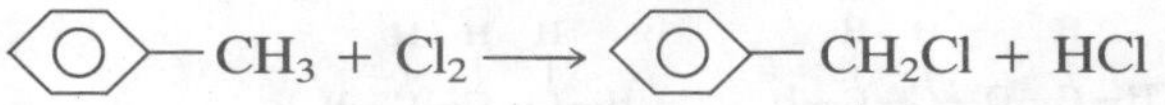

**15** $CH_3COCH_3 + HCN \rightarrow CH_3CH(OH)(CN)CH_3$

**16**

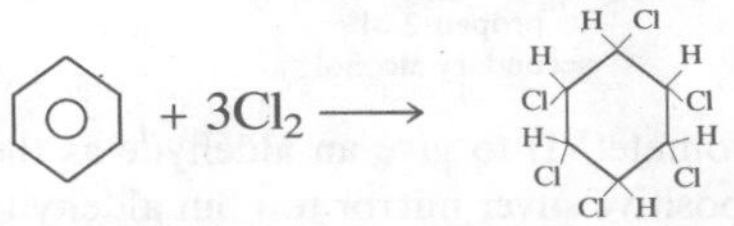

**Multiple completion questions**

Summary of combinations of responses:

| Combination | (i) (ii) (iii) | (i) (iii) | (ii) (iv) | (vi) | (i) (iv) |
|---|---|---|---|---|---|
| Response | A | B | C | D | E |

**17** Which of the following statements about the solution formed by mixing solutions of ammonia and ammonium chloride is (are) true?
 (i) the resulting mixture is a buffer solution
 (ii) most of the ammonium ions in the solution came from the ammonium chloride
 (iii) the ammonium ions act as an acid on the addition of $OH^-$ ions
 (iv) the pH of the resulting mixture is less than 7

**18** Using the following information:

$$Zn^{2+}(aq) + 2e^- \rightarrow Zn(s) \qquad E^0 = -0.76 \text{ V}$$
$$Ni^{2+}(aq) + 2e^- \rightarrow Ni(s) \qquad E^0 = -0.25 \text{ V}$$

Which of the following statements about the cell is (are) true?

$$Zn(s) \mid Zn^{2+}(aq) \vdots Ni^{2+}(aq) \mid Ni(s)$$

 (i) the electrons travel through the external circuit from the zinc to the nickel
 (ii) the e.m.f. of the cell is 0.51 V
 (iii) the mass of the zinc rod decreases during the operation of the cell
 (iv) the concentration of nickel ions increase during the operation of the cell

**19** Which of the following systems contain delocalized electrons?
 (i) benzene      (iii) sodium
 (ii) graphite      (iv) cyclohexene

**20** The bond angles are **less than** 109.5° (tetrahedral) in
(i) $H_2O$     (iii) $NH_3$
(ii) $BF_3$     (iv) $CH_4$

The answers to the preceding questions are:
1 C; 2 D; 3 B; 4 B; 5 D; 6 C; 7 D; 8 A; 9 D; 10 D;
11 E; 12 E; 13 C; 14 B; 15 E; 16 A; 17 A; 18 A; 19 A; 20 B.

### Structured questions

Each of these questions consists of a series of parts linked to a common theme. There is space on the paper for each answer and the amount of space given should be a guide to the length of answer required. Generally the earlier parts of each question will be easier than the later parts.

The following question is an example of a structured question.

A simple aliphatic compound, (**A**), has the empirical formula, $C_3H_8O$, and relative molecular mass, 60. [$A_r$ (H) = 1 ; (C) = 12 ; (O) = 16]

(a)  (i)  Give its molecular formula.                                                                                (1 line)
     (ii) Write down the possible structural formulae.                                                       (5 lines)
(b) State the conclusions which can be drawn from **each** of the following:
     (i)  Mild oxidation of (**A**) with acid dichromate(VI) gives a compound of formula $C_3H_6O$, which gives a positive silver mirror test.                                                                          (5 lines)
     (ii) $C_3H_6O$ is further oxidizable to $C_3H_6O_2$, which produces $CO_2$ on reaction with aqueous sodium carbonate.                                                                                      (3 lines)
(c)  (i)  Give the reaction of $C_3H_6O$ with 2,4-dinitrophenylhydrazine.                           (3 lines)
     (ii) Give the reaction of $C_3H_6O_2$ with $PCl_3$.                                                       (3 lines)
     (iii) Give the reaction of (**A**) with excess, hot, concentrated sulphuric acid, (or phosphoric acid).   (3 lines)
*(Welsh Joint Education Committee)*

*Sample answer*
(a)  (i) $C_3H_8O$  (ii)

propan-1-ol            methoxyethane            propan-2-ol
primary alcohol            ether            secondary alcohol

(b)  (i) **Primary alcohols are oxidized by warm, dilute acid dichromate(VI) to give an aldehyde as the first product. Secondary alcohols give a ketone. Ketones do not give a positive silver mirror test but aldehydes do, (A) must be propan-1-ol.**

$$CH_3CH_2CH_2OH + [O] \rightarrow CH_3CH_2CHO + H_2O$$
propanal

(ii) More vigorous oxidizing conditions of the propanal formed in (i) produces propanoic acid (no loss of carbon atoms). As an acid this evolves carbon dioxide with a carbonate.

$$CH_3CH_2CHO + [O] \rightarrow CH_3CH_2COOH$$

sodium propanoate

(c)(i)

propanal 2,4-dinitrophenylhydrazone

Condensation (or addition – elimination) reaction

(ii)

propanoyl chloride

Steamy fumes of hydrogen chloride produced.

(iii)
$$CH_3CH_2CH_2OH \rightarrow CH_3CH = CH_2 + H_2O$$
propene

Dehydration reaction. The use of phosphoric acid reduces charring. Excess acid prevents ether formation.

### Guided essays

It is important for you as an A-level candidate to write a complete and well-constructed answer to a question. You need to be able to select and use your knowledge to answer the question.

The following are sample essay titles.

**1**. Write an account of the chemistry of halogenoalkanes.

You should consider the reactions suitable for their preparation and their characteristic reactions. You should also refer to some industrial applications and to at least one of their reactions.     (*Nuffield*)

**2**. Describe the *chemistry* of THREE environmental problems that have arisen from the introduction of three products or processes by industry. Suitable examples include detergents, insecticides, aerosols, waste gases from factories and moter cars.     (*Nuffield*)

## Comprehension

The Associated Examining Board Syllabus 654 includes a compulsory comprehension question in Paper 3. Obviously, as in any comprehension exercise, it is essential that you read the passage thoroughly and understand it before attempting the questions. Usually the chemistry in the passage will be unfamiliar to you. A typical comprehension exercise is given below.

Read the following passage carefully and study the flow diagram, then answer questions (*a*) to (*k*).

Extract from **HYDROGEN PEROXIDE** by B. Haines

*Bulletin No 4    The Schools Information Centre on the Chemical Industry*

**Manufacture of hydrogen peroxide by the autoxidation process**

### Raw Materials

The following are required to produce hydrogen peroxide:

1. 2-alkylanthraquinone.

2. Hydrogen – obtained by reforming butane, naphtha or methane or as by-product hydrogen from electrolytic chlorine/alkali processes.

3. Oxygen – air is used.

4. Catalyst – this may be either

   (*a*) a precious metal, *e.g.* palladium or platinum, on a support of alumina or silica/alumina

   or (*b*) Raney nickel.

   (*a*) is expenisve but more resistant to physical and chemical attack than the cheaper Raney nickel, which is easily oxidised. A Raney nickel catalyst requires frequent regeneration and the reaction solution must be more stringently purified before it is recycled.

5. Solvents – (*a*) for the quinone. This is usually a high boiling (150-200°C range) mixture of aromatic and/or aliphatic hydrocarbons.

   (*b*) for the quinol. This is usually a hydrophilic compound, *e.g.* $C_8$–$C_9$ aliphatic or alicyclic alcohol, ketone, or a carboxylic or phosphoric acid ester.

### The Autoxidation Process

The process can be represented by the following equations, where R = ethyl, butyl, or pentyl.

2-alkylanthraquinone     2-alkylanthraquinol

### Process Description

The alkylanthraquinone solution passes from a storage tank (**1**), through a preheater into the hydrogenator (**2**), where the quinone is reduced to form the quinol.

The hydrogenation catalyst may be contained in a fixed-bed arrangement, and in this case the hydrogenator must be shut down periodically and the catalyst bed removed for regeneration. Alternatively, a fluidized-bed of finer catalyst particles may be employed. Here the catalyst is suspended by physical stirring and/or a stream of hydrogen, which is pumped through the reaction solution. Fresh catalyst is added to maintain the reaction rate, and a proportion of the catalyst is occasionally removed for regeneration or for recovery of the precious metal component. If a fluidized-bed hydrogenator is used, the solution is filtered (**3**) to remove any entrained catalyst and the catalyst is then recycled.

After cooling, the quinol enters the oxidizer (**4**) where the solution may flow counter or co-current with the air flow. If co-current flow is used the gas/liquid mixture enters a separator (**5**) which finally removes the gases present (oxygen-depleted air). The gases from this stage are cooled and pass through a solvent recovery plant and thence into the atmosphere. The liquid mixture – hydrogen peroxide and anthraquinone in the organic solvents – is cooled before passing into an extraction unit (**6**). The extraction unit may be a simple perforated plate or a column packed with ceramic material. In the latter case, the mixture is fed into the bottom of the column where it meets a stream of high purity, deionised water flowing down the unit.

De-ionized water is used, since any metallic impurities present could decompose the hydrogen peroxide.

As the organic solution (alkylanthraquinone + solvents) is less dense than water, this rises to the top and is drawn off, purified and returned to the storage tank (**1**). The dilute solution of hydrogen peroxide (20–40% w/w) is removed from the bottom of the column and passes into a storage tank (**7**).

Hydrogen peroxide is concentrated by distillation (**8**) under reduced pressure.

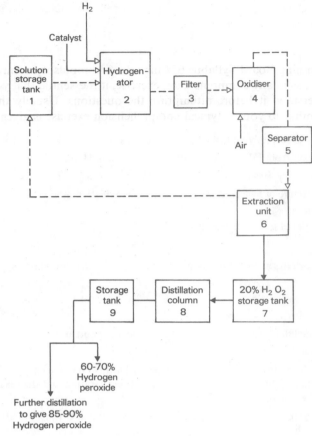

Production of hydrogen peroxide by the autoxidation process
      – – – organic solution
      ——— hydrogen peroxide (aqueous)

The solution is vapourised and the vapour, with excess liquid, enters a separator. Here the liquid is removed and recirculated, while the vapour passes into a fractionation column. De-ionized water is used as the refluxing agent. The concentrated hydrogen peroxide is then stored in a tank until required for use.

H.T.P. – high test peroxide (85–90% w/w) is produced by further distillation of the 60–70% w/w solution.

(*a*) State **two** methods by which hydrogen is obtained for the process. (2)
(*b*) List the catalysts available for the hydrogenation and discuss their relative merits. (8)
(*c*) State the nature and a property of the solvent required for (i) the 2-alkylanthraquinone and (ii) the 2-alkylanthraquinol. (4)
(*d*) Write an equation for the overall reaction of the autoxidation process. (2)
(*e*) Name the oxygen-containing functional groups present in (i) 2-alkylanthraquinone and (ii) 2-alkylanthraquinol. (2)
(*f*) Name and write structural formulae for (i) a $C_8$ aliphatic alcohol and (ii) a $C_8$ carboxylic acid ester (4)
(*g*) Explain what is meant by a *fluidized-bed hydrogenator* and compare its advantages and disadvantages with those of a fixed-bed hydrogenator. (6)
(*h*) State, and illustrate using arrows, the meanings of (i) co-current and (ii) counter-current flow. (2)
(*i*) What are the functions of the chemical plant labelled in the flow diagram as (i) storage tank (**1**), (ii) filter (**3**) and (iii) oxidiser (**4**)? (6)
(*j*) Describe what happens in the extraction unit (**6**) when it is packed with ceramic material. (5)
(*k*) Explain the term *distillation under reduced pressure*, and say why this process is preferred to concentrate an aqueous solution of hydrogen peroxide. (4)

*(Associated Examining Board)*

*Sample answer*

(*a*) **By reforming naphtha, methane or butane.**
**By-product from alkali/chlorine processes.**

(*b*) **Catalysts are palladium or platinum on a support of alumina or silica-alumina or Raney nickel.**

**Palladium or platinum is more resistant to chemical attack but more expensive. Raney nickel is easily oxidised, requires more regeneration and the solution must be purified before recycling.**

(c) (i) for quinone. High boiling point hydrocarbon.

(ii) for quinol. Hydrophilic compound. $C_8 - C_9$ aliphatic alcohol etc.

(d) $H_2 + O_2 \rightarrow H_2O_2$

(e) quinone – carbonyl or ketone group

quinol – phenol or enol group

(f) Octan-l-ol

Any suitable ester *e.g.* butyl butanoate

(g) Fluidised-bed hydrogenator has fine catalyst particles kept in suspension by a stream of hydrogen or by stirring.

Advantages: fresh catalyst can be added from time to time without shut down. Small amounts of catalyst can be removed from time to time. Uses finer particles.

Disadvantages: filter must be used.

(h) Co-current

solution ———→

air ———→

Counter-current

solution ———→

←——— air

(i) (i) storage tank contains a solution of alkylanthraquinone.

(ii) filter removes any catalyst from a fluidized bed so that it can be recycled.

(iii) oxidiser where the quinol is oxidised by air to the quinone.

(j) The mixture of organic solvents, quinone and hydrogen peroxide is fed into the bottom and meets a stream of de-ionized water. Organics removed from top for recycling. 20% hydrogen peroxide leaves bottom.

(k) Liquids boil when saturated vapour pressure reaches external pressure. Boiling point is lower if external pressure is reduced. Hydrogen peroxide decomposes on heating.

## Writing Summaries

Nuffield papers require students to write a summary of a passage of technical writing. The following is an example of this type of question.

**You must attempt this question and you are advised to spend about 40 minutes on it.**

Read the following passage carefully, and then make a summary of *not more than* 220 words. You should select what you consider to be the most important points and include them in your summary. Credit will be given for answers written in good English, using complete sentences.

Do not copy sentences or exceed 220 words; numbers count as one word, as do standard abbreviations and hyphenated words. Chemical equations do not count in the word total.

At the end of your summary, state clearly the number of words you have used.

### The Chemistry of Lubricating Oils

(Adapted from P.H. Harding and R. Oxtoby, *Education in Chemistry,* September 1970)

Lubricating oils are manufactured from the heavier fractions of crude oil after the lower-boiling gasoline, kerosine and gas oil have been removed by distillation.

A number of special processes, e.g. vacuum distillation, filtration, solvent extraction, sulphuric acid refining, contact treatment with activated clay, are employed to produce finished lubricating oil basestocks from the 'topped crude'. Commercial oils are prepared by blending one or more finished basestocks of the desired viscosity with the additives required for the expected service conditions. Without these additives, the oil is not usually able to withstand the stresses and temperatures imposed by modern engines and high-speed machinery.

Hydrocarbon mineral oils make good lubricants because they have high boiling points, are chemically inert, remain fluid at low temperatures and have viscosities which enable them to meet a wide range of operating requirements. They have the added advantages of being readily available and relatively cheap. They range from low-viscosity oils with relative molecular masses of about 250 to very viscous lubricants with molecular masses as high as 1000. Their properties depend quire markedly on the relative distribution of the different hydrocarbon types, i.e. alkane, cycloalkane and arene components. For example, for a given molecular size the alkanes possess high freezing points and low densities, viscosities and oxidation stabilities compared with other hydrocarbon types. Conventional separation techniques in combination with measurements of density, refractive index, relative molecular mass and spectroscopic characteristics may be used to determine the proportions of structural types present.

For a given application, a mineral oil will often be expected to satisfy a number of different requirements and it is not usually feasible to obtain such an oil economically by refining alone, even with the most advanced techniques. Using additives is an attractive method of achieving the required performance level economically.

Viscosity index improvers provide a cheap and effective way of controlling the viscosity/temperature characteristics of an oil. Commercially, the two most widely used additives are the polymethacrylates and the polyisobutylenes. Multigrade oils incorporating these polymers were introduced in the 1950s and have shown considerable advantages over conventional mineral oils. In solution, the polymers are believed to exist as random coils swollen by the basestock oil. As the temperature increases, the oil viscosity decreases but the polymer coil unravels and tends to counteract this effect. The improving power of the polymer is dependent on the relative molecular mass and chemical structure of the polymer.

The universal viscosity index improver, having excellent viscosity/temperature characteristics, high mechanical and oxidation stability and ready availability at low cost, has yet to be produced. Lubricating oils are frequently in contact with air and in use at high temperatures. Oxidation of the oil results in the formation of both oil soluble compounds and insoluble compounds that appear as resins, sludges and acidic materials. These have seriously adverse effects on the performance of the oil and the use of antioxidants offers the best means of achieving the desired oxidation stability.

Chain-breaking antioxidants, e.g. phenols and aromatic amines, act by interfering with the free-radical chain oxidation of hydrocarbons, believed to take place in the temperature range of operation of lubricating oils, at stages 2 and 3:

**1.**        Initiation:        $RH + O_2 \rightarrow RO_2\cdot + H\cdot$

**2.** ⎫
      ⎬    Propagation:    ⎧ $RO_2 + RH \rightarrow RO_2H + R\cdot$
**3.** ⎭                        ⎩ $R\cdot + O_2 \rightarrow RO_2\cdot$

These may be mixed with other types of antioxidant to give best overall protection.

Additives will continue to be used in increasing quantities and in wider varieties. One of the main difficulties for the chemist who attempts to correlate the chemical structure of additives with their effectiveness in practice is the problem of relating laboratory methods of examination to performance tests in an engine. Nevertheless, until we know more about the detailed chemistry of the action of certain types of additive, selection must be done largely on a trial and error basis and fundamental improvements will be far less likely.                                    (*Nuffield*)

## Sample answer

Lubricating oils are made from the heavier fractions of crude oil by processes of distillation, filtration, and refining. Commercial oils are blends of these oils and additives to produce the required viscosity for the conditions. High boiling points, low freezing points, chemical inertness, suitable viscosities and inexpensive availability make hydrocarbon mineral oils good lubricants.

Molecular masses range from 250 (low-viscosity) to 1000 (high-viscosity), with alkane, cycloalkane and arene components defining their properties, in proportions measured by spectroscopic characteristics. Alkanes increase freezing points, reduce viscosity, density and oxidation stability.

Using additives achieves performances unobtainable just by refining oils. Viscosity index improvers (e.g. polymethacrylates and polyisobutylenes) control viscosity temperature characteristics. As the temperature increases, these polymers uncoil and counteract the reduction in viscosity of the oil, their effectiveness being dependent on their relative molecular masses and structures. No low-cost viscosity index improver, with optimum viscosity temperature characteristics and oxidation stability, has yet been found. Working at high temperatures and in contact with the air causes oxidation, resulting in insoluble sludges, seriously affecting performance. Anti-oxidants improve oxidation stability. The best protection is achieved by mixtures of antioxidants including chain-breaking antioxidants (e.g. phenols, aromatic amines) which interfere with free radicle oxidation (2 & 3):

**1.**        Initiation:        $RH + O_2$        $RO_2\cdot + H\cdot$

**2.** ⎫
      ⎬    Propagation:    $RO_2\cdot + RH$        $RO_2H + R\cdot$
**3.** ⎭                    $R\cdot + O_2$        $RO_2\cdot$

Detailed chemical investigation of additives is essential if improvements in performance are to be made.

# The practical examination

Any A-level chemistry course contains a practical component. This may be examined by means of a practical examination or your teacher might carry out, on behalf of the Examination Board, an assessment of your practical work throughout the course. If the latter is the case, he or she will give you full information of how the assessment will be made.

If you are taking a practical examination there are a few points for you to remember.

(*i*) Candidates are often more worried at the thought of taking a practical examination than a theory examination. This should not be so. Very few candidates fail to satisfy the requirements of a practical examination and usually the practical examination adds a few marks to the overall marks of the candidates.

(*ii*) The practical examination will give you full instructions on what to do. You are being tested on your ability to follow these instructions accurately within the allotted time and making all the correct observations and readings. Accuracy and attention to detail in all your practical work is essential, so carry out instructions to the letter. Many marks are often lost through just not reading the question sufficiently carefully.

(*iii*) The practical examiner is not present when you take your practical examination. He or she can only mark what you write down on your answer paper. Do not assume anything. If you make observations or take readings, make sure they are written down. Some candidates write down all their readings in pencil and rub out the ones they later find incorrect. Examiners like to see all your results and you will not be penalised if you have made incorrect readings and then, at a second attempt, made correct readings.

(*iv*) In any practical examination, the time element is important. Plan the amount of time to be spent on each exercise and possibly the order of attempting the exercises at the start of the examination.

There are a great variety of exercises on A-level papers but they can be divided into two types.

## 1 Quantitative exercises

This could involve accurate weighings, titration readings, temperature measurements etc.

If you are asked to weigh out a sample of a chemical accurately within a certain range, it is important to ensure that the mass obtained is within the range given. Never weigh directly on the balance pan. Record your weighings as in the example below.

Mass of weighing bottle + solid = 46.121g
Mass of weighing bottle = 45.684g
Mass of solid = 0.437g

If you are preparing a standard solution it is important not to lose any of the compound during the dissolving and making up to the final volume. The standard solution should be thoroughly shaken before use.

When carrying out titrations, the first titration should be carried out quickly as a rough guide to the volume of solution required. Accurate titrations can then be carried out. Do not cheat by writing down results by guesswork. Your table of results should be as follows:

*Titration of sodium hydroxide with hydrochloric acid using screened methyl orange. Using 25cm³ samples of sodium hydroxide solution.*

|  | Rough | 1st | 2nd | 3rd | 4th |
|---|---|---|---|---|---|
| Final burette reading | 25.70cm³ | 25.75cm³ | 25.75cm³ | 24.95cm³ | 25.65cm³ |
| Initial burette reading | 0.00cm³ | 0.25cm³ | 0.20cm³ | 0.00cm³ | 0.05cm³ |
| Volume of acid added | 25.70cm³ | 25.50cm³ | 25.55cm³ | 24.95cm³ | 25.60cm³ |

*Average titration reading (average 1st, 2nd and 4th only) = 25.55cm³*

The following points should be noted:

(*i*) Record your burette readings to two decimal places. In the second place you should write a 0 if the bottom of the meniscus is exactly on a division. If the bottom of the meniscus is between two divisions, you should write a 5.

(*ii*) It is not necessary to start each titration with the reading exactly zero. It should be close to zero and you should read the burette accurately.

(*iii*) When averaging your results only average readings which are close to being correct. Don't use the rough reading where the end point has obviously been passed. In the example, titration 3 is suspect, and using it in the average would lead to inaccuracy.

(*iv*) You should aim to get at least three titrations within $0.10cm^3$. You may need to carry out further titrations before you can make a reliable average.

As a final piece of advice, do not panic. The examination is not intended to be an ordeal and the examiners are human.

## 2 Qualitative exercises

These exercises involve carrying out simple tests using inorganic or organic chemicals. The tests are given in full and you are required to make all the possible observations and draw all the correct conclusions from these tests. You may not be expected to identify the compounds being used. It is important to record exactly what you observe. If a precipitate is formed it is better to describe it as a pale yellow precipitate or a lemon yellow precipitate rather than a yellow precipitate. If a gas is evolved you should attempt to identify it with chemical tests.

It does not matter if you cannot draw all the correct inferences. You should have enough marks from your observations.

The tests, your observations and your inferences should be given in the form of a table. If the table is not given on the paper, the following table is an example of what is required.

| Tests | Observations | Inferences |
|-------|--------------|------------|
|       |              |            |

The following information may be useful when attempting qualitative exercises.

INORGANIC

### Appearance

The colour of an inorganic compound may suggest the presence of certain ions.
Coloured: probably contains a transition metal ion.

Blue: $Cu^{2+}$, $Ni^{2+}$            Green: $Cu^{2+}$, $Ni^{2+}$, $Fe^{2+}$, $Cr^{3+}$
Yellow: $Fe^{3+}$, $CrO_4^{2-}$       Orange: $Cr_2O_7^{2-}$              Purple: $MnO_4^-$

### Flame test
Orange-yellow: $Na^+$            Lilac: $K^+$                    Red: $Ca^{2+}$, $Sr^{2+}$. $Li^+$
Apple-green: $Ba^{2+}$           Blue-green: $Cu^{2+}$           Blue: $Pb^{2+}$

### Action of heat

Sublimation: $NH_4^+$ salts. Oxygen evolved (relights a glowing splint): certain oxides or nitrates. Carbon dioxide evolved (turns limewater milky): hydrogencarbonates or certain carbonates (not $Na_2CO_3$, $K_2CO_3$). Ammonia gas evolved (turns red litmus blue): decomposition of certain ammonium compounds. Water evolved (colourless liquid formed of the cooler part of test tube): hydrates, certain hydroxides or hydrogencarbonates. Brown gas evolved: $NO_2$ from certain nitrates (not $NaNO_3$ or $KNO_3$) or $Br_2$ from certain bromides.

### Solubilities of inorganic compounds in water at room temperature

All common salts of $Na^+$, $K^+$ and $NH_4^+$ are soluble.
All nitrates and ethanoates are soluble.
All sulphates are soluble except $Ba^{2+}$, $Sr^{2+}$ and $Pb^{2+}$ ($CaSO_4$ is only sparingly soluble).
All chlorides, bromides and iodides are soluble except $Ag^+$, $Cu^+$, $Pb^{2+}$, $Hg_2^{2+}$ (lead (II) halides are more soluble in hot water).
All carbonates and phosphates, are insoluble except those of $Na^+$, $K^+$ or $NH_4^+$.
All hydroxides are insoluble except those of $Na^+$, $K^+$, $NH_4^+$, $Sr^{2+}$ and $Ba^{2+}$.

*Precipitation of metal hydroxides with aqueous sodium hydroxide*

| Cation | Addition of sodium hydroxide solution | |
| --- | --- | --- |
| | A couple of drops | Excess |
| Potassium $K^+$ | no precipitate | no precipitate |
| Sodium $Na^+$ | no precipitate | no precipitate |
| Calcium $Ca^{2+}$ | white precipitate | precipitate insoluble |
| Magnesium $Mg^{2+}$ | white precipitate | precipitate insoluble |
| Aluminium $Al^{3+}$ | white precipitate | precipitate soluble—colourless solution |
| Zinc $Zn^{2+}$ | white precipitate | precipitate soluble—colourless solution |
| Iron(II) $Fe^{2+}$ | green precipitate | precipitate insoluble |
| Iron(III) $Fe^{3+}$ | red–brown precipitate | precipitate insoluble |
| Lead $Pb^{2+}$ | white precipitate | precipitate soluble—colourless solution |
| Copper(II) $Cu^{2+}$ | blue precipitate | precipitate insoluble |
| Silver $Ag^+$ | grey–brown precipitate | precipitate insoluble |

*Precipitation of metal hydroxides with ammonia solution*

| Cation | Addition of ammonia solution | |
| --- | --- | --- |
| | A couple of drops | Excess |
| Potassium $K^+$ | no precipitate | no precipitate |
| Sodium $Na^+$ | no precipitate | no precipitate |
| Calcium $Ca^{2+}$ | no precipitate | no precipitate |
| Magnesium $Mg^{2+}$ | white precipitate | precipitate insoluble |
| Aluminium $Al^{3+}$ | white precipitate | precipitate insoluble |
| Zinc $Zn^{2+}$ | white precipitate | precipitate soluble—colourless solution |
| Iron(II) $Fe^{2+}$ | green precipitate | precipitate insoluble |
| Iron(III) $Fe^{3+}$ | red–brown precipitate | precipitate insoluble |
| Lead $Pb^{2+}$ | white precipitate | precipitate insoluble |
| Copper $Cu^{2+}$ | blue precipitate | precipitate soluble—blue solution |
| Silver $Ag^+$ | brown precipitate | precipitate soluble |

*Tests for*

$CO_3^{2-}$. Carbon dioxide evolved when dilute hydrochloric acid added.

$SO_3^{2-}$. Sulphur dioxide evolved when dilute hydrochloric acid added and mixture heated.

$NO_2^-$. Brown $NO_2$ gas evolved when dilute hydrochloric acid added. Soln turns green.

$NO_3^-$. Brown ring or ammonia formed when sodium hydroxide solution and aluminium powder (or De Varda's alloy) heated with solution.

$SO_4^{2-}$. Add dilute hydrochloric acid and barium chloride solution to a solution. White ppt. of barium sulphate confirms $SO_4^{2-}$.

$Cl^-, Br^-, I^-$. Add dilute nitric acid and silver nitrate solution to solution. White ppt. of silver chloride turning purplish in sunlight and dissolving completely in ammonia solution confirms $Cl^-$. Cream ppt. of silver bromide turning green in sunlight and partially soluble in ammonia solution confirms $Br^-$. Yellow ppt. of silver iodide unaffected by sunlight and insoluble in ammonia solution confirms $I^-$.

$PO_4^{3-}$. Add ammonium molybdate solution and a few drops of conc. nitric acid to a solution. A yellow ppt. confirms $PO_4^{3-}$.

$CrO_4^{2-}$. Turns orange when dilute acid added to soln. Then add 1 drop of hydrogen peroxide and soln turns blue, then green and evolves oxygen.

ORGANIC

**Appearance**

The following are liquids at room temperature: alcohols, aliphatic carboxylic acids, esters, ethers, aldehydes, ketones, organohalogen compounds.

The following are solid at room temperature: aromatic carboxylic acids, phenols, amides, amino acids, carboxylic acid salts, amino salts, dicarboxylic acids.

**Smell**

Sharp, sour: carboxylic acids. Spirity: alcohols. Fruity: esters. Fishy: amines. Antiseptic: phenols.

**Solubility in water**

Soluble. Solids–carbohydrates, amino acids, urea, **phenols, carboxylic acids,** some amides.

Also salts of carboxylic acids and *amines*. Liquids. Some alcohols, aldhydes, ketones, **carboxylic acids, acid chlorides** and **amines.**

NB Substances in **bold type** form acidic solutions; substances in *italics* form alkaline solutions.

### Reaction with sodium hydroxide solution

Ammonia gas evolved when ammonium salts and amides are heated with sodium hydroxide solution.

### Reaction with neutral iron(III) chloride solution

Red colour: $HCOO^-$ or $CH_3COO^-$          Violet colour: phenols
Green colour: ethanedioate ion          Buff ppt: salt or aromatic acid

### Tests for unsaturation ($>C\equiv C<$ or $-C\equiv C-$)

1. Bromine water or solution of bromine in hexane—decolourized,
2. Alkaline solution of potassium manganate(VII)—decolourized.

### Tests for reducing agents

Acidified potassium manganate(VII) turns colourless if warmed with methanoates, ethanediates, aldehydes, primary alcohols and secondary alcohols.

Acidified potassium dichromate(VI) turns from orange to green on warming with methanoates, ethanoates, aldehydes, primary alcohols and secondary alcohols.

Fehling's solution (or Benedict's) forms an orange red ppt. when warmed with aldehydes.

Tollen's reagent forms a silver mirror on warming with aldehydes.

### Tests with 2,4–dinitrophenylhydrazine or similar

Yellow or orange ppt. formed with aldehydes or ketones.

Here is a qualitative exercise from a practical examination. My answer should give you some idea about the detail required.

You are provided with approximately 1.5 g of an organic compound of molecular formula $C_4H_4O_4$, labelled **Z**. Carry out the following tests, recording your observations and inferences, to determine the nature of **Z**. **Z** is sparingly soluble in water but soluble in ethanol.

*(Associated Examining Board)*

| TEST | OBSERVATIONS | INFERENCE |
|---|---|---|
| (*a*) Heat strongly (ignite) in a fume cupboard approximately 0.2 g of **Z** on a spatula, or crucible lid. | Burns (1 mark) with a non-smokey flame (1) | Not aromatic (1) |
| (*b*) Mix about 0.2 g of **Z** with about the same amount of sodium hydrogencarbonate. Add a few drops of water and observe carefully. | Effervescence (1) | **Z** is an acid (1) not a phenol (1) |
| (*c*) Pour 2 cm³ of bromine water (aqueous bromine) in each of two test tubes. To one add 0.2 g of **Z**. Stand both tubes in a beaker of hot (**not boiling**) water. | Bromine decolourized (1) in test tube containing **Z**. No change in other test tube (1) | **Z** contains $\diagdown C = C \diagup$ (1) |
| (*d*) To approximately 0.2 g of **Z** add 1 cm³ of aqueous potassium manganate(VII). Mix thoroughly. | Solution decolourized (1). Brown precipitate formed (1) | **Z** is unsaturated (1) |
| (*e*) Warm gently the remainder of **Z** with 5 cm³ of ethanol. Carry out the following tests using 1 cm³ **of the solution** for tests (i) and (ii). | | |
| (*i*) Test with damp Universal Indicator paper (or litmus paper). | Universal Indicator (or litmus) turns red (1) | Confirms **Z** is an acid (1) |
| (*ii*) Add 1 cm³ of water and allow to stand for 1 minute. Then add 1 cm³ of aqueous ammonia. | White precipitate (1) which disappears when ammonia is added (1) | **Z** is insoluble in water (1) /ammonium salt soluble in water (1) |
| (*iii*) To the remainder of the solution add 3 drops concentrated sulphuric acid, boil the mixture and pour into 10 cm³ of cold water. | Sweet smell or oily (1) layer. | Ester formed (1) |
| Summarize your inferences about the nature of **Z**. From the experiment, **Z** is an unsaturated (1) acid (1). From the formula, **Z** is possibly butenedioic acid $$\begin{array}{cc} H & H \\ \diagdown & \diagup \\ C = C & (2) \\ \diagup & \diagdown \\ HOOC & COOH \end{array}$$ | | |

## Summary of practical examination requirements

| | Oxford | Welsh | Oxford and Cambridge | AEB | London | Cambridge | JMB (Both Syllabuses) |
|---|---|---|---|---|---|---|---|
| *Alternative to practical exam* | None | None | None | None | Internal Assessment | None | Internal Assessment |
| Volumetric Analysis | ● | ● | ● | ● | ● | ● | ● |
| Acid – alkali | ● | ● | ● | ● | ● | ● | ● |
| Potassium manganate (VII) | ● | ● | ● | ● | ● | ● | ● |
| Sodium thiosulphate | ● | ● | ● | ● | ● | ● | ● |
| Silver nitrate | ● | ● | ● | ● | ● | ● | ● |
| Rate of reaction | ● | ● | ● | ● | ● | ● | ● |
| Melting and Boiling points | ● | ● | ● | | ● | | ● |
| Partition | | | ● | | ● | | |
| Solubility | | | ● | | ● | | |
| Systematic Qualitative | ● | | | | | | |
| Inorganic Analysis | | | | | | | |
| Tests for: | | | | | | | |
| $Na^+$ | ● | ● | ● | ● | | ● | |
| $K^+$ | ● | ● | ● | ● | | | |
| $NH_4^+$ | ● | ● | ● | ● | | | |
| $Li^+$ | | ● | | ● | | | |
| $Mg^{2+}$ | ● | ● | ● | ● | | ● | |
| $Ca^{2+}$ | ● | ● | ● | ● | | ● | |
| $Ba^{2+}$ | ● | ● | | ● | | ● | |
| $Sr^{2+}$ | | ● | | ● | | | |
| $Al^{3+}$ | ● | ● | ● | | | ● | ● |
| $Fe^{2+}/Fe^{3+}$ | ● | ● | ● | ● | | ● | ● |
| $Cr^{3+}$ | ● | | | ● | | ● | |
| $Mn^{2+}$ | ● | | ● | ● | | ● | |
| $Ni^{2+}$ | ● | ● | | | | ● | |
| $Cu^{2+}$ | ● | ● | | | | ● | ● |
| $Ag^+$ | ● | | | | | | ● |
| $Zn^{2+}$ | ● | ● | | | | ● | |
| $Pb^{2+}$ | ● | ● | ● | | | ● | |
| $Sn^{2+}$ | | ● | | | | | |
| $Co^{2+}$ | | | | | | | ● |
| $F^-$ | | | | ● | | | ● |
| $Cl^- Br^- I^-$ | ● | ● | ● | ● | | ● | ● |
| $SO_4^{2-}$, $CO_3^{2-}$ $NO_3^-$ | ● | | ● | ● | | ● | ● |
| $SO_3^{2-}$ | ● | | ● | ● | | ● | ● |
| $NO_2^-$ | ● | | ● | ● | | ● | ● |
| $PO_4^{3-}$ | ● | | | ● | | | |
| $S^{2-}$ | ● | | ● | | | ● | |
| $CH_3COO^-$ | ● | | ● | | | | |

(London column, through the ion-test rows: NOT SPECIFIED IN THE SYLLABUS)

Footnote: Practical notebooks must be submitted prior to the examination for Southern Universities.

# AS-level chemistry

For many years there has been criticism that students at A level were, of necessity, highly specialized. For example, a student studying Chemistry, Physics and Biology had no experience outside science. AS levels are new courses, designed to enable students to study more subjects to a high level and enabling them to keep higher education and career options open. They are offered in addition to existing A levels. They may be very suitable for students not going on to higher education and they have been enthusiastically received by universities, other higher education institutions and employers. A student studying two sciences at A level might take an AS humanity or modern language subject, for instance.

The first courses started in September 1987, for examination in summer 1989. They are intended as two-year courses, although there is no reason why a student could not complete them in one. The courses will take about half of the teaching and studying time of an A level.

AS courses will be graded in the same way as A level. AS passes will be graded A–E. The standards reached will be similar to those in A level but the syllabus is much shorter.

The table on page viii summarizes the sections required for new AS Chemistry syllabuses. The questions that will be asked will be similar to the A-level questions shown.

It will be interesting to see the take-up of these new courses in future years and the progress that successful candidates make. AS-level courses should certainly be considered seriously by students going on to post-16 studies. AS-level Chemistry would be a useful qualification for many. For mature students returning to study, an AS level might be a good introduction, as it is perhaps more achievable and could be followed by an A level.

## Examination boards

*AEB*
The Associated Examining Board
Stag Hill House, Guildford, Surrey GU2 5XJ

*Cambridge*
University of Cambridge Local Examinations Syndicate
Syndicate Buildings, 1 Hills Road, Cambridge CB1 2EU

*COSSEC (AS only)* As for Cambridge, Oxford and Cambridge or SUJB.

*JMB*
Joint Matriculation Board
Devas Street, Manchester M15 6EU

*London*
University of London Schools Examinations Board
Stewart House, 32 Russell Square, London WC1B 5DN

*NISEC*
Northern Ireland Schools Examinations Council
Beechill House, 42 Beechill Road, Belfast BT8 4RS

*Oxford*
Oxford Delegacy of Local Examinations
Ewert Place, Summertown, Oxford OX2 7BZ

*O and C*
Oxford and Cambridge Schools Examinations Board
10 Trumpington Street, Cambridge and Elsfield Way, Oxford

*SEB*
Scottish Examinations Board
Ironmills Road, Dalkeith, Midlothian EH22 1BR

*SUJB*
Southern Universities' Joint Board for School Examinations
Cotham Road, Bristol BS6 6DD

*WJEC*
Welsh Joint Education Committee
245 Western Avenue, Cardiff CF5 2YX

# 1 Atomic structure

## 1.1 INTRODUCTION

An elementary treatment of atomic structure is an essential component of any GCSE course. The treatment required at A level is significantly more detailed than that at GCSE.

The ability to work out the electronic structure of an atom is required throughout the course – in particular in Units 2, 3, 4, 5, 6 and 21–28. In particular it is important to understand the relationship between atomic structure and the Periodic Table (Unit 2).

## 1.2 PARTICLES IN AN ATOM

Atoms are composed of **protons, neutrons** and **electrons**. Table 1.1 compares the properties of these three particles.

**Table 1.1 Particles in an atom**

| Name of particle | Mass | Charge |
|---|---|---|
| proton p | 1 amu* | +1 |
| neutron n | 1 amu | 0 |
| electron e | negligible | −1 |

*amu — atomic mass unit

An atom is neutral and contains equal numbers of protons and electrons.

**Atomic number** (Z) – the number of protons in an atom.

**Mass number** (A) – the total number of protons plus neutrons in an atom.

*E.g.* $^{39}_{19}$K represents a potassium atom with mass number 39 and atomic number 19.

The number of protons, neutrons and electrons in this potassium atom are 19,20 (*i.e.* 39 – 19) and 19 respectively. It is possible to get other potassium atoms, namely $^{40}_{19}$K and $^{41}_{19}$K containing the same number of protons and electrons but different numbers of neutrons (21 and 22 respectively). These different atoms of the same element are called **isotopes**. A sample of potassium contains 93.1% potassium-39, 0.012% potassium-40 and 6.9% potassium-41.

The **relative atomic mass** (represented by $A_r$) is the mass of an atom based on a scale such that the $^{12}_{6}$C isotope has a mass of 12.00 units.

$$\text{relative atomic mass} = \frac{\text{mass of 1 atom of element} \times 12}{\text{mass of 1 atom of carbon-12}}$$

Relative atomic masses (formerly called atomic masses or atomic weights) will be given on your examination papers for you to use. *e.g.* ($A_r(O) = 16$ or O = 16)

## 1.3 EVIDENCE FOR PROTONS, NEUTRONS AND ELECTRONS

**Electrons**

J. J. Thomson discovered that a beam of rays was emitted from the cathode when an electric discharge was passed through a gas at a very low pressure. These rays were deflected by a magnetic field and behaved in the same way whichever gas was used. These rays consist of streams of **electrons**.

**Protons**

The passage of an electric discharge at very low pressures also produces a stream of particles from the anode. These particles are oppositely charged to electrons and were different for each gas used. The lightest particles were obtained when hydrogen was the gas used and these particles were assumed to be **protons**.

**Neutrons**

The neutron was more difficult to characterise than the electron or proton. Chadwick (1932) bombarded the element beryllium with α (alpha) particles and noticed that fast moving, highly penetrating particles were produced. These particles, now known to be **neutrons**, were not deflected by electric or magnetic fields. The following equation summarizes the change taking place.

$$^{9}_{4}\text{Be} + ^{4}_{2}\text{He} \rightarrow ^{12}_{6}\text{C} + ^{1}_{0}\text{n}$$

($^{4}_{2}$He represents an alpha particle).

### 1.4    THE NUCLEUS

The protons and neutrons in each atom are tightly packed in a positively charged **nucleus** and the electrons move around the nucleus. In chemical reactions the nucleus remains unchanged.

Radioactivity (Unit 20) is the breaking down of the nucleus with the emission of $\alpha$, $\beta$ or $\gamma$ rays. Radioactivity results in the formation of a new element.

Geiger and Marsden bombarded a thin gold foil with a beam of $\alpha$ particles. Most of the particles passed through the foil without deflection and were detected by a flash of light when the $\alpha$ particle struck a zinc sulphide screen. A few were deflected and some of these were deflected at angles greater than 90° suggesting they had been repelled by large positive charges within the foil – nuclei of atoms of gold.

### 1.5    ARRANGEMENT OF ELECTRONS AROUND THE NUCLEUS

From GCSE work you should be familiar with a simple model of the arrangement of electrons around the nucleus. The electrons are in certain **energy levels** and each energy level can hold only up to a certain maximum number of electrons. This is summarized in Table 1.2.

**Table 1.2 Energy levels**

| Energy level or shell | Maximum number of electrons |
| --- | --- |
| 1st or K shell | 2 |
| 2nd or L shell | 8 |
| 3rd or M shell | 18 |
| 4th or N shell | 32 |
| 5th or O shell etc | 50 |

A sodium atom containing 11 electrons has an electronic arrangement of 2,8,1 – the 1st and 2nd shells being full. This is sometimes represented in a simple Bohr diagram (Fig. 1.1).

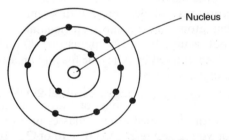

**Fig. 1.1** Simple representation of sodium atom

These ideas of electron arrangement, although useful, are very simplified and it is necessary for you to have a more refined model at A level. It has been found, by spectroscopic means, that all the electrons in the 2nd energy level are not exactly identical in energy. It is possible to break down this energy level into sub shells.

Electrons, partly because of their very small size, are impossible to locate exactly at any particular time. It is, however possible to indicate a region or volume where the electron is most likely to be. This region is called an **orbital**. Each orbital is capable of holding a maximum of two electrons. Orbitals can be divided into *s*, *p*, *d* and *f* types. Each type has its own characteristic shape. The shapes of *s* and *p* orbitals are shown in Fig. 1.2 but it should be remembered that these orbitals are not flat but are three-dimensional.

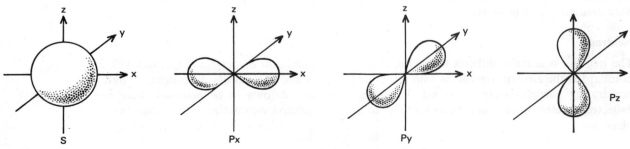

**Fig. 1.2** The *s* and *p* orbitals

The first energy level holds a maximum of two electrons in one *s*-type orbital (called 1*s*). There are no *p*, *d* or *f* orbitals available.

The second energy level consists of one *s* type orbital and three types of *p* orbitals – 2*s*, 2*p_x*, 2*p_y*, 2*p_z*.

Note: There are three *p* orbitals of identical energy – one along the *x* axis, one along the *y* axis and one along the *z* axis.

These four orbitals can hold a total of eight electrons (*i.e.* 2 electrons each). There are no 2*d* or 2*f* orbitals.

The third energy level consists of one *s* type orbital, three *p* type orbitals and five *d* type orbitals. These nine orbitals can hold a maximum of 18 electrons altogether (two electrons each). You are not expected to know the shapes of *d* or *f* orbitals.

When filling the available orbitals with electrons two important principles should be remembered:

(*i*) Electrons fill the lowest energy orbitals first and other orbitals in order of ascending energy. As will be seen with *d* block elements (Unit 28), it is incorrect to assume that an energy level is always completely filled before electrons enter the next energy level. The order of filling orbitals is shown in Fig. 1.3.

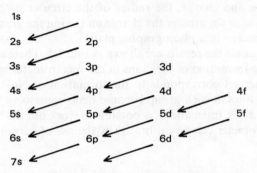

**Fig. 1.3** The order of filling orbitals

The order of filling orbitals is 1*s*, 2*s*, 2*p*, 3*s*, 3*p*, 4*s*, 3*d*, 4*p*, 5*s*, 4*d*, 5*p*, 6*s*, 4*f*, 5*d*, 6*p*.

(*ii*) Where there are several orbitals of exactly the same energy (*e.g.* three 2*p* orbitals), electrons will occupy different orbitals whenever possible (*e.g.* nitrogen is $1s^2 2s^2 2p_x^1 2p_y^1 2p_z^1$ and not $1s^2 2s^2 2p_x^2 2p_y^1$. When an orbital only contains one electron then this electron is said to be **unpaired**. This principle is called **Hund's rule**.

Table 1.3 summarizes the electron arrangements of the first 20 elements. For comparison the simple O level electron arrangements are given.

**Table 1.3 Electron arrangements of the first 20 elements**

| Element | O level | A level |
|---|---|---|
| Hydrogen | 1 | $1s^1$ |
| Helium | 2 | $1s^2$ |
| Lithium | 2,1 | $1s^2 2s^1$ |
| Beryllium | 2,2 | $1s^2 2s^2$ |
| Boron | 2,3 | $1s^2 2s^2 2p_x^1$ |
| Carbon | 2,4 | $1s^2 2s^2 2p_x^1 2p_y^1$ |
| Nitrogen | 2,5 | $1s^2 2s^2 2p_x^1 2p_y^1 2p_z^1$ |
| Oxygen | 2,6 | $1s^2 2s^2 2p_x^2 2p_y^1 2p_z^1$ |
| Fluorine | 2,7 | $1s^2 2s^2 2p_x^2 2p_y^2 2p_z^1$ |
| Neon | 2,8 | $1s^2 2s^2 2p_x^2 2p_y^2 2p_z^2$ |
| Sodium | 2,8,1 | $1s^2 2s^2 2p_x^2 2p_y^2 2p_z^2 3s^1$ |
| Magnesium | 2,8,2 | $1s^2 2s^2 2p_x^2 2p_y^2 2p_z^2 3s^2$ |
| Aluminium | 2,8,3 | $1s^2 2s^2 2p_x^2 2p_y^2 2p_z^2 3s^2 3p_x^1$ |
| Silicon | 2,8,4 | $1s^2 2s^2 2p_x^2 2p_y^2 2p_z^2 3s^2 3p_x^1 3p_y^1$ |
| Phosphorus | 2,8,5 | $1s^2 2s^2 2p_x^2 2p_y^2 2p_z^2 3s^2 3p_x^1 3p_y^1 3p_z^1$ |
| Sulphur | 2,8,6 | $1s^2 2s^2 2p_x^2 2p_y^2 2p_z^2 3s^2 3p_x^2 3p_y^1 3p_z^1$ |
| Chlorine | 2,8,7 | $1s^2 2s^2 2p_x^2 2p_y^2 2p_z^2 3s^2 3p_x^2 3p_y^2 3p_z^1$ |
| Argon | 2,8,8 | $1s^2 2s^2 2p_x^2 2p_y^2 2p_z^2 3s^2 3p_x^2 3p_y^2 3p_z^2$ |
| Potassium | 2,8,8,1 | $1s^2 2s^2 2p_x^2 2p_y^2 2p_z^2 3s^2 3p_x^2 3p_y^2 3p_z^2 4s^1$ |
| Calcium | 2,8,8,2 | $1s^2 2s^2 2p_x^2 2p_y^2 2p_z^2 3s^2 3p_x^2 3p_y^2 3p_z^2 4s^2$ |

Note: sometimes electronic structures are shown in a slightly condensed form *e.g.* Calcium $1s^2 2s^2 2p^6 3s^2 3p^6 4s^2$

*Notes on Table 1.3*

(*i*) The small number above the orbital refers to the number of electrons in the orbital *e.g.* $1s^2$ means two electrons in a $1s$ orbital.

(*ii*) The noble gases (helium, neon, argon, krypton, radon and xenon) contain filled *s* and, in all cases except helium, completely filled *p* orbitals. This $s^2p^6$ electron arrangement is very difficult to break down.

(*iii*) The electronic arrangements are sometimes abbreviated. For example, the electronic arrangement of calcium may be written as $1s^2\,2s^2\,2p^6\,3s^2\,3p^6\,4s^2$. It is important to remember, however, that the six electrons in $2p$ and $3p$ orbitals are in three separate orbitals each holding two electrons.

## 1.6    Mass spectrometer

The mass spectrometer (Fig. 1.4) is an instrument used for accurately measuring atomic mass and for finding the number of isotopes of an element present. The sample being tested is introduced into the instrument and is then ionized by heating, electrical discharge or electron bombardment. The ions produced are accelerated and passed through a slit to give a fine beam of ions all moving with the same velocity. The individual ions in the beam will differ slightly in mass and charge. The beam of ions is then subjected to a magnetic field which bends the beam into a circular path. Depending upon the mass and charge, the radius of the circular path for each ion is slightly different. The lighter the ion or the greater the charge on the ion the greater will be the deflection. The ions are detected by means of a photographic plate.

If neon is used as the sample the results are shown in Fig. 1.5. There are two sets of three lines. One set corresponds to the formation of $Ne^+$ ions in the spectrometer by the loss of one electron from each atom. The other set corresponds to the formation of $Ne^{2+}$ ions by the loss of two electrons. There are three lines in each group because neon consists of three isotopes – neon-20, neon-21 and neon-22. From the intensities it is possible to work out the relative abundance of the three isotopes. The spectrometer can be calibrated by the use of carbon-12.

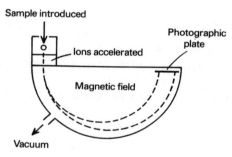

**Fig. 1.4** Mass spectrometer

**Fig. 1.5** The mass spectrum of neon

## 1.7    Atomic spectrum of hydrogen

When an electrical discharge is passed through a sample of hydrogen gas at low pressure, a pinkish glow is observed. This glow can be examined using a simple spectroscope – a number of separate lines can be observed and there are also lines in the ultraviolet and infrared regions (Fig. 1.6).

Lines in the spectrum correspond to electronic changes within the atoms. The fact that only certain lines are observed indicates that only certain energy changes are possible and this is evidence for the existence of distinct energy levels within the hydrogen atom.

When the electrical discharge is passed through hydrogen gas, energy is absorbed and the electron in the first energy level is promoted to a higher energy level. This electron may then drop

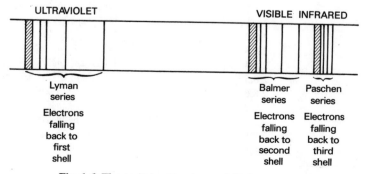

**Fig. 1.6** The emission spectrum of the hydrogen atom

back into a lower energy level giving out energy in the form of electromagnetic radiation of a definite frequency. This energy change is related to the frequency of the radiation by the formula

$$\Delta E = h\nu$$

where $\Delta E$ is the energy change

$h$ is a constant (Planck's constant)

$\nu$ is the frequency of the radiation (light etc.)

It is important to realize that each line in the spectrum corresponds to a transition between the separate energy levels.

When the electron drops back into the first energy level this results in one of the lines in the **Lyman series**. The **Balmer series** corresponds to falling back into the second shell from a higher shell, while the **Paschen series** corresponds to electrons falling back to the third shell.

## QUESTIONS

You may frequently be expected to work out the electronic structure of an atom of a particular element. This may be in a multiple choice question or in the early part of a structured question.

1 Give the atomic number of nitrogen and the electronic arrangement of its isolated atom.

From the Periodic Table (page 21) you can find that nitrogen has an atomic number of 7. The nitrogen atom contains 7 electrons and has an electron arrangement $1s^2 2s^2 2p_x^1 2p_y^1 2p_z^1$.

2 The element **Q** has the electronic configuration $1s^2 2s^2 2p^6 3s^2 3p^6 4s^2 3d^{10} 4p^1$.

(a) Write down the atomic number of **Q**.

(b) **Q** occurs naturally as a mixture of $^{69}$**Q** and $^{71}$**Q**. Explain the significance of the numbers 69 and 71, and say what these two components in the natural element are called.

(c) If $^{69}$**Q** and $^{71}$**Q** occur in the proportions 60% and 40% respectively, calculate the relative atomic mass of **Q**.

*(Southern Universities Joint Board)*

(a) **Q** contains 31 electrons and must also contain 31 protons. The atomic number (*i.e.* number of protons) of **Q** is 31. Many candidates will give an answer of between 69 and 71 as they are confused between the terms atomic number and mass number.

(b) The numbers 69 and 71 are mass numbers *i.e.* numbers of protons plus neutrons. Different atoms of the same element containing different numbers of neutrons are called isotopes.

(c) A natural mixture of **Q** contains 60% of isotope $^{69}$**Q** and 40% of isotope $^{71}$**Q**. The relative atomic mass of **Q** is an average of the two mass numbers but with regard for the relative amounts of the two isotopes.

$$\text{Relative atomic mass of } \mathbf{Q} = (69 \times \frac{60}{100}) + (71 \times \frac{40}{100})$$
$$= 41.4 + 28.4$$
$$= 69.8$$

If the two isotopes were present in equal amounts, the relative atomic mass of **Q** would be exactly 70. Having done your calculation you should check that your answer is reasonable before passing on to the next question. Because there is more $^{69}$**Q** present in the sample than $^{71}$**Q**, the relative atomic mass should be between 69 and 71 but closer to 69. Checking that your answer is reasonable can save many marks in examinations.

3 The element gallium consists of 60.4% of an isotope of atomic mass 68.93 and 39.6% of an isotope of atomic mass 70.92. Calculate, to three significant figures, the relative atomic mass of gallium.

This is a similar question to **3**, but the arithmetic is more complex.

Because the isotope with atomic mass 68.93 is present in larger amounts, the relative atomic mass should be closer to 68.93 than 70.92. If you work out your answer in a similar way to that in **3** you should obtain an answer of 69.72.

4 The diagram (Fig. 1.6) represents the atomic emission spectrum of hydrogen.

(*i*) Explain why the spectrum is composed of lines rather than being continuous.
(*ii*) What does each line indicate?
(*iii*) Explain why the lines become progressively closer to each other.

*(Associated Examining Board)*

(*i*) Only certain energy changes are possible within a hydrogen atom because the electrons can only exist in separate (or discrete) energy levels.
(*ii*) Each line represents the falling back of an electron from a higher energy level to a lower energy level.

(*iii*) This is best answered with the aid of a diagram (Fig. 1.7).

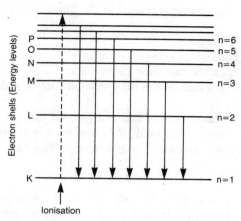

**Fig. 1.7** To show why the emission lines of hydrogen become progressively closer together

The relative increase in energy given out when an electron falls back from say $n = 4$ to $n = 1$ is less than when it falls from $n = 3$ to $n = 1$.

5 The diagram (Fig. 1.8) shows the mass spectrum of chlorine.

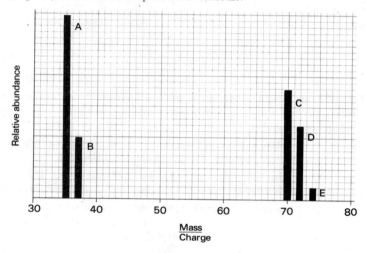

**Fig. 1.8** The mass spectrum of chlorine

If peak A represents the ion $^{35}Cl^+$, suggest possible ions which would give the peaks B and D.

*(Scottish Higher)*

Peaks B and D correspond to $^{37}Cl^+$ and $(^{35}Cl - ^{37}Cl)^+$.

Incidently peaks C and E correspond to $(^{35}Cl - ^{35}Cl)^+$ and $(^{37}Cl - ^{37}Cl)^+$.

# 2 Periodicity and the Periodic Table

## 2.1 INTRODUCTION

Most GCSE syllabuses include a brief treatment of the **Periodic Table** but the full extent of its usefulness may not be obvious.

At A level the use of the Periodic Table is extremely important. Many Boards provide a copy of the Periodic Table for you to use in the examination. Units 21–28 in particular make frequent reference to the Periodic Table and they should not be attempted until this unit is thoroughly understood. This unit is closely linked with Unit 1.

In this unit, a revision of the Periodic Table will be made. In addition various properties of elements will be considered including **electronegativity**, **ionization energy** and **electron affinity**.

**Transition elements**

**d block**

**f block**

Fig. 2.1 The Periodic Table of elements

KEY:

Atomic Mass
Symbol
Name
Atomic Number

| I | II | | | | III | IV | V | VI | VII | O |
|---|----|--|--|--|-----|----|---|----|-----|---|
| | | | | | | | | | | 4 **He** Helium 2 |
| 7 **Li** Lithium 3 | 9 **Be** Beryllium 4 | | | | 11 **B** Boron 5 | 12 **C** Carbon 6 | 14 **N** Nitrogen 7 | 16 **O** Oxygen 8 | 19 **F** Fluorine 9 | 20 **Ne** Neon 10 |
| 23 **Na** Sodium 11 | 24 **Mg** Magnesium 12 | | | | 27 **Al** Aluminium 13 | 28 **Si** Silicon 14 | 31 **P** Phosphorus 15 | 32 **S** Sulphur 16 | 35.5 **Cl** Chlorine 17 | 40 **Ar** Argon 18 |
| 39 **K** Potassium 19 | 40 **Ca** Calcium 20 | | | | 70 **Ga** Gallium 31 | 73 **Ge** Germanium 32 | 75 **As** Arsenic 33 | 79 **Se** Selenium 34 | 80 **Br** Bromine 35 | 84 **Kr** Krypton 36 |
| 85.5 **Rb** Rubidium 37 | 88 **Sr** Strontium 38 | | | | 115 **In** Indium 49 | 119 **Sn** Tin 50 | 122 **Sb** Antimony 51 | 128 **Te** Tellurium 52 | 127 **I** Iodine 53 | 131 **Xe** Xenon 54 |
| 133 **Cs** Caesium 55 | 137 **Ba** Barium 56 | | | | 204 **Tl** Thallium 81 | 207 **Pb** Lead 82 | 209 **Bi** Bismuth 83 | 210 **Po** Polonium 84 | 210 **At** Astatine 85 | 222 **Rn** Radon 86 |
| 223 **Fr** Francium 87 | 226 **Ra** Radium 88 | | | | | | | | | |

1 **H** Hydrogen 1

**d block transition elements:**

| 45 **Sc** Scandium 21 | 48 **Ti** Titanium 22 | 51 **V** Vanadium 23 | 52 **Cr** Chromium 24 | 55 **Mn** Manganese 25 | 56 **Fe** Iron 26 | 59 **Co** Cobalt 27 | 59 **Ni** Nickel 28 | 64 **Cu** Copper 29 | 65 **Zn** Zinc 30 |
|---|---|---|---|---|---|---|---|---|---|
| 89 **Y** Yttrium 39 | 91 **Zr** Zirconium 40 | 93 **Nb** Niobium 41 | 96 **Mo** Molybdenum 42 | 98 **Tc** Technetium 43 | 101 **Ru** Ruthenium 44 | 103 **Rh** Rhodium 45 | 106 **Pd** Palladium 46 | 108 **Ag** Silver 47 | 112 **Cd** Cadmium 48 |
| 139 **La** Lanthanum 57 | 178.5 **Hf** Hafnium 72 | 181 **Ta** Tantalum 73 | 184 **W** Tungsten 74 | 186 **Re** Rhenium 75 | 190 **Os** Osmium 76 | 192 **Ir** Iridium 77 | 195 **Pt** Platinum 78 | 197 **Au** Gold 79 | 201 **Hg** Mercury 80 |
| 227 **Ac** Actinium 89 | | | | | | | | | |

**f block:**

| 140 **Ce** Cerium 58 | 141 **Pr** Praseodymium 59 | 144 **Nd** Neodymium 60 | 147 **Pm** Promethium 61 | 150 **Sm** Samarium 62 | 152 **Eu** Europium 63 | 157 **Gd** Gadolinium 64 | 159 **Tb** Terbium 65 | 162.5 **Dy** Dysprosium 66 | 165 **Ho** Holmium 67 | 167 **Er** Erbium 68 | 169 **Tm** Thulium 69 | 173 **Yb** Ytterbium 70 | 175 **Lu** Lutetium 71 |
|---|---|---|---|---|---|---|---|---|---|---|---|---|---|
| 232 **Th** Thorium 90 | 231 **Pa** Protactinium 91 | 238 **U** Uranium 92 | 237 **Np** Neptunium 93 | 242 **Pu** Plutonium 94 | 243 **Am** Americium 95 | 247 **Cm** Curium 96 | 247 **Bk** Berkelium 97 | 251 **Cf** Californium 98 | 254 **Es** Einsteinium 99 | 253 **Fm** Fermium 100 | 256 **Md** Mendelevium 101 | 254 **No** Nobelium 102 | 257 **Lw** Lawrencium 103 |

The way in which these properties change through the Periodic Table will be examined. This leads to the concept of **periodicity**. Two other topics for consideration are **diagonal relationship** and the properties of the Group 0 elements (noble gases).

## 2.2    THE PERIODIC TABLE

The Periodic Table is an arrangement of all the chemical elements in order of increasing atomic number – elements with similar properties (*i.e.* the same chemical family) are placed in the same vertical column.

The Periodic Table was devised in 1869 by Mendeléef. At this time a number of elements that we now know had not been discovered, but Mendeléef left gaps in his table and even predicted the properties of some of these undiscovered elements. Fig. 2.1 shows the modern Periodic Table.

The main difference between Mendeléef's table and the modern table, apart from the extra elements, is that the latter arranges the elements in order of increasing atomic number rather than atomic mass.

The vertical columns are called **Groups**. A Group contains elements with similar properties and similar outer electronic arrangements. The Groups are given Roman numbers *e.g.* Group IV. The elements in the alkali metal family (Group I) all have a single electron in the highest energy level (see Unit 22).

The outer electron arrangements of elements in the different Groups are shown in Table 2.1

**Table 2.1  Arrangements of outer electrons**

| Group | Outer electron arrangement |
|-------|---------------------------|
| I | $s^1$ |
| II | $s^2$ |
| III | $s^2 p^1$ |
| IV | $s^2 p^2$ |
| V | $s^2 p^3$ |
| VI | $s^2 p^4$ |
| VII | $s^2 p^5$ |
| 0 | $s^2 p^6$ * |

*Helium (the first element in Group 0) has an electron arrangement of $1s^2$.

The horizontal rows in the Periodic Table are called **Periods**.

The elements shaded in Fig. 2.1 are called the *s*- and *p*-block elements. Most of the chemistry in A level involves the study of these main block elements. The elements between the two shaded portions are called the **d-block** or **transition elements**. A study of the first row of *d*-block elements Sc–Zn will be found in Unit 28. The two rows of elements at the bottom of the table are called ***f*-block** elements.

The bold line in Fig. 2.1 is an attempt to divide elements into metals and non metals. Elements on the right hand side of this line are non metals and on the left hand side are metals; but elements close to the line often show both metallic and non-metallic properties, particularly if they exist in allotropic (polymorphic) forms.

## 2.3    TRENDS WITHIN THE PERIODIC TABLE

Physical properties such as **ionisation energy, electron affinity, melting** and **boiling points** are related to electron arrangements. If one of these properties is plotted on a graph for each element, *e.g.* a graph of the property against atomic number, the graph has a characteristic shape. The shape consists of a series of peaks and troughs. An example is shown in Fig. 2.2.

You will notice that the element at the top of each peak is a noble gas (Group 0) and the element at the bottom is an alkali metal (Group I). Elements in the same Group are found to occur at similar positions on the different peaks.

In the following sections some of these physical properties will be discussed in greater detail.

## 2.4    IONIZATION ENERGY (IONIZATION POTENTIAL)

Ionisation energy is the energy absorbed when a mole of electrons is removed from a mole of atoms of an element in the gaseous state to form a mole of positively charged ions.

$$i.e.\ M(g) \rightarrow M^+(g) + e^-$$

This is the **first ionisation energy**. The energy is usually quoted in units of kilojoules per mole (kJ mol$^{-1}$). Energy is required to remove an electron from any atom because there is an attractive force between the nucleus and the electron being removed which has to be overcome. The value of the first ionisation energy depends upon:

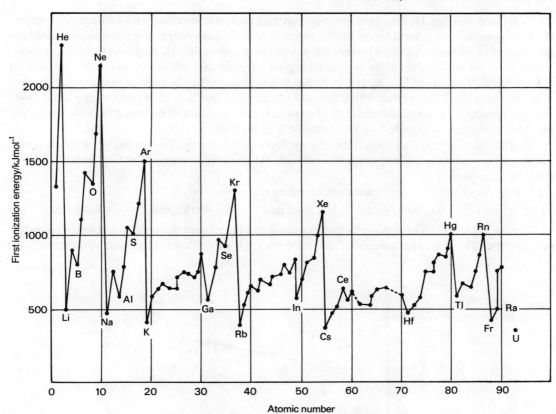

**Fig. 2.2** The first ionization energies of the elements

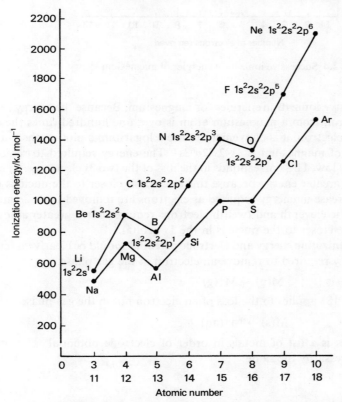

**Fig. 2.3** The variation of first ionization energies for the second and third Periods

(*i*) the effective nuclear charge,

(*ii*) the distance between the electron and the nucleus, and

(*iii*) the 'shielding' produced by lower energy orbitals.

Shielding involves the repulsion between electrons in inner filled orbitals and the electron being removed from the outer orbital.

Fig. 2.3 shows the variation of first ionization energy within the second and third Periods. Across each Period there is an increase in ionization energy.

Beryllium (Group II) has an extra electron and proton compared with lithium. The extra electron goes into the same $2s$ orbital. The increase in ionization energy here can be attributed to the increased nuclear charge. The two breaks in the graphs (Be→B and N→O in the second period) can be explained by the increased repulsion produced when two electrons are in the same orbital. The ionization energy of boron is less than that of beryllium because, in boron, there is a complete $2s^2$ orbital. The increased shielding of the $2s$ orbital reduces the ionization energy. Similary, the ionization energy of oxygen is less than that of nitrogen because the extra electron in oxygen is shielded by the half-filled $2p$ orbitals.

Within a Group the first ionization energy decreases down the Group as the outer electron becomes progressively further from the nucleus. Also, there is more shielding because of the extra filled orbitals.

*E.g.* for group I the first ionization energies are

Li 520 kJ mol$^{-1}$     Na 500 kJ mol$^{-1}$     K 420 kJ mol$^{-1}$     Rb 400 kJ mol$^{-1}$     Cs 380 kJ mol$^{-1}$

The energy required to remove three electrons is not three times the energy required to remove one electron. The values for successive ionization energies are related to electronic structures.

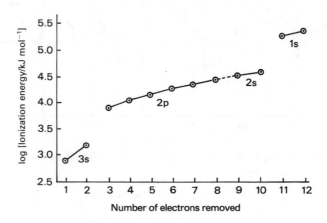

**Fig. 2.4** Successive ionization energies of magnesium

Fig. 2.4 shows the successive ionization energies for magnesium. Because the energy required to remove the eleventh electron from a magnesium atom is over one hundred times the energy required to remove just one electron, it is convenient to use a logarithmic plot on the $y$-axis.

The electronic structure of magnesium is $1s^22s^22p^63s^2$. The energy required to remove the first two electrons is relatively low. This corresponds to the loss of the two $3s$ electrons. To remove a third electron needs much greater energy because this electron is closer to the nucleus in a $2p$ orbital. There is a steady increase in energy required as electrons are removed from $2p$ and then $2s$ orbitals. The removal of the eleventh and twelfth electrons requires much greater amounts of energy, as these electrons are closer to the nucleus in the $1s$ orbital.

The distinction between **ionization energy** and **electrode potential** should be clearly understood. **Ionization energy** is the energy required to remove an electron in the gas phase.

$$M(g) \rightarrow M^+(g) + e^-$$

The **electrode potential** (Unit 18) applies to the loss of an electron not in the gas phase.

$$M(s) \rightarrow M^+(aq) + e^-$$

The **electrochemical series** is a list of metals in order of electrode potential. The order of ionization energies is not necessarily the same.

## *2.5   ELECTRON AFFINITY

Electron affinity is the energy change when one mole of isolated gaseous atoms gains one mole of electrons to form a mole of negatively charged ions.

$$X(g) + e^- \rightarrow X^-(g)$$

It is the atoms of elements on the right hand side of a Period which readily accept electrons to form negative ions. In these atoms there are greater nuclear charges to attract electrons. The electron affinities of chlorine, bromine and iodine are $-364$, $-342$ and $-295$ kJ mol$^{-1}$ respectively. In each case the value is negative as energy is released.

## 2.6    ELECTRONEGATIVITY

The tendency of an atom to attract an electron to itself is called its **electronegativity**. The electronegativity of an element can be regarded as related to ionization energy and electron affinity, and can be evaluated in various ways. In a covalent bond such as between two chlorine atoms, the two atoms attract the pair of electrons in the bond equally. However in hydrogen fluoride, because fluorine is more electronegative than hydrogen, the pair of electrons is attracted more by the fluorine than the hydrogen. This results in a slight separation of charge.

$$\overset{\delta^+}{H} - \overset{\delta^-}{F}$$

Fig. 2.5 shows the variation of electronegativity (according to values obtained by Pauling) with atomic number for the first 60 elements.

In any Group of the Periodic Table there is a decrease in electronegativity down the Group. Across a Period there is an increase in electronegativity from left to right. Elements in the top right hand corner of the Periodic Table are therefore most electronegative.

In bond formation, the difference in electronegativity between two atoms will determine whether the bonding is predominately ionic or covalent (Unit 4).

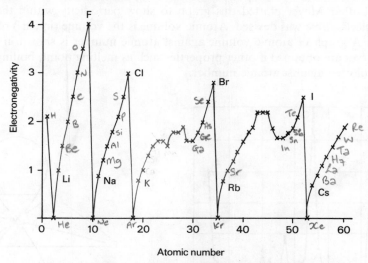

**Fig. 2.5** Variation of electronegativity with atomic number

## 2.7    ATOMIC AND IONIC RADII

The atomic radius can be determined by X-ray or electron diffraction. In any period there is a **decrease** in atomic radius from left to right. This surprises many students, because from sodium to chlorine there is an addition of 6 protons, 6 electrons and a number of neutrons and in spite of this the atomic radii fall from 0.157nm to 0.099nm. The explanation is that the extra electrons are being added to the same 3s and 3p orbitals (therefore there is no increase in size) but the extra nuclear charge attracts these electrons drawing them closer to the nucleus.

For the elements in a Group of the Periodic Table, the atomic radii increase down the Group. Down the Group, extra electrons are added to additional orbitals. Despite the extra nuclear charge and, because of increased screening, there is a net increase in atomic radius down the Group.

Fig. 2.6 shows a graph of atomic radius against atomic number. The trends in ionic radii down a Group are similar to those for atomic radii.

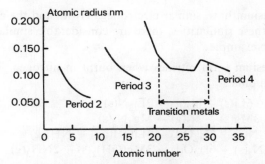

**Fig. 2.6** Graph of atomic radius against atomic number

**Cations** (*i.e.* positive ions) have a smaller radius than the corresponding atoms.

*E.g.* Na 0.157nm    Na$^+$ 0.098nm

The removal of one electron from a sodium atom empties the $3s$ orbital. The outer electron is therefore in a $2p$ orbital.

**Anions** have a larger radius than the corresponding atom because of repulsion between the electrons. There is no extra nuclear charge.

*E.g.* F 0.064nm   F$^-$ 0.133nm

The three ions Na$^+$, Mg$^{2+}$ and Al$^{3+}$ are isoelectronic, *i.e.* they contain the same number of electrons in an arrangement $1s^2 2s^2 2p^6$. The ionic radii of these ions are 0.098, 0.065 and 0.045nm respectively. The decreasing radius is due to the additional nuclear charge – Na$^+$ contains 11 protons, Mg$^{2+}$ 12 and Al$^{3+}$ 13.

## 2.8 OTHER PROPERTIES

**Oxidation states** vary regularly through the Periodic Table. The variation in oxidation state will be discussed in Unit 6.

**Atomic volume** is another property which varies periodically through the Periodic Table. Historically, Lothar Meyer plotted this graph to show periodicity within the Periodic Table before Mendeléef's table was devised. Atomic volume is the volume (in cm$^3$) of one mole of the solid element. A graph of atomic volume against atomic number is shown in Fig. 2.7.

Similar curves are obtained if other properties such as melting point, boiling point, enthalpy of fusion are plotted against atomic number.

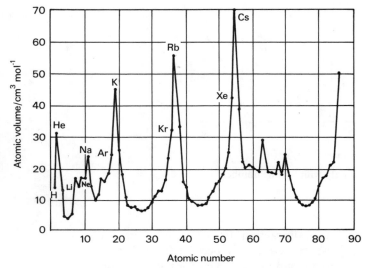

**Fig. 2.7** Graph of atomic volume against atomic number

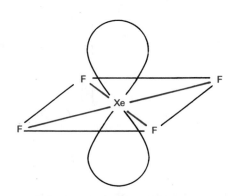

**Fig. 2.8** Structure of xenon tetrafluoride

## 2.9 DIAGONAL RELATIONSHIPS

We will see obvious relationships between elements in the same Group of the Periodic Table. There are also trends within any Period.

In addition to these, there are similarities between elements diagonally related in the Periodic Table. The best example of this is probably the relationship between lithium and magnesium. The diagonal relationship is, however, limited to elements in the top left hand corner of the Periodic Table.

Lithium and magnesium have similar electronegativities and the atomic and ionic radii are similar. As a result of these similarities, there are considerable similarities in the chemistry of these two elements. For example:

(*i*) Lithium and magnesium form nitrides when burnt in nitrogen. These solid nitrides react with water to form ammonia.

$$6Li(s) + N_2(g) \rightarrow 2Li_3N(s)$$
$$3Mg(s) + N_2(g) \rightarrow Mg_3N_2(s)$$
$$Li_3N(s) + 3H_2O(l) \rightarrow 3LiOH(aq) + NH_3(g)$$
$$Mg_3N_2(s) + 6H_2O(l) \rightarrow 3Mg(OH)_2(s) + 2NH_3(g)$$

(*ii*) Many of the salts of lithium and magnesium are insoluble in water. In this lithium differs from other alkali metals. Lithium chloride and magnesium chloride are soluble in organic solvents showing a tendency to covalent bonding.

Other common examples of diagonal relationship are given by beryllium and aluminium and boron and silicon.

### 2.10  GROUP 0 ELEMENTS

The noble gas elements were discovered after the Periodic Table had been devised. Their lack of reactivity was the reason for the delay in their discovery. The gases are helium He, neon Ne, argon Ar, krypton Kr, xenon Xe and radon Rn, and all are present in air.

The gases are all monatomic and sparingly soluble in water. The melting and boiling points increase regularly with increasing atomic number.

Compounds of noble gases are difficult to produce. The best known compound is xenon tetrafluoride which is prepared by heating xenon and fluorine together at 400°C under a pressure of 13 atmospheres.

$$Xe(g) + 2F_2(g) \rightarrow XeF_4(s)$$

Xenon tetrafluoride is a colourless crystalline solid. It is hydrolysed by water to produce xenon(VI) oxide.

$$6XeF_4(s) + 12H_2O(l) \rightarrow 4Xe(g) + 3O_2(g) + 24HF(g) + 2XeO_3(s)$$

The structure of xenon tetrafluoride is shown in Fig. 2.8.

### QUESTIONS

1 The table below gives the ionisation energies of lithium, sodium and potassium.

| Ionisation energy kJ mol$^{-1}$ | 1st | 2nd | 3rd | 4th |
|---|---|---|---|---|
| lithium | 526 | 7310 | 11 800 | —— |
| sodium | 502 | 4560 | 6920 | 9540 |
| potassium | 425 | 3060 | 4440 | 5880 |

(*a*) Why does the first ionization energy decrease in the order lithium → sodium → potassium?

(*b*) Why is the second ionization energy for each element much greater than the first ionization energy?

(*c*) Why is there no value given for the fourth ionization energy of lithium?

*(Scottish Higher)*

(*a*) The electronic arrangements of the three elements are:

Li    $1s^22s^1$

Na   $1s^22s^22p^63s^1$

K    $1s^22s^22p^63s^23p^64s^1$

In each case, there is a single electron in the highest energy level. This electron is lost more easily from potassium because the potassium atom is larger and the electron is further from the positively charged nucleus. Also, the filled lower energy orbitals provide shielding of the nucleus. These two factors compensate for the extra protons in the nucleus.

(*b*) The second ionization energy is much larger in each case because the second electron comes from a lower energy level. For lithium, the first electron lost comes from the $2s$ orbital but the second comes from the $1s$ orbital. More energy is required to remove this.

(*c*) There is no fourth ionisation energy because a lithium atom only contains three electrons.

2 The table below shows the first eight ionization energies of four different elements labelled A–D. (These are not chemical symbols). For each element, state the Group of the Periodic Table to which it belongs. Give a brief explanation of your answer.

| | Ionisation energy/kJ mol$^{-1}$ | | | | | | | |
|---|---|---|---|---|---|---|---|---|
| | 1st | 2nd | 3rd | 4th | 5th | 6th | 7th | 8th |
| A | 418 | 3070 | 4600 | 5860 | 7990 | 9620 | 11 400 | 14 900 |
| B | 786 | 1580 | 3230 | 4360 | 16 000 | 20 000 | 23 600 | 29 100 |
| C | 1680 | 3400 | 6000 | 8400 | 11 000 | 15 200 | 17 900 | 92 000 |
| D | 2080 | 3950 | 6150 | 9290 | 12 100 | 15 200 | 20 000 | 23 000 |

**A** This element is in Group I. The first electron can be removed easily, but it is comparatively difficult to remove the second electron.

**B** This element is in Group IV. Four electrons can be removed ($s^2p^2$) before the sharp rise in ionization energy occurs.

**C** Group VII. Seven electrons can be removed before the sharp rise. The element has an electronic arrangement $s^2p^5$.

**D** Group 0. There is only a gradual change in ionization energy with no sharp break.

3 Explain why the first electron affinity of a sulphur atom has a negative value ($-199.5$kJ mol$^{-1}$) but the second electron affinity is positive ($+648.5$kJ mol$^{-1}$).

First electron affinity of sulphur atom:

$$S(g) + e^- \rightarrow S^-(g)$$

Second electron affinity:

$$S^-(g) + e^- \rightarrow S^{2-}(g)$$

The second process is endothermic because it involves the repulsion of a negatively charged electron by a negatively charged ion. This repulsion makes the process unlikely. There is no repulsion in the first process.

4 This question concerns the following elements whose electronegativities are listed.

Al $= 1.5$    C $= 2.5$    H $= 2.1$    N $= 3.0$    P $= 2.1$
B $= 2.0$    Cl $= 3.0$    Li $= 1.0$    Na $= 0.9$    S $= 2.5$
Be $= 1.5$    F $= 4.0$    Mg $= 1.2$    O $= 3.5$    Si $= 1.8$

(*a*) What is meant by the term electronegativity? What factors determine the electronegativity of an element?

(*b*) Arrange the elements according to their positions in the Periodic Table.

(*c*) What is the relationship between electronegativity and the position of elements (*i*) in a Period (*ii*) in a Group? In each case give a brief explanation.

(*d*) In what major respect do electronegativity and electron affinity differ?

(*e*) Arrange the following substances in order of increasing ionic character (*i.e.* putting the least ionic first).

$$CO_2, \; LiCl, \; MgF_2, \; NaCl, \; NH_3, \; S_2Cl_2$$

(*a*) Electronegativity of an atom is the ability of an atom to attract electrons to itself. The factors which affect the electronegativity of the element are (*i*) atomic radius; (*ii*) nuclear charge and (*iii*) screening by lower orbitals.

(*b*) In this examination, no copy of the Periodic Table was supplied. This part had to be attempted from memory.

| | | | | | | | |
|------|------|---|------|------|------|------|------|
| | | | | H | | | |
| | | | | 2.1 | | | |
| Li | Be | | B | C | N | O | F |
| 1.0 | 1.5 | | 2.0 | 2.5 | 3.0 | 3.5 | 4.0 |
| Na | Mg | | Al | Si | P | S | Cl |
| 0.9 | 1.2 | | 1.5 | 1.8 | 2.1 | 2.5 | 3.0 |

(*c*) (*i*) Across a Period there is an increase in electronegativity. This can be explained by increasing nuclear charge and decreasing atomic radius.

(*ii*) Down a Group there is a decrease in electronegativity. This can be explained by increasing atomic radius and increased screening. This overcomes the additional nuclear charge.

(*d*) Electronegativity is only a tendency to attract electrons and it is measured in arbitrary units. Electron affinity is the energy liberated or absorbed when an electron is **added**.

$$X(g) + e^- \rightarrow X^-(g)$$

It has units of kJ mol$^{-1}$.

(*e*) See Unit 4. The type of bonding depends upon the difference in electronegativity between the two atoms. The bigger the difference in electronegativity the more ionic the bonding between two atoms.

$$CO_2 \; 1.0, \quad LiCl \; 2.0, \quad MgF_2 \; 2.8, \quad NaCl \; 2.1, \quad NH_3 \; 0.9, \quad S_2Cl_2 \; 0.5$$

The order is therefore:

$$S_2Cl_2, \; NH_3, \; CO_2, \; LiCl, \; NaCl, \; MgF_2$$

least ionic $\longrightarrow$ most ionic

5 Which one of the following has the greatest first ionization energy?

     A   H
     B   Ne
     C   Na
     D   C
     E   Cl

Answer B. Neon is a member of the noble gas family (Group 0).

# 3 Symbols, formulae and equations

## 3.1 INTRODUCTION

In order to obtain a grade C at GCSE, you should have used chemical symbols and formulae. You will also have written chemical equations and used them in simple calculations of quantities required for reaction.

I make no apology for covering this work again and in detail. When marking A-level scripts it is surprising how many elementary mistakes involving symbols, formulae and equations are seen.

## 3.2 SYMBOLS AND FORMULAE

Each **element** is represented by a symbol – either a capital letter alone or a capital letter and a small letter. The complete list of symbols can be obtained by reference to the Periodic Table (Fig. 2.1 on page 21).

Each **compound** is represented by a formula which gives the proportions of the different elements it contains. For example, 24 g of magnesium combines with 16 g of oxygen in every 40 g sample of magnesium oxide. The formula is MgO, showing that 1 mole of magnesium atoms (24 g) combines with 1 mole of oxygen atoms (16 g).

The formula of many compounds can be found by use of the list of ions in Table 3.1.

**Table 3.1 Common ions**

| Positive ions | Negative ions |
|---|---|
| lithium $Li^+$ | fluoride $F^-$ |
| sodium $Na^+$ | chloride $Cl^-$ |
| potassium $K^+$ | bromide $Br^-$ |
| silver $Ag^+$ | iodide $I^-$ |
| copper(II) $Cu^{2+}$ | hydroxide $OH^-$ |
| lead(II) $Pb^{2+}$ | *nitrate $NO_3^-$ (nitrate(V)) |
| magnesium $Mg^{2+}$ | *nitrite $NO_2^-$ (nitrate(III)) |
| calcium $Ca^{2+}$ | carbonate $CO_3^{2-}$ |
| strontium $Sr^{2+}$ | hydrogencarbonate $HCO_3^-$ |
| barium $Ba^{2+}$ | *sulphate $SO_4^{2-}$ (sulphate(VI)) |
| zinc $Zn^{2+}$ | hydrogensulphate $HSO_4^-$ |
| aluminium $Al^{3+}$ | *sulphite $SO_3^{2-}$ (sulphate(IV)) |
| iron(II) $Fe^{2+}$ | oxide $O^{2-}$ |
| iron(III) $Fe^{3+}$ | sulphide $S^{2-}$ |
| chromium(III) $Cr^{3+}$ | *phosphate $PO_4^{3-}$ (phosphate(V)) |
| manganese(II) $Mn^{2+}$ | manganate(VII) $MnO_4^-$ |
| cobalt(II) $Co^{2+}$ | ethanedioate $C_2O_4^{2-}$ |
| nickel(II) $Ni^{2+}$ | chlorate(V) $ClO_3^-$ |
| ammonium $NH_4^+$ | chlorate(I) $OCl^-$ |
| **hydrogen $H^+$ | chromate(VI) $CrO_4^{2-}$ |
| | dichromate(VI) $Cr_2O_7^{2-}$ |

*Notes* *Ions marked can be named in alternative ways. The systematic name is given in brackets, but the first name given is still preferred by IUPAC, national authorities and most Examination Boards. There are alternative names also for the parent acids
*e.g.* sulphuric acid (sulphuric(VI) acid)
    sulphurous acid (sulphuric(IV) acid)
    nitric acid (nitric(V) acid)
    nitrous acid (nitric(III) acid).

**The hydrogen ion may be written as $H^+$ or in the hydrated form as $H_3O^+$ (oxonium ion) or $H^+(aq)$.

In forming the compound, the number of ions used is such that the number of positive charges equals the number of negative charges. For example, sodium chloride is made up from $Na^+$ and $Cl^-$ ions. Since a sodium ion has a single positive charge and a chloride ion has a single negative charge, the formula of sodium chloride is NaCl. Sodium sulphate is made up from $Na^+$ and $SO_4^{2-}$ ions. Twice as many sodium ions as sulphate ions are necessary in order to have equal numbers of positive and negative charges. The formula is $Na_2SO_4$. Chromium(III) sulphate is made up from $Cr^{3+}$ and $SO_4^{2-}$ ions and its formula is $Cr_2(SO_4)_3$. Table 3.2 contains further examples.

**Table 3.2 Finding the formula**

| Compound | Ions present | | Formula |
|---|---|---|---|
| copper(II) oxide | $Cu^{2+}$ | $O^{2-}$ | $CuO$ |
| ammonium sulphate | $NH_4^+$ | $SO_4^{2-}$ | $(NH_4)_2SO_4$ |
| potassium manganate(VII) | $K^+$ | $MnO_4^-$ | $KMnO_4$ |
| calcium ethanedioate | $Ca^{2+}$ | $C_2O_4^{2-}$ | $CaC_2O_4$ |
| chromium(III) oxide | $Cr^{3+}$ | $O^{2-}$ | $Cr_2O_3$ |
| hydrochloric acid | $H^+$ | $Cl^-$ | $HCl$ |
| nitric acid | $H^+$ | $NO_3^-$ | $HNO_3$ |
| sulphuric acid | $H^+$ | $SO_4^{2-}$ | $H_2SO_4$ |

NB  *(i)* All acids contain hydrogen ions (but see also Unit 16).

*(ii)* A small number after a bracket multiplies everything inside the bracket, *e.g.* $(NH_4)_2SO_4$ is composed of three ions–two $NH_4$ ions and one $SO_4^-$ ion. Overall there are two nitrogen atoms, eight hydrogen atoms, one sulphur atom and four oxygen atoms.

All of the compounds above are composed of ions. Many compounds are not ionized. The formulae of some of these compounds are shown in Table 3.3.

**Table 3.3 Some compounds not composed of ions**

| Compound | Formula | Compound | Formula |
|---|---|---|---|
| Water | $H_2O$ | Sulphur dioxide | $SO_2$ |
| Carbon dioxide | $CO_2$ | Sulphur(VI) oxide | $SO_3$ |
| Carbon monoxide | $CO$ | Ammonia | $NH_3$ |
| Nitrogen monoxide | $NO$ | Hydrogen chloride | $HCl$ |
| Nitrogen dioxide | $NO_2$ | Methane | $CH_4$ |

### 3.3    CHEMICAL EQUATIONS

Chemical equations are widely used in textbooks and examination papers. An equation is a useful summary of a chemical reaction, and it is always theoretically possible to obtain the equation from the results of an experiment. It is advisable to be able to write equations in your answers to A level questions. You should write any relevant equation even if you are not directly asked to do so.

The steps in writing a chemical equation are as follows:

*(i)* Write down the equation as a word equation using either the information given or your memory. Include all reacting substances (reactants) and all products (do not forget small molecules such as water) *e.g.*

calcium hydroxide + hydrochloric acid → calcium chloride + water

*(ii)* Fill in the correct formulae for all the reacting substances and products.

$$Ca(OH)_2 + HCl \rightarrow CaCl_2 + H_2O$$

*(iii)* Now balance the equation. During a chemical reaction, atoms cannot be created or destroyed (Law of conservation of mass). There must be the same total numbers of the different atoms before and after the reaction. When balancing an equation only the *proportions* of the reacting substances and products can be altered – not the formulae.

$$Ca(OH)_2 + 2HCl \rightarrow CaCl_2 + 2H_2O$$

In organic chemistry the equations can be more complex. Frequently, in the interest of clarity, the formulae are drawn out as structural formulae to show clearly the changes that have taken place. Also, less emphasis may be given to balancing the equation. You should however balance the equation if possible.

For example, the reaction between bromoethane and a solution of potassium hydroxide in ethanol can be written.

$$CH_3CH_2Br \xrightarrow{-HBr} C_2H_4$$

It is better to write

$$H-\underset{\underset{H}{|}}{\overset{\overset{H}{|}}{C}}-\underset{\underset{H}{|}}{\overset{\overset{H}{|}}{C}}-Br + KOH \rightarrow \underset{\underset{H}{\diagdown}}{\overset{\overset{H}{\diagup}}{C}}=\underset{\underset{H}{\diagup}}{\overset{\overset{H}{\diagdown}}{C}} + KBr + H_2O$$

bromoethane          ethene

This equation is balanced and shows clearly the change taking place *i.e.* elimination of HBr.

(*iv*) Finally, the states of reacting substances and products can be included in small brackets after the formulae.

(s)  for solid (sometimes (c) is seen for crystalline solid)
(l)  for liquid
(g)  for gas
(aq) for a solution with water as solvent

These state symbols may not be required and may not give you extra credit from the examiner but they do help your thinking considerably and give a good impression.

$$Ca(OH)_2(aq) + 2HCl(aq) \rightarrow CaCl_2(aq) + 2H_2O(l)$$

Sometimes, state symbols are specifically requested. Then, of course, they must be given. In the interests of simplicity of this text, state symbols are not included throughout the book, unless considered important.

### 3.4  INFORMATION PROVIDED BY AN EQUATION

$$CaCO_3(s) + 2HCl(aq) \rightarrow CaCl_2(aq) + CO_2(g) + H_2O(l)$$

This equation gives the following information: 1 mole of calcium carbonate reacts with 2 moles of hydrochloric acid to produce 1 mole of calcium chloride, 1 mole of carbon dioxide and 1 mole of water. The equation also gives the states of all reactants and products.

In general, a chemical equation gives:
(*i*) the chemicals reacting together and the chemicals produced,
(*ii*) the physical states of reactants and products, and
(*iii*) the quantities of chemicals reacting together and produced.

The equation does not, however, give information about energy changes, the feasibility of reaction, the rate of reaction or the conditions necessary for reaction to take place.

At O level you will have attempted calculations of quantities of chemicals reacting. For example, using the above equation and the relative atomic masses ($A_r(H)=1$, $A_r(C)=12$, $A_r(O)=16$, $A_r(Cl)=35.5$, $A_r(Ca)=40$), calculate the mass of solid calcium chloride which could be produced from 20g of calcium carbonate.

From the equation, 1 mole of calcium carbonate ($CaCO_3$) produces 1 mole of calcium chloride ($CaCl_2$)

$$40 + 12 + (3 \times 16)g\ CaCO_3\ \text{produces}\ 40 + (35.5 \times 2)g\ CaCl_2$$

$$100g\ CaCO_3\ \text{produces}\ 111g\ CaCl_2$$

$$20g\ CaCO_3\ \text{produces}\ \frac{111}{100} \times 20g\ CaCl_2$$

$$= 22.2g\ \text{of}\ CaCl_2$$

You will attempt similar questions at A level. In some cases, for example when considering carbon dioxide as a product, it is more useful to measure the volume of gas produced. This can then be converted into numbers of moles of carbon dioxide by using the information (given on the examination paper) that 1 mole of any gas occupies $22\,400cm^3$ at standard temperature and pressure, stp, (see also Unit 8.4). If the volume is measured at conditions other than stp, and it usually is, it is necessary to convert the volume measured to the volume the gas would occupy at stp (273K and 101.3kPa) using the equation:

$$\frac{P_1 V_1}{T_1} = \frac{P_2 V_2}{T_2} \qquad\qquad \text{(Unit 8.4)}$$

For example, what mass of calcium carbonate would react with dilute hydrochloric acid to produce $133.5cm^3$ of carbon dioxide at 27°C and 100kPa.

Volume of carbon dioxide at stp $= V_1 cm^3$

$$\frac{V_1 \times 101.3}{273} = \frac{133.5 \times 100}{(273 + 27)}$$

$$V_1 = \frac{133.5 \times 100 \times 273}{101.3 \times 300}$$

$$= 120cm^3$$

From the equation,

100g of calcium carbonate (1 mole) produces 22 400cm³ $CO_2$ at stp

$\dfrac{100 \times 120}{22\,400}$g of calcium carbonate produces 120cm³ $CO_2$ at stp

Mass of calcium carbonate = 0.54g

You will also use equations in calculations in Unit 19.

### 3.5 CALCULATION OF PERCENTAGE YIELD

In organic chemistry, a 100% conversion of reactants into products is rarely achieved. From an equation it is possible to calculate a theoretical yield, *i.e.* the mass that would be produced if a 100% conversion were achieved. If you are given the actual mass produced, it is possible to calculate the percentage yield by:

$$\text{Percentage yield} = \frac{\text{mass obtained}}{\text{theoretical yield}} \times 100$$

*Example:* Propan-2-ol reacts with iodine and sodium hydroxide on warming to produce triiodomethane. In an experiment a student obtained 39g of triiodomethane from 10g of propan-2-ol. What was the percentage yield? ($A_r(H)=1$, $A_r(C)=12$, $A_r(O)=16$, $A_r(Na)=23$, $A_r(I)=127$).

$$CH_3CH(OH)CH_3 + 4I_2 + 6NaOH \rightarrow CHI_3 + CH_3COONa + 5NaI + 5H_2O$$

Mass of 1 mole of propan-2-ol = $(12 \times 3) + 8 + 16g = 60g$
Mass of 1 mole of triiodomethane = $12 + 1 + (3 \times 127) = 394g$
From 60g of propan-2-ol, 394g of triiodomethane would be produced if 100% conversion.

Theoretical mass of triiodomethane from 10g propan-2-ol = $\dfrac{394}{6}$g = 65.66g

$$\text{Percentage yield} = \frac{39}{65.66} \times 100$$
$$= 59.4\%$$

### 3.6 IONIC EQUATIONS

You may well have written ionic equations at O level but they become much more significant at A level. Ionic equations are useful because they emphasize the important changes taking place in a chemical reaction.

Taking the reaction between acidified potassium manganate(VII) and iron(II) sulphate as an example. The full equation is:

$$2\,KMnO_4 + 8H_2SO_4 + 10FeSO_4 \rightarrow 2MnSO_4 + 8H_2O + 5Fe_2(SO_4)_3 + K_2SO_4$$

In this equation there are a number of ions which take no part in the chemistry; they are present before and after the reaction. These tend to confuse the change taking place and can often be ignored.
Expanding the equation to show the ions present.

$$2(K^+MnO_4^-) + 8(2H^+SO_4^{2-}) + 10(Fe^{2+}SO_4^{2-}) \rightarrow 2(Mn^{2+}SO_4^{2-}) + 8H_2O + 5(2Fe^{3+}3SO_4^{2-})$$
$$+ 2K^+SO_4^{2-}$$

Cross out ions appearing on both sides of the equation

$$2MnO_4^- + 16H^+ + 10Fe^{2+} \rightarrow 2Mn^{2+} + 8H_2O + 10Fe^{3+}$$

The equation can be divided through by 2 to give the simplest equation.

$$MnO_4^- + 8H^+ + 5Fe^{2+} \rightarrow Mn^{2+} + 4H_2O + 5Fe^{3+}$$

This represents the simplest ionic equation for this reaction. The equation must be balanced in the usual way with equal numbers of each type of atom on left and right hand sides. In addition, the algebraic sum of the charges on the left hand side equals that of the right hand side.
In the example,

$$\begin{aligned} LHS \quad &(-) + 8(+) + 5(2+) = 17(+) \\ RHS \quad &(2+) + 4(0) + 5(3+) = 17(+) \end{aligned}$$

The ionic equation can be broken down into 2 half equations which added together produce the overall equation.

(*i*) $\qquad MnO_4^- + 8H^+ + 5e^- \rightarrow Mn^{2+} + 4H_2O$
(*ii*) $\qquad\qquad\qquad\quad 5Fe^{2+} \rightarrow 5Fe^{3+} + 5e^-$

---

$$MnO_4^- + 8H^+ + 5Fe^{2+} \rightarrow Mn^{2+} + 4H_2O + 5Fe^{3+}$$

Equation (*i*) represents the reduction of manganese(VII) to manganese(II). Equation (*ii*) represents the oxidation of iron(II) to iron(III). A list of common half equations and their use in building up ionic equations is given in Unit 6.

QUESTIONS

1 Write down the correct chemical formulae for the following compounds:

(*i*) sodium carbonate; (*ii*) lead(II) nitrate; (*iii*) potassium hydrogencarbonate; (*iv*) silver nitrate; (*v*) ammonium chloride; (*vi*) calcium hydroxide; (*vii*) iron(III) hydroxide; (*viii*) cobalt(II) chloride; (*ix*) potassium dichromate(VI); (*x*) sodium chlorate(I).

Answers: (*i*) $Na_2CO_3$; (*ii*) $Pb(NO_3)_2$; (*iii*) $KHCO_3$; (*iv*) $AgNO_3$; (*v*) $NH_4Cl$; (*vi*) $Ca(OH)_2$; (*vii*) $Fe(OH)_3$; (*viii*) $CoCl_2$; (*ix*) $K_2Cr_2O_7$; (*x*) NaOCl.

2 Balance the equations in the following reactions:

(*i*) $\quad\quad\quad\quad\quad\quad\quad\quad Zn(s) + Fe^{3+}(aq) \rightarrow Zn^{2+}(aq) + Fe^{2+}(aq)$

(*ii*) $\quad\quad\quad\quad\quad\quad\quad Sn(s) + HNO_3(l) \rightarrow SnO_2(s) + NO_2(g) + H_2O(l)$

(*iii*) $\quad\quad\quad\quad\quad\quad\quad Cu^{2+}(aq) + I^-(aq) \rightarrow Cu_2I_2(s) + I_2(aq)$

(*iv*) $\quad\quad\quad SO_2(aq) + Br_2(aq) + 2H_2O(l) \rightarrow H^+(aq) + SO_4^{2-}(aq) + Br^-(aq)$

Despite the practice candidates should have had in balancing equations before they take an A level examination, a surprising number of candidates will make mistakes. Remember in (*i*), (*iii*) and (*iv*) where the equations are ionic equations, the charges must also balance.

(*i*) $\quad\quad\quad\quad\quad\quad\quad Zn(s) + 2Fe^{3+}(aq) \rightarrow Zn^{2+}(aq) + 2Fe^{2+}(aq)$

(*ii*) $\quad\quad\quad\quad\quad\quad Sn(s) + 4HNO_3(l) \rightarrow SnO_2(s) + 4NO_2(g) + 2H_2O(l)$

(*iii*) $\quad\quad\quad\quad\quad\quad 2Cu^{2+}(aq) + 4I^-(aq) \rightarrow Cu_2I_2(s) + I_2(aq)$

(*iv*) $\quad\quad SO_2(aq) + Br_2(aq) + 2H_2O(l) \rightarrow 4H^+(aq) + SO_4^{2-}(aq) + 2Br^-(aq)$

3 A compound $PBr_x$ contains 88.6% by mass of bromine. Deduce the formula of $PBr_x$.

$$(A_r(P) = 31, \ A_r(Br) = 80)$$

*(Oxford and Cambridge)*

This is similar to questions which appear on O level papers.

In a 100g sample of the bromide of phosphorus,

88.6g of bromine combine with (100 − 88.6)g of phosphorus.

88.6g of bromine combine with 11.4g of phosphorus.

31g of phosphorus combine with $\dfrac{88.6}{11.4} \times 31$g of bromine

$$= 240.9\text{g of bromine}$$

Dividing 240.9g by the relative atomic mass of bromine to find the number of moles of bromine atoms = 3 (to nearest whole number).

The phosphorus bromide has a formula $PBr_3$.

# 4 Chemical bonding

## 4.1 INTRODUCTION

In the GCSE course you will have encountered two methods of bonding or joining atoms together—**ionic** and **covalent** bonding. These are only two of the ways in which atoms can be held together. In this unit they will be considered in further detail and, in addition, other forces which hold particles together will be considered.

There are six types of bonding considered in this unit—**ionic, covalent, coordinate, hydrogen bonding, dipole-dipole forces** and **metallic bonding**. The nature of the forces of attraction between particles will have a considerable bearing on the properties of the substance and this relationship is particularly important at A level.

In any atom it is the electrons in the **highest energy orbitals** that are involved in the process of bonding. These electrons are sometimes called **valency electrons**. In Unit 1 electronic configurations of elements are discussed. The electronic configurations of the noble gases are as follows:

| | | | |
|---|---|---|---|
| Helium | $1s^2$ | Krypton | $1s^2\,2s^2\,2p^6\,3s^2\,3p^6\,4s^2\,3d^{10}\,4p^6$ |
| Neon | $1s^2\,2s^2\,2p^6$ | Xenon | $1s^2\,2s^2\,2p^6\,3s^2\,3p^6\,4s^2\,3d^{10}\,4p^6\,5s^2\,4d^{10}\,5p^6$ |
| Argon | $1s^2\,2s^2\,2p^6\,3s^2\,3p^6$ | | etc. |

These electronic configurations are regarded as particularly stable and are not easily broken down. The electronic theory of bonding states that, on bonding, atoms attempt to achieve noble gas electronic configurations.

This unit on bonding is closely linked with Units 1, 3 and 14.

### 4.2 IONIC (OR ELECTROVALENT) BONDING

Atoms can achieve noble gas electronic configurations by loss or gain of electrons to form ions. Metals (with low electronegativities) lose electrons to form positive ions and non-metals gain electrons.

The most frequently quoted example of ionic bonding is sodium chloride. A complete transfer of one electron from a sodium atom to a chlorine atom leads to the formation of a positive sodium ion and a negative chloride ion. These ions are held together by strong electrostatic forces. Both sodium and chloride ions have electronic configurations identical to noble gas atoms. The changes are summarized in Table 4.1.

**Table 4.1**

| Na | $\xrightarrow{-e^-}$ | Na$^+$ | Cl | $\xrightarrow{+e^-}$ | Cl$^-$ |
|---|---|---|---|---|---|
| $1s^2 2s^2 2p^6 3s^1$ | | $1s^2 2s^2 2p^6$ | $1s^2 2s^2 2p^6 3s^2 3p^5$ | | $1s^2 2s^2 2p^6 3s^2 3p^6$ |

Another example of ionic bonding is magnesium oxide. In this case each magnesium atom loses two electrons to form an ion with a 2 + charge. Each oxygen gains two electrons to form an ion with a 2 − charge. This is is summarized in Table 4.2.

**Table 4.2**

| Mg | $\xrightarrow{-2e^-}$ | Mg$^{2+}$ | O | $\xrightarrow{+2e^-}$ | O$^{2-}$ |
|---|---|---|---|---|---|
| $1s^2 2s^2 2p^6 3s^2$ | | $1s^2 2s^2 2p^6$ | $1s^2 2s^2 2p^4$ | | $1s^2 2s^2 2p^6$ |

A detailed account of the energy changes involved in the formation of an ionic compound will be found in Unit 13.

### 4.3 PROPERTIES OF COMPOUNDS WITH IONIC BONDS

It is incorrect to speak of, for example, a molecule of sodium chloride (NaCl) as this suggests that one particular sodium ion is associated with one particular chloride ion. Sodium chloride consists of a regular three dimensional arrangement of sodium and chloride ions called a **lattice** (Fig. 10.5). Within the sodium chloride lattice, each sodium ion is surrounded by six chloride ions and each chloride ion by six sodium ions.

The electrostatic forces holding the ions together in the lattice are very strong, and a large amount of energy (called the **lattice energy**) is required to break up the lattice. As a consequence of the high lattice energies of ionic compounds, these compounds have high melting and boiling points. The lattice energies of sodium chloride and magnesium oxide are 771 and 3889 kJ mol$^{-1}$ respectively. The much higher value for magnesium oxide can be explained because the ions have 2+ and 2− charges and therefore the forces of attraction are greater. The melting points of these two compounds are approximately 800°C and 1400°C.

Although different crystal lattices are possible, magnesium oxide and sodium chloride have similar lattice structures. Caesium chloride (CsCl) has a different crystal structure in which each caseium ion is surrounded by eight chloride ions (Fig. 10.6). A caesium ion is considerably larger than a sodium ion and it is geometrically possible to pack eight chloride ions around a caesium ion but not a sodium ion.

The particular lattice structure for an ionic compound is determined by the **relative radii** of the two ions (Unit 10).

Water and other polar liquids are very good solvents for compounds containing ionic bonds. This process is explained using water as the solvent in the simplified diagrams (Fig. 4.1). These are only two-dimensional representations and they cannot show the movement of the particles.

The energy released when ions are surrounded by polar solvent molecules is called **solvation energy**. If the solvent is water this energy may be called the **hydration energy**.

The solvation energy released when the ions are surrounded by solvent molecules provides the energy required to break up the lattice (lattice energy). In a compound with a very high lattice energy (*e.g.* calcium fluoride) the solvation energy may not be sufficient to break down the lattice and, in this case, the compound will be insoluble.

Compounds containing ionic bonding are electrolytes: when molten or in solution they conduct electricity and undergo electrolysis. The electricity is carried by ions released and able to move when the compound melts or dissolves.

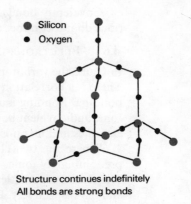

Fig. 4.1 Dissolving an ionic crystal    Fig. 4.2 Silicon(IV) oxide structure

Ionic compounds do not dissolve in non-polar liquids such as methylbenzene and hexane. Polar liquids contain slight positive and negative charges called **dipoles** (Unit 4.6) but these charges are not present to any degree in non polar liquids. They cannot, therefore, break down ionic lattices in the same way.

### 4.4    COVALENT BONDING

The simplest case of covalent bonding is within a hydrogen molecule ($H_2$). Each hydrogen atom has a single electron in a $1s$ orbital. It is highly unlikely that one hydrogen atom will lose an electron and the other atom gain an electron to produce ions. Instead each atom donates its electron to form a **shared pair** of electrons in a **molecular orbital** between the two hydrogen atoms. This is called a **covalent bond** and each hydrogen atom has a share of two electrons. (Remember a helium atom has an electronic configuration $1s^2$). The bonding can be represented by the simplified diagrams:

$$H \overset{\times}{\cdot} H \qquad H—H$$

The small crosses represent valency electrons and the single stroke represents a covalent bond. Other simple examples of covalent bonding are shown in Table 4.3.

**Table 4.3**

|  methane  |  water  |  carbon dioxide  |  chlorine  |

|  ethane  |  ethene  |  ethyne  |

In all the above examples covalent bonding leads to the formation of small molecules. It is also, however, possible for covalent bonding to produce an extremely large arrangement of atoms called a **giant structure**. Silicon(IV) oxide (silicon dioxide) is an example of this and can be represented in a simplified form (Fig. 4.2).

### 4.5    PROPERTIES OF COVALENT COMPOUNDS

Substances containing covalent bonding, unless they are giant structures, are relatively small molecules and the forces between the molecules (Unit 4.9) are weak. As a result these substances have low melting and boiling points. At room temperature and pressure they are likely to be gases, volatile liquids or low melting point solids.

Solubility in water is usually low although some covalently bonded compounds are hydrolysed by water (Unit 26). The bonding in dry hydrogen chloride changes from covalent to ionic when dissolved in water.

$$H - Cl(g) \rightarrow H^+(aq) + Cl^-(aq)$$

The energy required to break the covalent bond is provided by the hydration energy of the ions produced.

Covalently bonded substances usually dissolve in non-polar or organic liquids. These solutions do not conduct electricity.

## 4.6 ELECTRONEGATIVITY AND ITS RELATIONSHIP TO BOND TYPE

In Unit 2 the variation of electronegativity in the main block of the Periodic Table was discussed. There is a correlation between the electronegativities of the combining elements and the type of bonding. Bonding is rarely entirely ionic or covalent but it is usually some mixture of the two. Ionic and covalent bonding are extremes of bonding. In the same way black and white paints are extremes and it is possible to obtain all shades of grey by mixing these extremes.

Where the two elements combined differ greatly in electronegativity the bonding will be predominately ionic. If the electronegativities are similar the bonding will be largely covalent. The electronegativities (according to Pauling) of potassium, aluminium and bromine are 0.8, 1.5 and 2.8 respectively. Potassium bromide (difference in electronegativity between potassium and bromine = 2.0) is predominately ionic bonding (about 63% ionic) while aluminium bromide (electronegativity difference = 1.3) is predominately covalent (about 66% covalent).

Fajans (1923) produced rules which enabled predictions to be made about which ionic compounds would have appreciably covalent tendencies, *i.e.* show **polarization**. The factors which produce polarization are:

(*i*) if either the positive or negative ion is highly charged,
(*ii*) if the positive ion is small, or
(*iii*) if the negative ion is large.

Consider three chlorides of the alkali metals – lithium chloride LiCl, sodium chloride NaCl and potassium chloride KCl. The greatest covalent character will occur in lithium chloride because the $Li^+$ ion is smaller than the $Na^+$ or $K^+$.

In a covalent bond where the two atoms combined have equal or very similar electronegativities it would be reasonable to assume that the shared pair of electrons in the bond is evenly distributed between the two atoms. Where the two atoms differ greatly in electronegativity it is reasonable to assume that the pair of electrons in the covalent bond is more closely associated with the more electronegative atom. This slight shift of electrons within the covalent bond produces slight positive and negative charges (represented by $\delta+$ and $\delta-$). Examples of this include

$$\underset{\delta+H \qquad\qquad H\delta+}{O^{\delta-}} \quad \text{and} \quad \underset{\delta+ \quad \delta-}{H-F}$$

The slight charges in these molecules can lead to the formation of **hydrogen bonds** (Unit 4.8).

## 4.7 COORDINATE BONDING (OR DATIVE COVALENCY)

**Coordinate bonding** is a type of covalent bonding. In this case, a pair of electrons is shared between two atoms to form a bond but one atom supplies both electrons and the other atom supplies none.

An example of coordinate bonding is the compound formed between boron trifluoride $BF_3$ and ammonia $NH_3$. Both electrons in the N and B bond are donated by the nitrogen atom. The molecule produced is

$$\begin{array}{ccc} H & F \\ | & | \\ H-N & \to B-F \\ | & | \\ H & F \end{array}$$

As a result of the coordinate bond formation there is a slight positive charge on the nitrogen atom and a slight negative charge on the boron atom.

Other examples of coordinate bonding are:

(*i*) Ammonium ion

$$\left[ \begin{array}{c} H \\ | \\ H-N \to H \\ | \\ H \end{array} \right]^+ \quad \text{Lone pair of electrons on nitrogen donated to form coordinate bond.}$$

(*ii*) Oxonium ion

$$\left[ \begin{array}{c} H \\ \diagdown \\ O \to H \\ \diagup \\ H \end{array} \right]^+ \quad \text{Lone pair of electrons on oxygen donated to form the coordinate bond.}$$

(*iii*) Aluminium chloride (see Unit 25).

There are many examples of coordinate bonding with transition metal ions (Unit 30).

## 4.8 HYDROGEN BONDING

Hydrogen bonding is more commonly encountered *between* molecules (**intermolecular**) rather than *within* molecules (**intramolecular**). It occurs in compounds whose molecules consist of a hydrogen atom covalently bonded to an electronegative atom—usually fluorine, oxygen or nitrogen. There are slight charges within the molecules caused by a slight shift of electrons in the covalent bond (Unit 4.6). As a result of these charges, weak electrostatic attractions exist between molecules and these are called **hydrogen bonds**

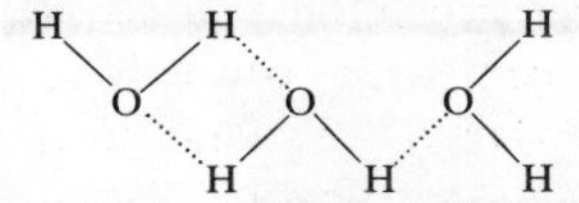

$$\underset{H}{\overset{H}{|}} \quad \underset{H}{\overset{H}{|}} \quad \underset{H}{\overset{H}{|}}$$

This association of molecules has no effect on chemical properties but alters physical properties. For example, the boiling point is increased as extra energy is required to break these bonds before the molecules can escape from the liquid and boiling take place. These extra bonds also increase the viscosity of the liquid.

Organic acids such as ethanoic acid and benzenecarboxylic acid exist as dimers when in non-polar solvents such as methylbenzene

## 4.9 DIPOLE-DIPOLE ATTRACTIONS

There are permanent forces of attraction in polar molecules which contain permanent dipoles. For example, in propanone.

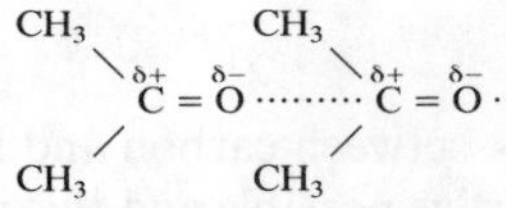

$$\underset{CH_3}{\overset{CH_3}{\diagdown}} \overset{\delta+}{C} = \overset{\delta-}{O} \cdots\cdots\cdots \underset{CH_3}{\overset{CH_3}{\diagdown}} \overset{\delta+}{C} = \overset{\delta-}{O} \cdots$$

The weakest forms of attraction between molecules are found in alkanes, noble gases etc. There are no permanent charges in these molecules. However, when two atoms or molecules approach one another, slight charges are momentarily induced due to a slight movement of electrons. There are forces of attraction between these particles. These induced forces are called Van der Waals' forces. They become greater as the number of electrons involved increases.

## 4.10 METALLIC BONDING

The atoms in a metal are closely packed together in a metallic lattice and the bonding within the lattice is very strong. The bonding between the atoms in a metal is rather complicated and cannot be explained by any of the types of bonding discussed.

The metallic lattice can be considered as a regular arrangement of metal ions. The outer valency electrons have been lost and form a 'sea' occupying the spaces between the ions, binding them together. The electrons are in orbitals delocalized throughout the lattice. This bonding accounts for the good electrical conductivity of heat and electricity in metals. A study of possible metal structures will be found in Unit 10.

**1** Pentane (boiling point 36°C) and 2,2-dimethylpropane (boiling point 9°C) are isomers. Explain why there is this difference in boiling points.

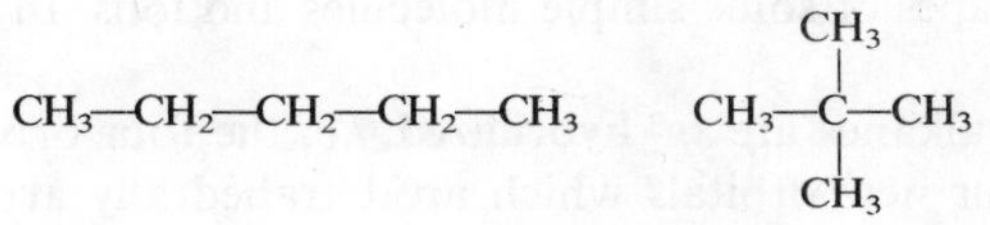

$$CH_3-CH_2-CH_2-CH_2-CH_3 \qquad \underset{CH_3}{\overset{CH_3}{\underset{|}{\overset{|}{CH_3-C-CH_3}}}}$$

Pentane is a long linear molecule but 2,2-dimethylpropane, because of branching, is approximately spherical. 2,2-Dimethylpropane molecules have a much smaller surface area to come into contact with other molecules. The van der Waals' forces are stronger between pentane molecules than between 2,2-dimethylpropane molecules. Pentane therefore has a higher boiling point.

# 5 Shapes of molecules

## 5.1 INTRODUCTION

One of the differences between GCSE and A level is the greater emphasis on the shapes of molecules or the spatial arrangements of atoms in a molecule. It is difficult sometimes to represent the three dimensional molecule on a piece of paper, and it is important that you practise drawing these simple molecules before taking the examination.

## 5.2 ELECTRON PAIR REPULSION THEORY

This theory assumes that the electron pairs in the valency shells of the atoms will repel one another and try to get as far apart as possible.

In methane

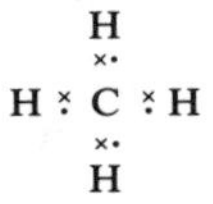

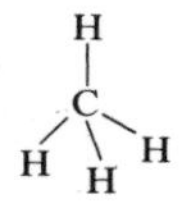

tetrahedral

there are four electron pairs in bonds between carbon and hydrogen atoms. These four electron pairs distribute themselves as far apart as possible and this produces a tetrahedral configuration. The angle between the bonds is 109°28′.

Ammonia contains three nitrogen–hydrogen bonds each containing a pair of electrons and a pair of electrons in a non-bonding orbital.

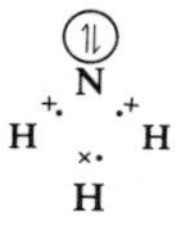

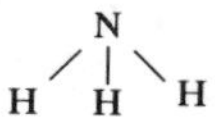

pyramidal

Again there are four electron pairs and the spatial arrangement is approximately tetrahedral. The non-bonding orbital occupies more space than the bonding orbitals. As a result the $H\hat{N}H$ is reduced to approximately 107°.

Water contains two oxygen–hydrogen bonds each containing a pair of electrons and two pairs of electrons in non-bonding orbitals.

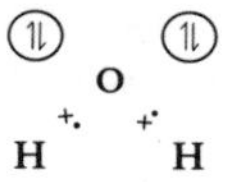

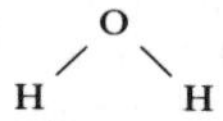

bent

There are still four pairs of electrons and the spatial arrangement is again approximately tetrahedral. Because of the extra space occupied by the non-bonding orbitals, the $H\hat{O}H$ is reduced to approximately 104.5°.

Table 5.1 gives the shapes of some simple molecules and ions. In Unit 28 shapes of *d*-block complexes are given.

The carbon atoms in alkanes are $sp^3$ **hybridized**, *i.e.* the four orbitals ($2s$, $2p_x$, $2p_y$, $2p_z$) are combined to produce four new orbitals which are tetrahedrally arranged around the carbon atom.

In an alkene such as ethene, the carbon atoms are $sp^2$ hybridized ($2s$, $2p_x$, $2p_y$) to produce three new hybrid orbitals all in the same plane with the angles between these orbitals 120°. The $p_z$ orbitals on two adjacent atoms overlap to produce a weak additional bond. The ethene molecule is planar.

In an alkyne such as ethyne, the carbon atoms are $sp$ hybridized ($2s$, $2p_x$) to produce two new hybrid orbitals. The $p_y$ and $p_z$ orbitals on two adjacent carbon atoms can overlap to produce two additional bonds. The molecule is linear.

**Table 5.1**

| Shape | | Examples |
|---|---|---|

Linear — 180° — Cl—Be—Cl   O=C=O   H—C≡N
H—C≡C—H

Bent — $H_2S$ (92°)   $NO_2^-$ (115°)   $ClO_2^-$ (110.5°)
O
H 104.5° H

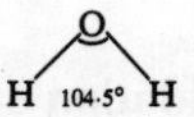

Trigonal planar — 120° — $NO_3^-$   $CO_3^{2-}$   $SO_3$   $C_2H_4$
F
B
F   F

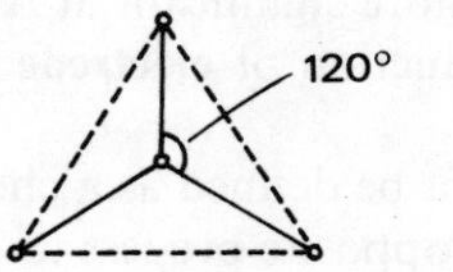

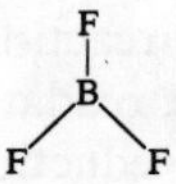

Pyramidal — $PH_3$ (93°)   $ClO_3^-$   $SO_3^{2-}$
N
H 107° H
H

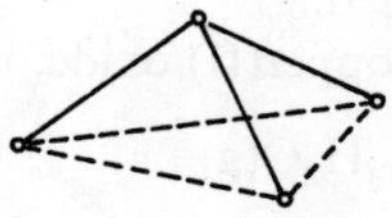

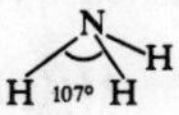

T-shape — Cl—F
F
F

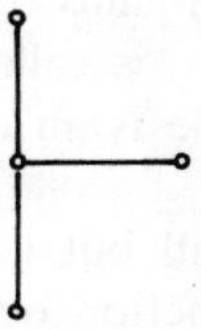

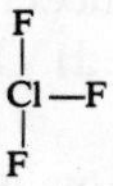

Tetrahedral — 109° 28' — $NH_4^+$   $ClO_4^-$   $SO_4^{2-}$   $PO_4^{3-}$
H
C
H   H
H

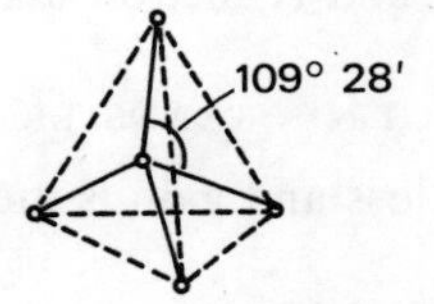

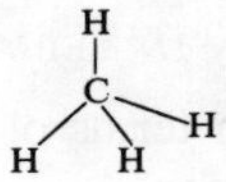

Square planar — $ICl_4^-$
F   F
Xe
F   F

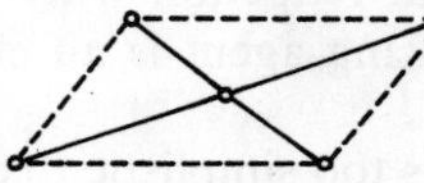

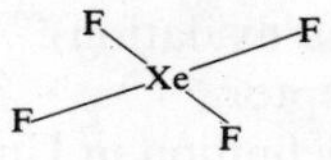

Trigonal bipyramidal — 90°   120° — 
Cl
Cl—P—Cl
Cl
Cl

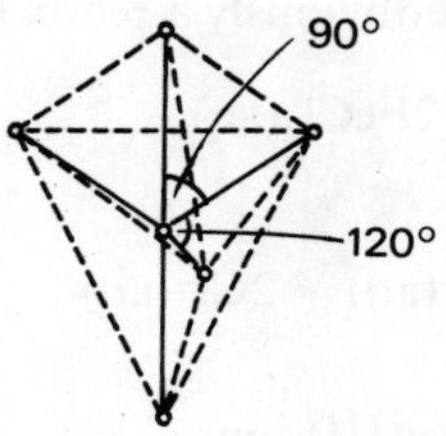

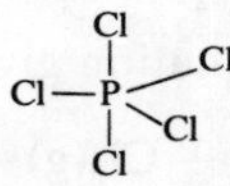

Octahedral — 90° — 
F   F
F   S   F
F   F

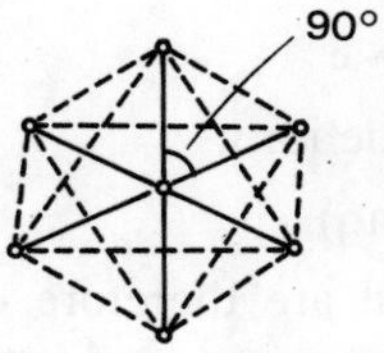

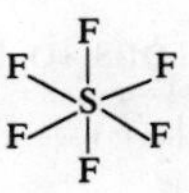

Pentagonal bipyramidal — 90°   72° — 
F   F
F   I   F
F   F
F

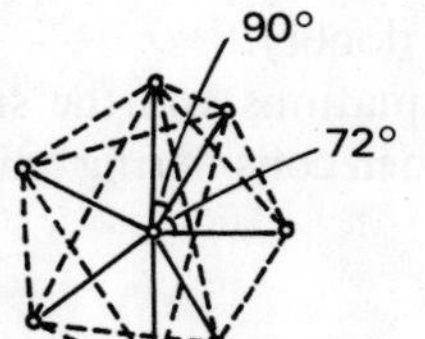

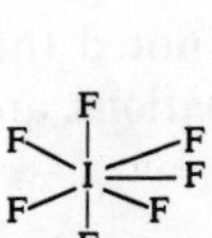

QUESTIONS

**1** Write notes on the shapes of $BF_3$, $NH_3$ and $H_2O$.

See Table 5.1.

(*Welsh Joint Education Committee*)

**2** Which one of the following gives the correct bond angles in a molecule of $NH_3$?

A  90°     D  120°
B  104°    E  180°
C  107°

The correct answer is C–tetrahedral angle, slightly reduced.

# 6 Oxidation and reduction

## 6.1 INTRODUCTION

During a GCSE course the terms **oxidation** and **reduction** should have been used. In this Unit these terms will be clearly defined. They become far more significant at A level. In Unit 18, the subject will be covered quantitatively with the introduction of **electrode potentials**, which can be used to predict the likelihood of a reaction.

In an elementary chemistry course, oxidation might be defined as a chemical reaction where oxygen is gained or hydrogen is lost. Reduction is the opposite process to oxidation. It is, therefore, a reaction where oxygen is lost or hydrogen is gained.

For example when hydrogen is passed over heated copper(II) oxide, the following reaction takes place.

$$CuO(s) + H_2(g) \rightarrow Cu(s) + H_2O(g)$$

The copper(II) oxide loses oxygen and is reduced while hydrogen gains oxygen and is oxidized. The hydrogen, which is required to reduce the copper(II) oxide, is called the reducing agent (remember the reducing agent is oxidized). The copper(II) oxide is an oxidizing agent and is itself reduced.

The elementary treatment of oxidation is easy to understand but it does not include all possible oxidation and reduction reactions. Oxidation and reduction occur together and the term **redox reaction** means a reaction where oxidation and reduction take place.

## 6.2 OXIDATION AND REDUCTION IN TERMS OF ELECTRON TRANSFER

Defining oxidation and reduction in terms of electron loss and gain is more useful, for not all reactions involve oxygen and hydrogen.

Oxidation is a process where electrons are lost and reduction where electrons are gained (Remember: **leo** – loss of electrons oxidation). A reducing agent is an electron donor and an oxidising agent is an electron acceptor.

For example, the elementary definition in Unit 6.1 is too simplistic to cope with the reaction of chlorine with iron(II) chloride solution which is not obviously a redox reaction. The equation is

$$2FeCl_2(aq) + Cl_2(g) \rightarrow 2FeCl_3(aq)$$

Writing this equation as an ionic equation gives

$$2Fe^{2+}(aq) + Cl_2(g) \rightarrow 2Fe^{3+}(aq) + 2Cl^-(aq)$$

There are two processes taking place here:
(*i*) Iron(II) ions $Fe^{2+}$ are losing electrons to form iron(III) ions

$$Fe^{2+}(aq) \rightarrow Fe^{3+}(aq) + e^-$$

(*ii*) Chlorine atoms are gaining electrons to form chloride ions.

$$Cl_2(g) + 2e^- \rightarrow 2Cl^-(aq)$$

During the reaction, iron(II) ions lose electrons and are therefore oxidized and chlorine which gains electrons is reduced. Chlorine is the electron acceptor and is therefore the oxidizing agent and the iron(II) ions the reducing agent (electron donor).

From this example it should be noted that ionic equations and the simple ionic equations (like (*i*) and (*ii*) above), called **half equations**, are useful when considering oxidation and reduction.

## 6.3 COMMON OXIDIZING AGENTS

In this section the common oxidizing agents are listed. It is a good idea to learn the half equation which accompanies each oxidizing agent. In all of these equations electrons are shown on the left hand side since all oxidizing agents are electron acceptors. (NB For the sake of simplicity the ion $H_3O^+$ is represented in these equations by $H^+(aq)$.)

| | |
|---|---|
| oxygen | $O_2(g) + 4e^- \rightarrow 2O^{2-}(s)$ |
| chlorine | $Cl_2(g) + 2e^- \rightarrow 2Cl^-(s)$ |
| bromine | $Br_2(g) + 2e^- \rightarrow 2Br^-(s)$ |
| iodine | $I_2(aq) + 2e^- \rightarrow 2I^-(aq)$ |

manganate(VII) in acid solution

$$MnO_4^-(aq) + 8H^+(aq) + 5e^- \rightarrow Mn^{2+}(aq) + 4H_2O(l)$$

manganate(VII) in alkaline solution

$$MnO_4^-(aq) + 2H_2O(l) + 3e^- \rightarrow MnO_2(s) + 4OH^-(aq)$$

dichromate(VI) in acid solution

$$Cr_2O_7^{2-}(aq) + 14H^+(aq) + 6e^- \rightarrow 2Cr^{3+}(aq) + 7H_2O(l)$$

iron(III) salts

$$Fe^{3+}(aq) + e^- \rightarrow Fe^{2+}(aq)$$

hydrogen ions

$$2H^+(aq) + 2e^- \rightarrow H_2(g)$$

hydrogen peroxide (in the absence of another oxidizing agent)

$$H_2O_2(aq) + 2H^+(aq) + 2e^- \rightarrow 2H_2O(l)$$

manganese(IV) oxide (in the presence of acid)

$$MnO_2(s) + 4H^+(aq) + 2e^- \rightarrow Mn^{2+}(aq) + 2H_2O(l)$$

concentrated sulphuric acid

$$2H_2SO_4(l) + 2e^- \rightarrow SO_4^{2-}(aq) + 2H_2O(l) + SO_2(g)$$

You will find other examples elsewhere in the book.

## 6.4    COMMON REDUCING AGENTS

In a similar way, a list of reducing agents can be drawn up. In all of these half equations, electrons appear on the right hand side of the equation as reducing agents are electron donors.

metals $\quad M(s) \rightarrow M^{n+}(aq) + ne^-$

$\quad\quad e.g.\ Zn(s) \rightarrow Zn^{2+}(aq) + 2e^-$

iron(II) salts

$$Fe^{2+}(aq) \rightarrow Fe^{3+}(aq) + e^-$$

acidified potassium iodide

$$2I^-(aq) \rightarrow I_2(aq) + 2e^-$$

thiosulphate

$$2S_2O_3^{2-}(aq) \rightarrow S_4O_6^{2-}(aq) + 2e^-$$

ethanedioic acid and ethanedioates (oxalic acid and oxalates)

$$C_2O_4^{2-}(aq) \rightarrow 2CO_2(g) + 2e^-$$

hydrogen sulphide and sulphides

$$S^{2-}(aq) \rightarrow S(s) + 2e^-$$

sulphurous acid (sulphuric(IV) acid)

$$SO_3^{2-}(aq) + H_2O(l) \rightarrow SO_4^{2-}(aq) + 2H^+(aq) + 2e^-$$

hydrogen peroxide (in the presence of a strong oxidizing agent)

$$H_2O_2(aq) \rightarrow O_2(g) + 2H^+(aq) + 2e^-$$

tin(II) ions in hydrochloric acid

$$Sn^{2+}(aq) \rightarrow Sn^{4+}(aq) + 2e^-$$

hydrogen $\quad H_2(g) + O^{2-}(s) \rightarrow H_2O(l) + 2e^-$

carbon $\quad\quad C(s) + O^{2-}(s) \rightarrow CO(g) + 2e^-$

carbon monoxide

$$CO(g) + O^{2-}(s) \rightarrow CO_2(g) + 2e^-$$

## 6.5    USING IONIC HALF EQUATIONS TO WRITE FULL IONIC EQUATIONS

Combination of two ionic half equations, one representing an oxidizing action and one representing a reducing action, will produce a full ionic equation. This could be used, if required, to produce a molecular equation. Remember, however, that it is possible to write equations for reactions which do not in fact take place. Whether a reaction takes place depends upon the standard electrode potentials (Unit 18).

*E.g.* reaction of acidified potassium manganate(VII) with iron(II) sulphate solution. Write out the relevant half equations.

OA $\quad\quad MnO_4^-(aq) + 8H^+(aq) + 5e^- \rightarrow Mn^{2+}(aq) + 4H_2O(l)$

RA $\quad\quad\quad\quad\quad\quad\quad\quad Fe^{2+}(aq) \rightarrow Fe^{3+}(aq) + e^-$

On combining these equations the electrons must cancel out. Multiply the RA equation by 5 and add

$$MnO_4^-(aq) + 8H^+(aq) + 5Fe^{2+}(aq) \rightarrow Mn^{2+}(aq) + 4H_2O(l) + 5Fe^{3+}(aq)$$

In this equation, manganate(VII) acts as the oxidizing agent and iron(II) ions as the reducing agent.

The reaction of acidified potassium manganate(VII) with acidified potassium iodide. Write the relevant half equations.

OA $\quad\quad MnO_4^-(aq) + 8H^+(aq) + 5e^- \rightarrow Mn^{2+}(aq) + 4H_2O(l)$

RA $\quad\quad\quad\quad\quad\quad 2I^-(aq) \rightarrow I_2(aq) + 2e^-$

In order to remove the electrons, multiply the OA equation by 2 and RA by 5 and add

$$2MnO_4^-(aq) + 16H^+(aq) + 10I^-(aq) \rightarrow 2Mn^{2+}(aq) + 8H_2O(l) + 5I_2(aq)$$

**6.6     RECOGNIZING OXIDATION AND REDUCTION FROM AN IONIC EQUATION**

If you are given a full ionic equation it is possible to split it up into its constituent half equations.

*E.g.* $\quad\quad\quad\quad Zn(s) + 2H^+(aq) \rightarrow Zn^{2+}(aq) + H_2(g)$

The half equations are:

OA $\quad\quad\quad\quad 2H^+(aq) + 2e^- \rightarrow H_2(g)$

RA $\quad\quad\quad\quad Zn(s) \rightarrow Zn^{2+}(aq) + 2e^-$

In this reaction $H^+(aq)$ ions are reduced and zinc atoms are oxidised.

**6.7     HYDROGEN PEROXIDE AS OXIDIZING AND REDUCING AGENT**

Hydrogen peroxide can act as an oxidizing agent or a reducing agent according to the conditions.

In the absence of a stronger oxidizing agent, hydrogen peroxide acts as an oxidizing agent.
*E.g.* hydrogen peroxide and acidified potassium iodide

OA $\quad\quad\quad H_2O_2(aq) + 2H^+(aq) + 2e^- \rightarrow 2H_2O(l)$

RA $\quad\quad\quad\quad\quad\quad 2I^-(aq) \rightarrow I_2(aq) + 2e^-$

Add $\quad\quad H_2O_2(aq) + 2H^+(aq) + 2I^-(aq) \rightarrow 2H_2O(l) + I_2(aq)$

When a stronger oxidizing agent is present, hydrogen peroxide then acts as a reducing agent.

OA $\quad\quad\quad MnO_4^-(aq) + 8H^+(aq) + 5e^- \rightarrow Mn^{2+}(aq) + 4H_2O(l)$

RA $\quad\quad\quad\quad H_2O_2(aq) \rightarrow O_2(g) + 2H^+(aq) + 2e^-$

Multiply OA equation by 2 and the RA equation by 5 and add:

$$2MnO_4^-(aq) + 16H^+(aq) + 5H_2O_2(aq) \rightarrow 2Mn^{2+}(aq) + 8H_2O(l) + 5O_2(g) + 10H^+(aq)$$

This equation can be slightly simplified. There are $16H^+(aq)$ on the left hand side and $10H^+(aq)$ on the right hand side. An equation should show change and $10H^+(aq)$ can be subtracted from both sides.

$$2MnO_4^-(aq) + 6H^+(aq) + 5H_2O_2(aq) \rightarrow 2Mn^{2+}(aq) + 8H_2O(l) + 5O_2(g)$$

If the two equations for hydrogen peroxide are written down and added together we get

OA $\quad\quad\quad H_2O_2(aq) + 2H^+(aq) + 2e^- \rightarrow 2H_2O(l)$

RA $\quad\quad\quad\quad H_2O_2(aq) \rightarrow O_2(g) + 2H^+(aq) + 2e^-$

Add $\quad\quad\quad 2H_2O_2(aq) \rightarrow 2H_2O(l) + O_2(g)$

This equation represents the decomposition of hydrogen peroxide into water and oxygen. The reaction can be regarded as the disproportionation (Unit 26) of hydrogen peroxide with hydrogen peroxide acting simultaneously as oxidizing and reducing agent.

**6.8     OXIDATION STATE (OR OXIDATION NUMBER)**

The system of **oxidation states** (or **oxidation numbers**) has been devised to give a guide to the extent of oxidation or reduction in a species. The system is without direct chemical foundation, but is extremely useful, being appropriate to both ionic and covalently bonded species.

The oxidation state can be defined simply as the number of electrons which must be added to a positive ion to get a neutral atom or removed from a negative ion to get a neutral atom.

*E.g.* $Fe^{2+}(aq)$ – two electrons have to be added and the oxidation state is $+2$.

$Cl^-(aq)$ – one electron has to be removed and the oxidation state is $-1$.

It is relatively easy to understand for simple ions because the electrons are definitely associated with certain ions.

For covalent species it is assumed that the electrons in the covalent bond actually go to the atom which is most electronegative. For example in ammonia, $NH_3$, nitrogen is the more electronegative element and it is assumed that the three electrons (one from each hydrogen atom) are associated with the nitrogen atom. Nitrogen, therefore (like nitrogen in $N^{3-}$) has an oxidation state of $-3$ and hydrogen $+1$.

The system will be clear with practice but the following rules are worth remembering.

(*i*) The oxidation state of all elements uncombined is zero. Therefore the oxidation state of oxygen in oxygen gas is zero.

(*ii*) The algebraic sum of the oxidation states of the elements in a compound is always zero.

*E.g.* in $NH_3$

| | | |
|---|---|---|
| N | OS | −3 |
| H | OS | +1 |
| H | OS | +1 |
| H | OS | +1 |
| | | 0 |

(*iii*) The algebraic sum of the oxidation states of the elements in an ion is equal to the charge on the ion.

*E.g.* in $CO_3^{2-}$

| | | |
|---|---|---|
| C | OS | +4 |
| O | OS | −2 |
| O | OS | −2 |
| O | OS | −2 |
| | | −2 |

(*iv*) The oxidation state of oxygen is −2 (except in oxygen gas and peroxides)

(*v*) The oxidation state of hydrogen is +1 (except when combined with Group I or II metals as hydrides).

In chemical names the oxidation state of a particular species may be shown in Roman numerals in brackets if there is the chance of any uncertainty. In these cases the sign is not given.

*E.g.* iron(III) chloride – iron is in oxidation state +3

tetracarbonylnickel(0) – nickel is in oxidation state zero.

If during a chemical reaction a species changes its oxidation state, then oxidation and reduction are taking place. An increase in oxidation state corresponds to oxidation and a decrease to reduction.

*E.g.* the reaction of chlorine with hydrogen sulphide

$$Cl_2(g) + H_2S(g) \rightarrow 2HCl(g) + S(s)$$

This could be written in two ionic half equations

OA $\qquad\qquad Cl_2(g) + 2e^- \rightarrow 2Cl^-(g)$

RA $\qquad\qquad S^{2-}(g) \rightarrow S(s) + 2e^-$

Add together $\qquad Cl_2(g) + S^{2-}(g) \rightarrow 2Cl^-(g) + S(s)$

The oxidation state of chlorine in $Cl_2$ is zero (because it is an element) and in HCl is −1. (Remember H is usually in OS +1 and the sum must be zero.)

The chlorine therefore is reduced because the oxidation state is reduced from 0 to −1. Similarly, the oxidation state increases from −2 in $H_2S$ to zero in S. The sulphur is oxidised as there is an increase in oxidation state.

With *d*-block elements (Unit 28), a variety of oxidation states is possible for each element.

QUESTIONS

1 Chlorine can exist in all oxidation states between +7 and −1 except +2. Complete the following table by inserting the following compounds in the appropriate place in Table 6.1.

HCl, HOCl, $ClF_3$, $Cl_2$, $ClO_2$, $KClO_4$, $Cl_2O_6$, $KClO_3$, $Cl_2O_6$, $CCl_4$

**Table 6.1**

| +7 | +6 | +5 | +4 | +3 | +1 | 0 | −1 |
|---|---|---|---|---|---|---|---|
| | | | | | | | |

HCl $\quad$ oxidation state of chlorine is −1 (H = +1 and the sum is zero)

HOCl $\quad$ oxidation state of chlorine is +1 (H = +1, O = −2 and the sum is zero)

$ClF_3$ $\quad$ oxidation state of chlorine is +3 (F is more electronegative than Cl F = −1).

$Cl_2$ $\quad$ oxidation state of chlorine is zero (element)

$ClO_2$ $\quad$ oxidation state of chlorine is +4 (O = −2)

$KClO_4$ oxidation state is +7 (K = +1, O = −2; (+1) +4(−2) = −7)

$Cl_2O_6$ $\quad$ oxidation state of chlorine is +6 (6 O each −2 = −12; divided between two chlorines)

$CCl_4$ $\quad$ oxidation state of chlorine is −1 (chlorine more electronegative than C).

$KClO^3$ oxidation state of chlorine is + 5

**2** In the following molecular equations, work out the oxidation states of the element in bold type in each case. Then decide whether this element is oxidized, reduced or neither oxidized nor reduced in that particular reaction.

(*a*) $2\textbf{Cu}SO_4(aq) + 4KCN(aq) \rightarrow 2\textbf{Cu}CN(s) + C_2N_2(g) + K_2SO_4(aq)$

(*b*) $10\textbf{Fe}SO_4(aq) + 2KMnO_4(aq) + 8H_2SO_4(aq) \rightarrow 5\textbf{Fe}_2(SO_4)_3(aq) + K_2SO_4(aq) + 2MnSO_4(aq) +$
$$8H_2O(l)$$

(*c*) $K\textbf{I}(aq) + AgNO_3(aq) \rightarrow Ag\textbf{I}(s) + KNO_3(aq)$

(*d*) $CuSO_4(aq) + \textbf{Zn}(s) \rightarrow Cu(s) + \textbf{Zn}SO_4(aq)$

(*e*) $\textbf{Sn}Cl_2(aq) + HgCl_2(aq) \rightarrow Hg(l) + \textbf{Sn}Cl_4(aq)$

(*a*) Oxidation state of Cu in $CuSO_4 = +2$
in $CuCN = +1$
Copper is reduced during the reaction $(+2 \rightarrow +1)$

(*b*) Oxidation state of Fe in $FeSO_4 = +2$
in $Fe_2(SO_4)_3 = +3$
Iron is oxidized during the reaction $(+2 \rightarrow +3)$

(*c*) Iodine is in oxidation state $-1$ in KI and AgI – neither oxidation or reduction.

(*d*) Oxidation state of zinc changes from 0 to $+2$ – it is therefore oxidized.

(*e*) Oxidation state of tin changes from $+2$ to $+4$. Tin is oxidized.

**3** (*a*) State the observations which could be recorded in a practical notebook as each of the following reactions is performed in a test tube.

(*i*) $Br_2 + SO_2 + 2H_2O \rightarrow SO_4^{2-} + 4H^+ + 2Br^-$

(*ii*) $Cu_2O + 2H^+ \rightarrow Cu^{2+} + Cu + H_2O$

(*iii*) $Zn(OH)_2 + 2OH^- \rightarrow Zn(OH)_4^{2-}$

(*iv*) $2CrO_4^{2-} + 2H^+ \rightarrow Cr_2O_7^{2-} + H_2O$

(*b*) Identify from the following list one entity which undergoes oxidation, one which undergoes reduction and one which undergoes disproportionation in the above-mentioned reactions. Give brief explanations of your choices.
$$Br_2, SO_2, Cu_2O, Zn(OH)_2, CrO_4^{2-}$$

(*c*) Balance the half equations (*i*) and (*ii*) below and so construct the redox equation for the action of chlorine on $MnO_4^{2-}$

(*i*) $\qquad\qquad\qquad MnO_4^{2-}(aq) \rightarrow MnO_4^-(aq)$

(*ii*) $\qquad\qquad\qquad Cl_2(g) \rightarrow Cl^-(aq)$

(*d*) Repeat (*c*) for the incomplete half equations (*i*) and (*ii*) and so construct the redox equation for the action of zinc on the ion $VO_2^+(aq)$.

(*i*) $\qquad\qquad\qquad VO_2^+(aq) \rightarrow V^{2+}(aq)$

(*ii*) $\qquad\qquad\qquad Zn(s) \rightarrow Zn^{2+}(aq)$

*(Southern Universities Joint Board)*

(*a*) (*i*) When colourless sulphur dioxide gas is passed through a red-brown solution of bromine, the solution is decolourized.

(*ii*) Red-brown copper(I) oxide is added to a dilute acid solution (*e.g.* dilute sulphuric acid). The solution turns blue and a brown precipitate of copper is produced.

(*iii*) A colourless solution is produced when sodium hydroxide is added to the white zinc hydroxide precipitate.

(*iv*) On addition of dilute acid, the yellow chromate(VI) solution turns to an orange dichromate(VII) solution.

(*b*) Bromine is reduced to bromide ions. A bromine atom gains an electron to form a $Br^-$ ion. The oxidation state of bromine changes from 0 to $-1$. Since oxidation and reduction occur together, sulphur dioxide is oxidised.

Copper(I) oxide undergoes disproportionation. Copper(I) oxide contains copper in oxidation state $+1$. On reaction some is oxidized to copper(II) and some copper(I) is reduced to copper (oxidation state 0).

(*c*) OA $\quad MnO_4^-(aq) + 8H^+(aq) + 5e^- \rightarrow Mn^{2+}(aq) + 4H_2O(l)$
RA $\quad Cl_2(g) + 2e^- \rightarrow 2Cl^-(aq)$
$$2MnO_4^-(aq) + 16H^+(aq) + 5Cl_2(g) \rightarrow 2Mn^{2+}(aq) + 8H_2O(l) + 10Cl^-(aq)$$

(*d*) (*i*) OA $\quad VO_2^+(aq) + 4H^+(aq) + 3e^- \rightarrow V^{2+}(aq) + 2H_2O(l)$

(*ii*) RA $\quad Zn(s) \rightarrow Zn^{2+}(aq) + 2e^-$
$$2VO_2^+(aq) + 8H^+(aq) + 3Zn(s) \rightarrow 2V^{2+}(aq) + 4H_2O(l) + 3Zn^{2+}(aq)$$

# 7 Introduction to organic chemistry

## 7.1 Introduction

Organic chemistry is the study of the compounds of carbon, but excluding simple compounds such as carbon monoxide, carbon dioxide, carbonates, hydrogencarbonates and carbides. The elements present in organic compounds are usually restricted to carbon, hydrogen, oxygen, nitrogen, sulphur, fluorine, chlorine, bromine, iodine and metals.

In this book Units 7 and 29–39 contain the necessary organic chemistry. Most GCSE courses contain a small section of organic chemistry – usually a brief introduction to hydrocarbons (Unit 29) and possibly the relationships between alcohols and acids. An elementary treatment of polymers (Unit 36) might also be given.

Organic chemistry is usually popular with A-level students as it is very systematic. This unit is designed to give an introduction to organic chemistry. Units 29–36 detail chemistry of different types of organic compound. Unit 37 is particularly important as it stresses the relationships between organic compounds.

There are a very large number of organic compounds because carbon forms four very stable covalent bonds. As a result long chains of carbon atoms, branched chains and rings can be formed.

octane

2-methylpropane

cyclohexane
(sometimes abbreviated to )

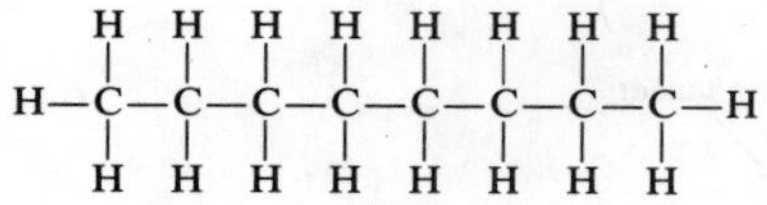

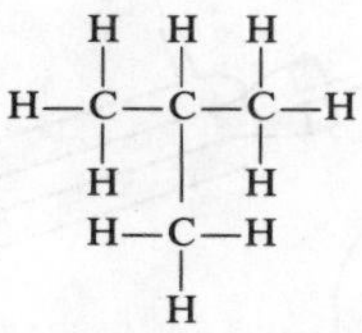

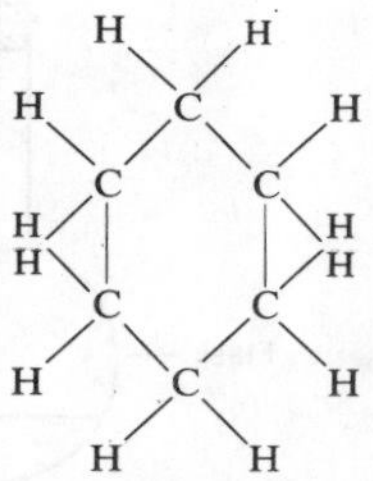

Unlike the other similar elements in Group IV (Unit 24) of the Periodic Table, carbon can also form double and triple bonds.

propanone

H—C≡C—H
ethyne

benzene
(sometimes abbreviated to  or )

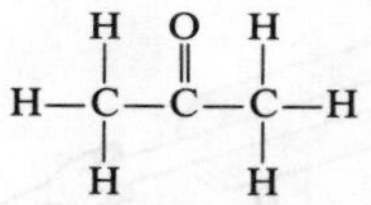

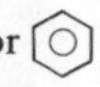

Compounds containing a benzene ring or similar are called **aromatic** compounds. Other organic compounds are called **aliphatic** compounds.

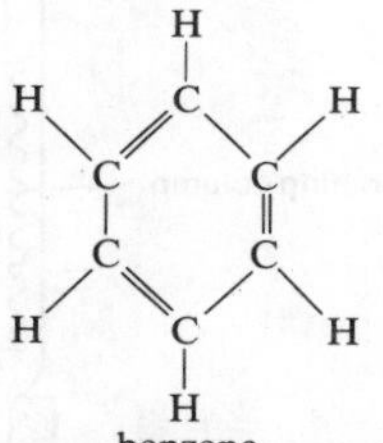

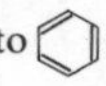

## 7.2 Purification of organic compounds

Provided it is not chemically changed by the heating process, a pure compound has a melting point at a definite temperature. If impurities are present the melting point is reduced and the compound will melt over a range of temperature.

There are various methods of purifying organic compounds including solvent extraction (or partition), using drying agents, simple distillation, fractional distillation, steam distillation, crystallization, reduced-pressure filtration and chromatography.

## 7.3 Solvent extraction

Most organic compounds are more soluble in non-polar solvents than in aqueous solutions. (Water is a polar solvent and dissolves most ionic compounds well but only a few organic compounds.) When ethoxyethane (diethyl ether) is shaken with water in a tap funnel, two layers are produced. If an organic compound is initially in the water, most of it will have transferred to the ethoxyethane after shaking.

### 7.4    DRYING SOLUTIONS WITH DRYING AGENTS

There are various chemicals that can be used to remove the last traces of water from a pure liquid. These are summarized in Table 7.1

**Table 7.1 Drying agents**

| Drying agent | Use |
|---|---|
| Anhydrous calcium chloride | Not suitable for alcohols, phenols, amines and acidic compounds |
| Anhydrous sodium sulphate | For most substances but it is very slow |
| Potassium hydroxide | For amines. Not suitable for acids, phenols and esters. |
| Anhydrous magnesium sulphate | For most substances – faster than sodium sulphate |
| Sodium wire | For ethers or hydrocarbons |

### 7.5    DISTILLATION

The apparatus in Fig. 7.1 can be used for simple distillation where a volatile solvent is to be removed from a non-volatile solid. An example is the removal of ethoxyethane from a solution of a non-volatile organic compound in ethoxyethane.

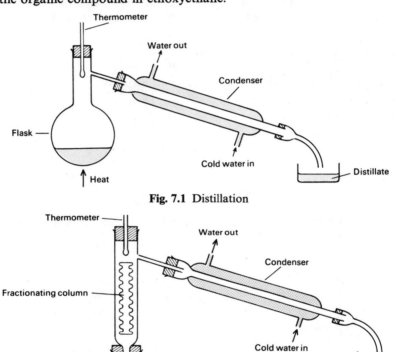

**Fig. 7.1** Distillation

**Fig. 7.2** Fractional distillation

If the boiling point of the liquid distilling off is above about 140°C, an air condenser is used. An air condenser is simply a glass tube. The air around the tube cools the contents of the tube.

If the liquid being distilled off decomposes at a temperature close to its boiling point, it is possible to distil it under reduced pressure or by **steam distillation**. Mixtures of volatile liquids can be separated by **fractional distillation** (Unit 11.3) using the apparatus in Fig. 7.2.

### 7.6    CRYSTALLIZATION

The impure solid is added to a small volume of hot solvent in which it dissolves. On cooling, the pure solid crystallizes out and the impurities, which are present in small amounts, remain dissolved in the solvent. It is necessary to choose the solvent carefully. The pure solid should be readily soluble in the hot solvent and much less soluble in the cold solvent. The pure crystals can be removed by filtration under reduced pressure using a Buchner funnel and flask (Fig. 7.3). Finally the crystals can be washed with cold solvent and dried by pressing between filter papers.

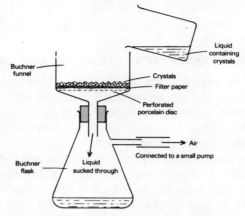

**Fig. 7.3** Filtration under reduced pressure

Benzenecarboxylic acid (benzoic acid) can be recrystallized using water as solvent.

### 7.7    CHROMATOGRAPHY

**Paper chromatography** can be used to show whether a substance is pure or contains other substances. If a small spot of the substance in solution is put on a strip of filter paper and the end of the filter paper dipped into a suitable solvent a separation will occur if the substance is impure. In Fig. 7.4 the chromatograms of a pure and impure substance are compared.

This technique requires only a very small amount of material, and is not used to purify compounds except on the minutest scale.

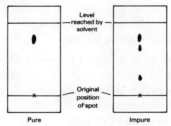

**Fig. 7.4** Chromatography of pure and impure substances

Two-way paper chromatography is a useful technique for separating complex mixtures of similar compounds. In Fig. 7.5 two amino acids were separated by using (*i*) a mixture of butan-1-ol and ethanoic acid and (*ii*) phenol as solvents in turn. After (*i*) the chromatography paper was turned through 90°. The $R_f$ values for amino acid A are 0.27 and 0.34, respectively. What are the $R_f$ values for amino acid B?

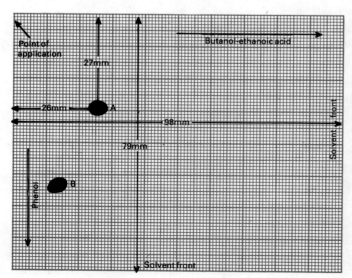

**Fig. 7.5** Two-way paper chromatogram

Answer: 0.14 and 0.65

NB Amino acids are colourless and the chromatogram has to be developed by spraying with a solution of ninhydrin and heating. A red-to-blue spot is obtained. Ninhydrin is a test for —$NH_2$ groups.

P.47.

Gas–liquid chromatography (GLC) is a most useful technique for analysis of complex mixtures. A minute quantity of the mixture to be analysed is injected into the apparatus and vaporised. The vapour is swept along by a 'carrier' gas through a long heated column packed with a suitable packing material. The different components pass through the column at different rates and the exit of each component is detected and registered graphically. GLC is used to detect and measure ethanol in breath in new breath-testing machinery.

**7.8    FINDING THE ELEMENTS PRESENT IN AN ORGANIC COMPOUND**

When a mixture of an organic compound and dry copper(II) oxide is heated, carbon dioxide and water are produced. The carbon dioxide is produced when carbon in the organic compound is oxidized. It can be detected by testing with limewater which turns milky. Water vapour is produced by oxidation of hydrogen in the original compound and can be detected with cobalt(II) chloride paper which turns from blue to pink.

The elements nitrogen, sulphur, chlorine, bromine and iodine are detected by means of the **Lassaigne's test**. The organic compound is fused with molten sodium which breaks up the compound. Any nitrogen, sulphur, chlorine, bromine or iodine in the compound are converted into sodium cyanide, sodium sulphide, sodium chloride, sodium bromide and sodium iodide respectively. The residue is added to water and the solution filtered. The following tests for the sodium compounds are then carried out.

### Sodium cyanide

Fresh iron(II) sulphate solution and sodium hydroxide solution are added to the extract and the solution is heated. The solution is then acidified and iron(III) chloride solution added. A finely divided dark-blue precipitate confirms the presence of nitrogen (Unit 28).

### Sodium sulphide

A freshly prepared solution of sodium pentacyano nitrosylferrate(II) $Na_4Fe(CN)_5NO_2$ (sodium nitroprusside) is added. A purple colour confirms the presence of sulphur.

### Sodium chloride, sodium bromide and sodium iodide

The extract is acidified with dilute nitric acid and the solution is boiled to remove any sulphide that might be present  Silver nitrate solution is added. A white precipitate of silver chloride confirms chloride, a cream coloured precipitate of silver bromide confirms bromide and a yellow precipitate of silver iodide confirms iodide (Unit 27).

It is not usual to test for the presence of oxygen.

**7.9    FINDING THE MOLECULAR FORMULA OF AN ORGANIC COMPOUND**

Having found the elements that are present by qualitative analysis, it is usual to carry out experiments to find the percentage of each element (apart from oxygen) by quantitative analysis. You are not required to study details of these methods, but the following calculation will illustrate how the percentage of carbon and hydrogen can be found.

**Sample calculation**

0.152g of an organic compound produced 0.223g of carbon dioxide and 0.091g of water on complete combustion. Calculate the percentage of carbon and hydrogen in the compound. ($A_r(H)=1$, $A_r(C)=12$, $A_r(O)=16$).

*Carbon dioxide $CO_2$*
Mass of 1 mole of carbon dioxide molecules = $12 + (2 \times 16) = 44g$.
1 mole of carbon dioxide molecules contain 12g of carbon.

0.223g of carbon dioxide contains $0.223 \times \dfrac{12}{44}$g of carbon = 0.061g

This carbon must all have come from the organic compound.
  percentage of carbon in organic compound = $\dfrac{0.061}{0.152} \times 100\% = 40.1\%$

*Water $H_2O$*
Mass of 1 mole of water molecules = $(2 \times 1) + 16 = 18g$.
1 mole of water molecules contains 2g of hydrogen.

0.091g of water contains $0.091 \times \dfrac{2}{18}$g of hydrogen = 0.010g

$\therefore$ Percentage of hydrogen in organic compound = $\dfrac{0.010}{0.152} \times 100\% = 6.6\%$

Assuming no other elements have been found to be present by qualitative analysis, the percentage of oxygen can be found:

Percentage of oxygen = 100 − (40.1 + 6.6)% = 53.3%

Having found the percentage of each element present it is possible to calculate the **empirical formula**, or simplest formula.

| Elements present | C | H | O |
|---|---|---|---|
| Percentage | 40.1% | 6.6% | 53.3% |
| Relative atomic mass | 12 | 1 | 16 |
| Percentage ÷ relative atomic mass | 3.34 | 6.6 | 3.33 |
| Divide by the smallest *i.e.* 3.33 | 1 | 2 | 1 |

Empirical formula $C_1H_2O_1$ or $CH_2O$

The empirical formula contains the elements in the correct proportions, but the actual or molecular formula may be the same as the empirical formula or a multiple of it *e.g.* in this case $C_2H_4O_2$, $C_3H_6O_3$, $C_4H_8O_4$ etc.

In order to find the molecular formula the relative molecular mass is found by experiment (Units 8 and 12). In this example if the relative molecular mass was found to be approximately 60, the molecular formula would be $C_2H_4O_2$.

## 7.10 STRUCTURAL FORMULAE

The **molecular formula** gives the number of atoms of each element in a molecule of the compound. It does not give any information about how the atoms are joined together. This information is given by the **structural formula**.

In order to obtain the structural formula it is necessary to identify which **functional groups** are present. A functional group is a group of atoms within a compound which gives characteristic properties. The common functional groups are given later in Table 7.3.

The functional groups can be identified by specific chemical tests in Units 29–35 or by physical means including infrared (ir) spectroscopy.

Infrared radiation is light with wavelengths of between 2500nm and 25 000nm (longer wavelengths than visible light). When ir radiation is passed through a compound, certain wavelengths are absorbed. The absorbed energy is used to make certain bonds within the molecule vibrate more vigorously. These vibrations involve stretching or bending. The wavelengths absorbed and the extent of absorption can be recorded on instruments (see Fig. 7.6).

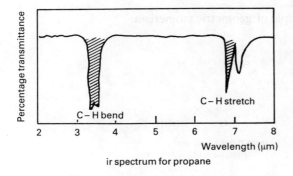

ir spectrum for propane

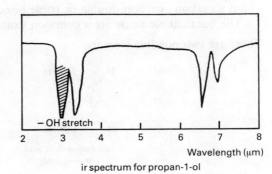

ir spectrum for propan-1-ol

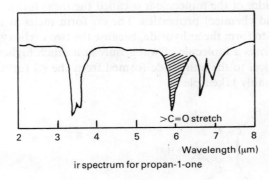

ir spectrum for propan-1-one

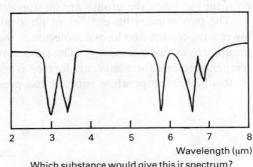

Which substance would give this ir spectrum?

**Answer: propanoic acid**

**Fig. 7.6**

It is sometimes possible to obtain two or more possible structural formulae from the same molecular formula. Examples include:

*Molecular formula*            *Structural formulae*

$C_4H_{10}$

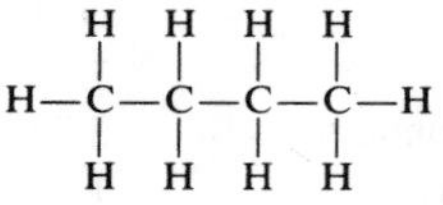

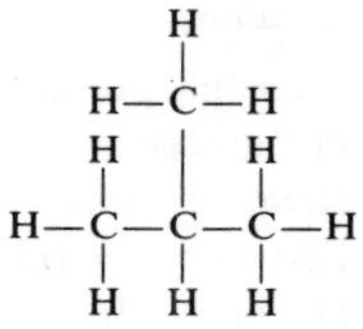

$C_3H_6O$

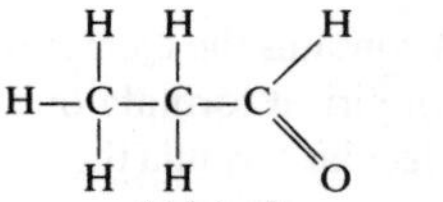

Aldehyde      Ketone

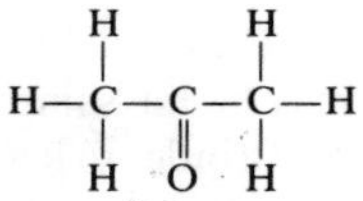

$C_2H_6O$

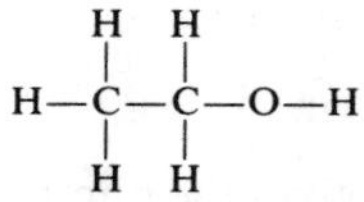

Alcohol      Ether

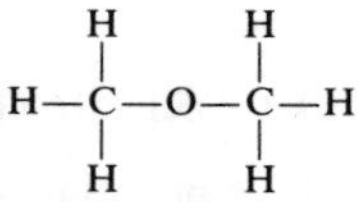

$C_2H_4O_2$

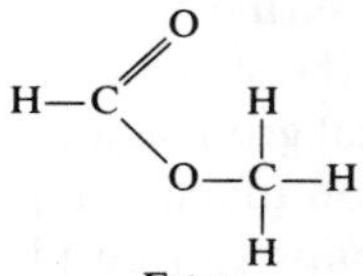

Ester      Acid

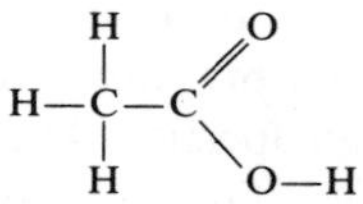

The existence of two or more forms of the same compound with the same structural formula but with different spatial arrangements is called **stereoisomerism**. There are two types – **geometric isomerism** and **optical isomerism**.

### 7.11 GEOMETRIC ISOMERISM

There is normally free rotation about a carbon–carbon single bond but there is no free rotation about a carbon–carbon double or triple bond.

The butanedioic acids are a common example of geometric isomerism.

There are two possible isomers.

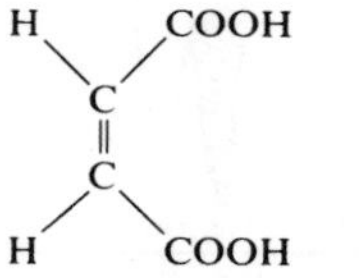

*cis*-butenedioic acid      *trans*-butenedioic acid
(maleic acid)          (fumaric acid)

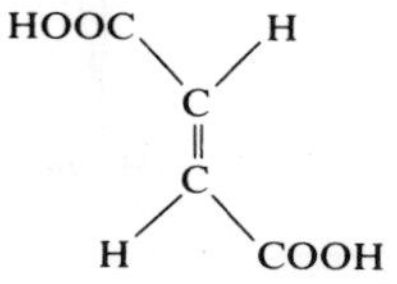

Lack of rotation in the carbon–carbon double bond explains the existence of the two isomers. The isomer with the two functional groups on the same side of the molecule is called the *cis* form but when the groups are on opposite sides of the molecule it is called the *trans* form.

The two isomers have different physical and chemical properties. The *cis* form melts at a lower temperature and loses a molecule of water to form the anhydride, because the two carboxyl groups are in close proximity. The *trans* form loses a molecule of water only at a much higher temperature and the anhydride formed is identical to the anhydride formed from the *cis* form. At this higher temperature rotation has presumably taken place.

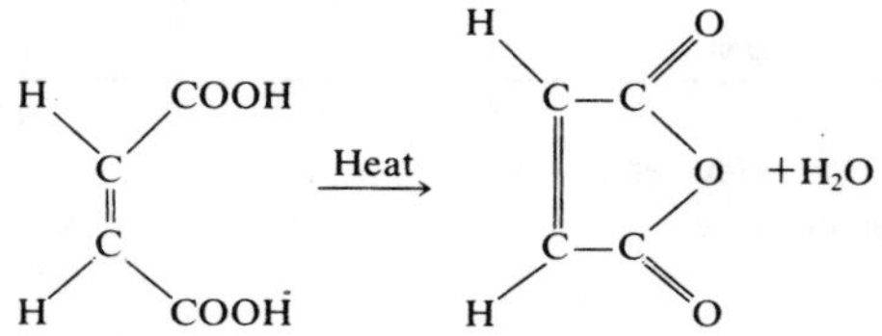

$+H_2O$

*cis*-butenedioic anhydride
(maleic anhydride)

### 7.12 OPTICAL ISOMERISM (OR CHIRALITY)

Optical isomerism in organic compounds exists because the four covalent bonds formed by a carbon atom are tetrahedrally arranged. When four different groups are attached to one carbon atom it is possible to produce two optical isomers that are mirror images and are not superimposable.

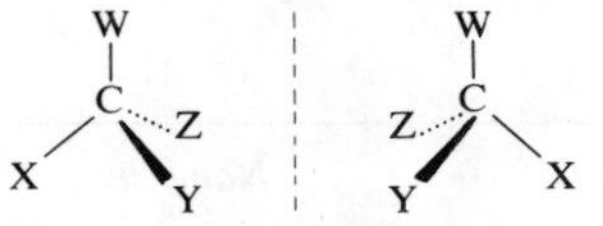

W, X, Y and Z represent different functional groups.
—— in the plane of paper
...... behind paper
▬▬ in front of paper

The carbon atom attached to four different groups is called an **asymmetric carbon atom** (or **chiral centre**).

The two optical isomers are called **enantiomers** or **enantiomorphs**. They have identical chemical properties and identical physical properties except for their effect on plane polarized light.

**Plane polarized light** is light in which the vibrations are all in one plane. The two forms rotate the plane of polarized light by equal angles but in opposite directions. The form which rotates polarized light to the right is said to be **dextrorotatory** (and this form is called the *d* or + form). The form which rotates to the left is said to be **laevorotatory** (and is called the *l* or − form)

A mixture of equal amounts of the *d* and *l* forms has no effect on the plane of polarized light is called a **racemic** (or ± or *dl*) **mixture**.

A common example of optical isomerism is 2-hydroxypropanoic acid

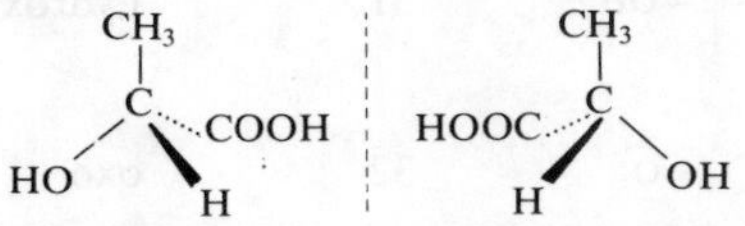

The separation of a racemic mixture into the separate *d* and *l* forms is called **resolution**. Because the physical properties are identical, it is not possible to resolve the mixture by distillation or crystallization. An optically active base can be used to resolve a racemic mixture of an acid, two salts are formed – one between the *d* form of the base and *d* form of the acid and the other between the *d* form of the base and the *l* form of the acid. These salts can be separated by fractional crystallization as they have different properties. The acid can be regenerated by hydrolysis.

### 7.13 NAMING ORGANIC COMPOUNDS

Many common organic compounds still retain a common name which is not related to the chemical composition, *e.g.* propanone is still frequently called acetone.

Since 1948, attempts have been made to systematize the names of all organic compounds.

**Aliphatic compounds**

The basis of systematic naming of aliphatic compounds is that every name consists of a **root**, one **suffix** and as many **prefixes** as necessary. The root is determined by the number of carbon atoms in the longest continuous chain. In the examples below the longest carbon chain in **A** is 3 and in **B** is 4.

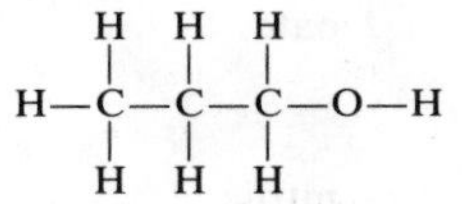

**A** 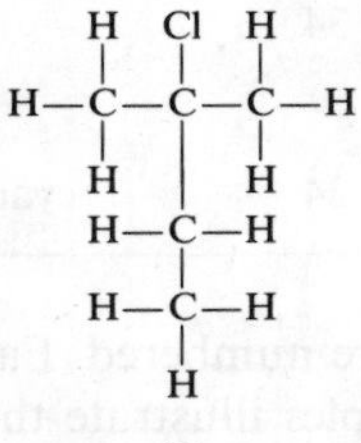      **B**

Table 7.2 lists the alkanes from which the roots are derived.

**Table 7.2 Alkane roots for aliphatic nomenclature**

| Number of carbon atoms in chain | Root |
|---|---|
| 1 | methane |
| 2 | ethane |
| 3 | propane |
| 4 | butane |
| 5 | pentane |
| 6 | hexane |
| 7 | heptane |
| 8 | octane |
| 9 | nonane |
| 10 | decane |

Any organic compound containing a continuous chain of five carbon atoms has a name based upon the alkane pentane. In **B** above, the name is based upon butane because the longest carbon chain contains four carbon atoms.

Having identified the longest carbon chain, it is necessary to identify the various functional groups and name them. Table 7.3 lists some of the common functional groups.

**Table 7.3  Common functional groups**

| Functional group | Structure | Unit | Name as prefix | Name as suffix |
|---|---|---|---|---|
| Double bond | $>C=C<$ | 29 | – | ene |
| Triple bond | $-C\equiv C-$ | 29 | – | yne |
| Halogen (X = Cl, Br, I) | $-C-X$ | 30 | chloro bromo iodo | chloride bromide iodide |
| Amine | $-C-NH_2$ | 35 | amino | amine |
| Hydroxyl | $-C-OH$ | 31 | hydroxy | ol |
| Carbonyl | $>C=O$ | 32 | oxo | al (in aldehydes) one (in ketones) |
| Carboxyl | $-C\overset{O}{\underset{O-H}{}}$ | 33 | carboxy | oic acid |
| Acid chloride | $-C\overset{O}{\underset{Cl}{}}$ | 34 | – | oyl chloride |
| Amide | $-C\overset{O}{\underset{NH_2}{}}$ | 34 | amido | amide |
| Acid anhydride | anhydride structure | 34 | – | oic anhydride |
| Ester | $-C\overset{O}{\underset{O-R}{}}$ | 34 | – | oate |
| Nitrile | $-C\equiv N$ | 34 | cyano | nitrile |

The carbon atoms in the longest chain are numbered. Each functional group is then named and added to the root. The following examples illustrate the system

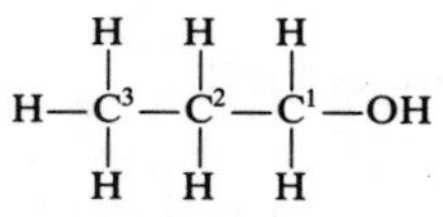

propan-1-ol

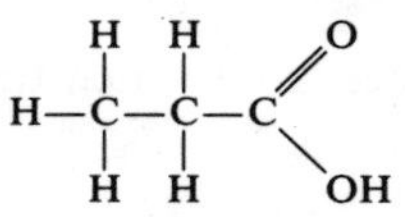

propanoic acid

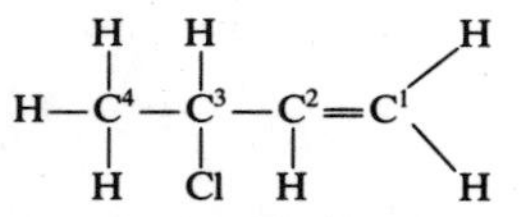

3-chlorobut-1-ene

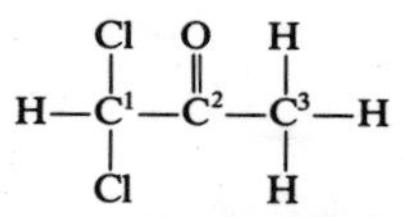

1,1-dichloropropanone

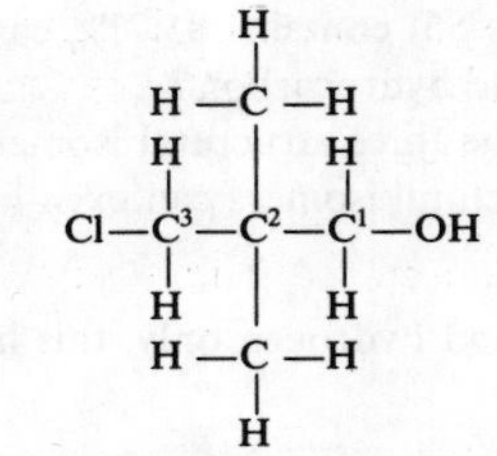

3-chloro-2,2-dimethylpropan-1-ol

NB (*i*) The chain is numbered to ensure that the *lowest* possible numbers appear in the name. Propan-1-ol could be named propan-3-ol if the numbering had been from the left hand end

*i.e.*

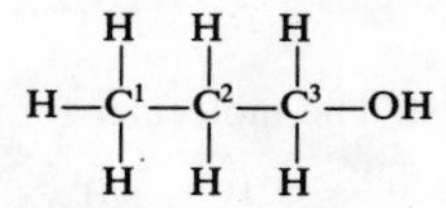

(*ii*) Where a name contains a number of prefixes the prefixes are arranged in alphabetical order. In 3-chloro-2,2-dimethylpropan-1-ol, the chloro prefix is placed before the methyl prefix (ignore di- and tri-).

(*ii*) With practice you should encounter no problems in using the system. Practice will be obtained as you work through Units 29–39.

### Aromatic compounds

The names of aromatic compounds are derived from benzene or similar aromatic hydrocarbons. The carbon atoms in the benzene ring are numbered from the carbon atom to which the principal group is attached.

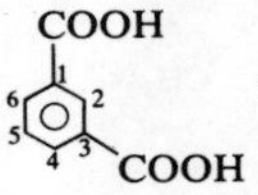

benzene-1,3-dicarboxylic acid

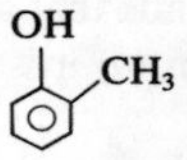

1-hydroxy-2-methylbenzene

You may still encounter a system for naming disubstituted benzene rings in which the prefixes *ortho-*, *meta-* and *para-* are used.

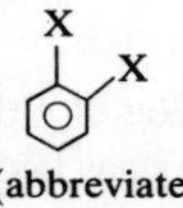

*ortho* (abbreviated *o*-) 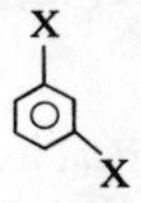   *meta-*(*m*-) 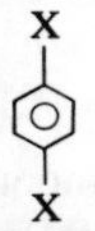   *para-*(*p*-)

substituents attached to adjacent carbon atoms

A fuller guide to the naming of organic (and inorganic) compounds will be found in the booklet: *Chemical nomenclature, symbols and terminology* (1979), The Association for Science Education, College Lane, Hatfield, Herts.

### QUESTIONS

1 Which one of the following compounds is NOT an isomer of heptane?

A 2-methylhexane
B 2,2-dimethylpentane
C 2,3-dimethylbutane
D 2,3-dimethylpentane

*(Scottish Higher)*

Draw out the structural formulae of the four compounds and find out which one does not have a molecular formula $C_7H_{16}$.

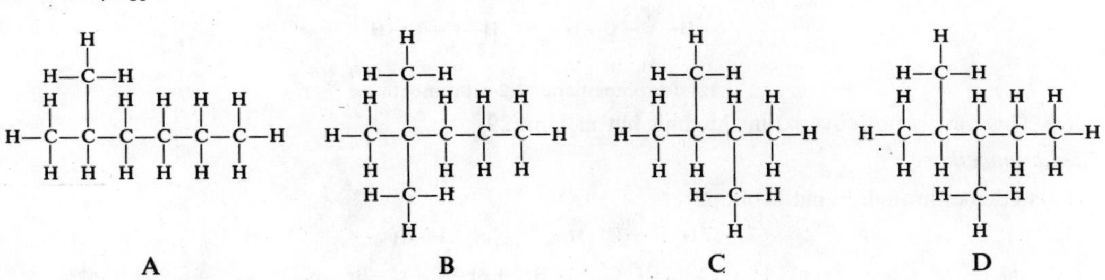

A          B          C          D

The correct answer is C.

2 A hydrocarbon (relative molecular mass 56) contains 85.71% carbon.
   (a) What is the molecular formula of the hydrocarbon?
   (b) Give three structural formulae for the three structural isomers with this molecular formula.
   (c) Explain why one of these three structural isomers can exist in two stereoisomeric forms.
   ($A_r(H) = 1$, $A_r(C) = 12$)

Since a hydrocarbon contains carbon and hydrogen only, this hydrocarbon must contain $100 - 85.71\%$ hydrogen *i.e.* 14.29%.

(a)
| Elements present | C | H |
|---|---|---|
| Percentage | 85.71 | 14.29 |
| Relative atomic mass | 12 | 1 |
| Percentage $\div A_r$ | 7.14 | 14.29 |
| Divide by smallest (7.14) | 1 | 2 |
| Empirical formula $= CH_2$ | | |

No compound with this formula can exist. The molecular formula must be $C_4H_8$. ($M_r = 56$).

(b)

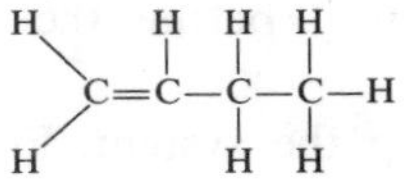

but-1-ene 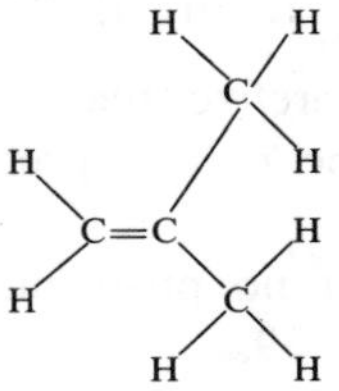          2-methylprop-1-ene 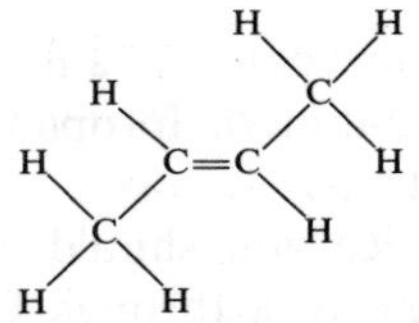          but-2-ene

(c) But-2-ene can exist in *cis* and *trans* forms (*i.e.* show geometric isomerism). This is because there is no rotation about the double bond.

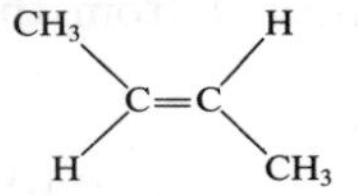

*trans* 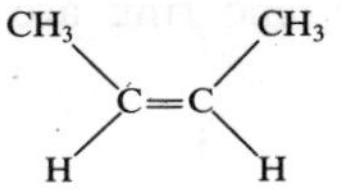                    *cis*

In this type of question the examiner often finds that a mistake in part (a) leads to an incorrect molecular formula. Even if you have made a mistake in (a) it is still possible to obtain marks in (b) and (c) if you use the molecular formula you obtain correctly.

You will probably lose marks if you are lazy and draw just 'stick' formulae *e.g.*

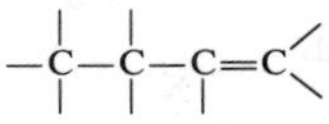

3 Two volatile isomeric compounds have the empirical formula $CH_2Br$. 1.26g of either isomer was found to occupy 150cm³ at stp ($A_r(H) = 1$. $A_r(C) = 12$, $A_r(Br) = 80$). 1 mole of any gas occupies 22 400cm³ at stp).
   (a) Calculate the relative molecular mass of these compounds.
   (b) Name and give the structural formulae of the two isomeric compounds.
   (c) How could each isomer be prepared, starting from a different named hydrocarbon in each case?
   (d) Outline a method whereby the presence of bromine could be confirmed in these compounds.

(a) 1.26g of compound occupies 150cm³ at stp

$\dfrac{1.26}{150}$ g of compound occupies 1cm³ at stp

$\dfrac{1.26}{150} \times 22\,400$ g of compound occupies 22 400cm³ at stp

Relative molecular mass $= 188.2$

(b) Empirical formula $= CH_2Br$
If this was the molecular formula also the relative molecular mass would be
$$12 + 2 + 80 = 94$$
Molecular formula $= C_2H_4Br_2$

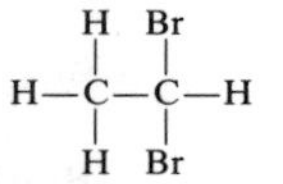

1,1-dibromoethane 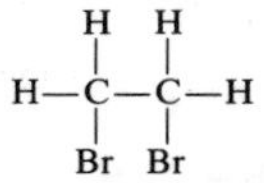   1,2-dibromoethane

(c) This part is not covered in this unit but in Unit 29.

*1,2-dibromoethane*

Pass ethene through liquid bromine

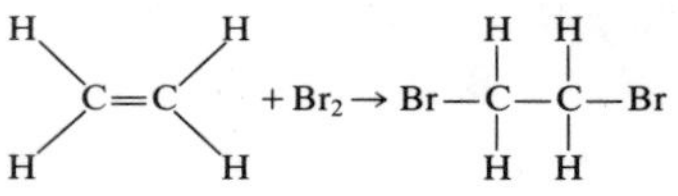

*1,1-dibromoethane*
Addition of hydrogen bromide to ethyne.

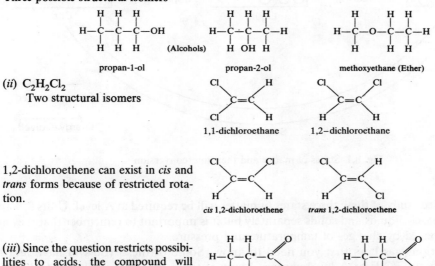

1-bromoethane

(*d*) Lassaigne's tests (Unit 7.8)

**4** Use the following compounds in a discussion of the various types of isomerism which occur in organic chemistry.

(*i*) $C_3H_8O$  (*ii*) $C_2H_2Cl_2$  (*iii*) $C_3H_6O_3$ (only acids)

Different types of isomerism are discussed in Units 7.10–7.13. It is reasonable to expect that the examiner is going to include all types of isomerism in the examples given

(*i*) $C_3H_8O$
Three possible structural isomers

propan-1-ol    (Alcohols)    propan-2-ol    methoxyethane (Ether)

(*ii*) $C_2H_2Cl_2$
Two structural isomers

1,1-dichloroethane    1,2–dichloroethane

1,2-dichloroethene can exist in *cis* and *trans* forms because of restricted rotation.

*cis* 1,2-dichloroethene    *trans* 1,2-dichloroethene

(*iii*) Since the question restricts possibilities to acids, the compound will contain a —COOH group

2-hydroxypropanoic acid    3-hydroxypropanoic acid

2-hydroxypropanoic acid and 3-hydroxypropanoic acid are structural isomers but 2-hydroxypropanoic acid contains an asymmetric carbon atom (marked *) and therefore exists in two optically active forms.

Many questions rely upon appreciating that an optically active compound contains an asymmetric carbon atom *i.e.* a carbon atom attached to four different groups.

*E.g.* An alcohol **X** with a molecular formula $C_4H_{10}O$ is optically active. Identify **X**.

The only alcohol with the formula $C_4H_{10}O$ that is optically active is

butan-2-ol

**5** (*a*) Write a systematic name for each of the following compounds.
A $CH_3CH(OH)CH_2CH_3$  B $(CH_3)_3CBr$  C $CH_3CH_2CH_2CHO$  D $CH_3CH_2CH{=}CH_2$
E $HO_2CCH_2CH_2CO_2H$
(*b*) Use the letters A to E to refer to the compounds (if any) which (*i*) exhibit optical isomerism, (*ii*) react with concentrated sulphuric acid to form an alkene, (*iii*) polymerize under appropriate conditions.
(*c*) Which two compounds, A to E, react to form (*i*) an ester, (*ii*) an acetal?
(*d*) Give the mechanism for the reaction of (*i*) D with bromine, (*ii*) B with aqueous alkali.

(*London University*)

(*a*) A butan-2-ol,  B 2-bromo 2-methylpropane,  C butanal,  D but-1-ene.  E butanedioic acid.
(*b*) (*i*) A (Unit 7.12). (*ii*) A (Units 31.4, 29.5). (*iii*) D (Unit 36.2).
(*c*) (*i*) A and E (Unit 34.4). (*ii*) Aldehydes react with alcohols in two stages in the presence of dry hydrogen chloride to form an acetal.

*e.g.*

ethanal    ethanol    1,1-diethoxyethane (an acetal)

Compounds reacting to form an acetal are A and C.
(*d*) (*i*) see 38.5 (page 222)
    (*ii*) see 38.6 (page 223)

# 8 Gases

## 8.1 INTRODUCTION

At O level you would have learnt that matter can exist in three physical states – solid, liquid and gas. This is summarized in Fig. 8.1.

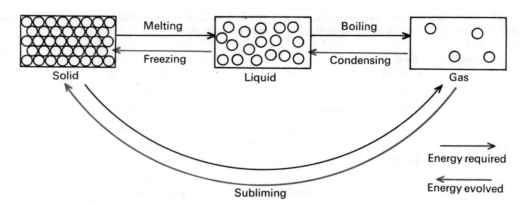

Fig. 8.1 States of matter and their interconversion

A more detailed study of these three states of matter will be required at A level. Units 8, 9 and 10 will consider gases, liquids and solids separately but it is important to remember that they are readily interconverted by changes of temperature and pressure.

There are various gas laws that you need to know. Some of these you will have met at GCSE or possibly in GCSE or A-level physics. Many of these laws only apply to **ideal gases**. Gases are never strictly ideal and it is necessary to understand why and how these laws can be applied to the non-ideal situation.

One difficulty that you might encounter with this subject is the difference in units of pressure used. Looking at A-level papers in recent years you will see:

(*i*) Pascal (Pa). This can be written as $Nm^{-2}$ (newtons per square metre) or $kg\,m^{-1}\,s^{-1}$. Standard pressure corresponds to 101 325Pa or 101.325kPa. (100kPa is sometimes called 1 bar and so standard pressure is approximately 1 bar.)

(*ii*) Millimetres of mercury (mm Hg). This is the height of a mercury column in millimetres supported by that pressure. Standard pressure corresponds to 760mm Hg.

(*iii*) Atmosphere (atm). This is useful for measuring high pressures. Standard pressure is 1 atmosphere.

The important thing to remember is that unless the question asks for specific units it does not matter to you or the examiner which units you use providing you use your units consistently.

## 8.2 GAS LAWS

Changes in temperature and pressure have little effect on the volume of a liquid or a solid, but a considerable effect on the volume of a gas. Various gas laws have been devised following experiments and they are considered individually in Units 8.3 – 8.7.

## 8.3 BOYLE'S LAW

**The volume of a fixed mass of gas is inversely proportional to pressure providing the temperature is kept constant.**

*i.e.* as the pressure ($P$) increases, the volume ($V$) decreases. This is expressed mathematically as

$$P = \text{constant} \times \frac{1}{V}$$

A graph of $P$ against $1/V$ is a straight line passing through the origin.

**8.4** CHARLES' LAW

**The volume of a fixed mass of gas, at constant pressure, is directly proportional to the absolute temperature.**

*i.e.* as the temperature ($T$) increases and the particles move faster, they move farther apart and the volume ($V$) increases.

Mathematically, $V = \text{constant} \times T$, where $T$ is temperature in degrees K (The temperature in degrees Kelvin (K) is found by adding 273° to the temperature Celsius *e.g.* 20°C = 20 + 273 = 293 K).

This law is based upon work done by Gay Lussac. He found that a gas expands or contracts by approximately 1/273 of its volume at 0°C for every °C that the temperature rises or falls.

273cm³ of a gas at 0°C will have a volume of 272cm³ at −1°C, 263cm³ at −10°C, and 293cm³ at 20°C. If the gas were cooled to −273°C (0 K) its volume would theoretically be zero. −273°C (0 K) is called the **absolute zero** and is the starting point of the Kelvin scale.

Boyle's law and Charles' law can be combined in a general gas equation.

$$\frac{P_1 V_1}{T_1} = \frac{P_2 V_2}{T_2}$$

This equation is useful for correcting the volume of a fixed mass of gas under one set of conditions of temperature and pressure to the volume of gas under a different set of conditions.

If a gas is at 0°C (273 K) and 101.3kPa, these conditions are called **standard temperature and pressure** (stp).

*Example*: A fixed mass of gas has a volume of 76cm³ at 27°C and 100kPa pressure. Find the volume that the gas would have at stp.

$V_1$ = volume before correction = 76cm³ | $V_2$ = volume after correction – unknown
$P_1$ = pressure before correction = 100kPa | $P_2$ = pressure after correction = 101.3kPa
$T_1$ = temperature before correction = 300 K | $T_2$ = temperature after correction = 273 K

$$\text{Substitute in } \frac{P_1 V_1}{T_1} = \frac{P_2 V_2}{T_2} \qquad \frac{100 \times 76}{300} = \frac{101.3}{273} \times V_2$$

$$V_2 = 68.3 \text{cm}^3$$

Since both the pressure has increased and the temperature has decreased, it is expected that the answer would be less than 76cm³.

**\*8.5** IDEAL GAS EQUATION

This equation applies only for ideal gases. The equation is

$$PV = nRT$$

where $P$ = pressure, $V$ = volume, $n$ = number of moles of gas, $R$ = molar gas constant and $T$ = absolute temperature.

The **molar gas constant** $R$ is the same for all gases. It can be calculated using the information that 1 mole of a gas at 0°C and 101 325Pa pressure occupies 0.0224m³.

Substituting in the equation $PV = nRT$

$$R = \frac{101300 \times 0.0224}{1 \times 273} = 8.31 \text{ J mol}^{-1}\text{K}^{-1}$$

The value for $R$ is given on the examination paper and it is important that the units you use are in agreement with the units for $R$.

**8.6** DALTON'S LAW OF PARTIAL PRESSURE

**In a mixture of gases, the total pressure is equal to the sum of the partial pressures of the individual gases.**

Mathematically, for three gases A, B and C in a mixture,

$$P = P_A + P_B + P_C$$

where $P$ is the total pressure of the mixture, $P_A$ is the partial pressure of A, $P_B$ is the partial pressure of B and $P_C$ is the partial pressure of C.

The **partial pressure** of a gas is the pressure that the gas would exert if it alone occupied the whole volume of the mixture at the same temperature. For A the partial pressure

$$P_A = \frac{\text{number of moles of A} \times \text{total pressure}}{\text{total number of moles in mixture}}$$

*Example:* Calculate the partial pressures of carbon dioxide and oxygen in a mixture of the two gases with total pressure 100kPa. The mixture consists of 4.4g. of carbon dioxide and 6.4g of oxygen (Relative atomic masses C = 12, O = 16)

$$\text{Number of moles of carbon dioxide} = \frac{\text{mass of carbon dioxide}}{\text{molar mass of carbon dioxide}}$$

$$= \frac{4.4}{12 + (2 \times 16)} = 0.1 \text{ moles}$$

$$\text{Number of moles of oxygen molecules} = \frac{\text{mass of oxygen}}{\text{relative molecular mass of oxygen}}$$

$$= \frac{6.4}{32} = 0.2 \text{ moles}$$

Total number of moles of carbon dioxide and oxygen = 0.1 + 0.2 = 0.3 mole.

$$\text{Partial pressure of carbon dioxide} = \frac{\text{no. of moles } CO_2 \times \text{total pressure}}{\text{total number of moles}}$$

$$= \frac{0.1}{0.3} \times 100 \text{ kPa} = 33.3 \text{ kPa}$$

$$\text{Partial pressure of oxygen} = \frac{\text{no. of moles oxygen} \times \text{total pressure}}{\text{total number of moles}}$$

$$= \frac{0.2}{0.3} \times 100 \text{ kPa} = 66.6 \text{ kPa}$$

(Check that the two answers add up to the total pressure)

### 8.7    AVOGADRO'S HYPOTHESIS AND GAY LUSSAC'S LAW

**Avogadro's hypothesis states that equal volumes of different gases under the same conditions of temperature and pressure, contain the same number of molecules.**

**Gay Lussac's law states that the volumes of gases reacting and the volumes of products if gaseous bear a simple numerical relation to one another, providing all measurements are made at the same temperature and pressure.**

For example, by experiment, it can be found that 10cm³ of nitrogen gas (1 volume) combines with 30cm³ of hydrogen gas (3 volumes) to produce 20cm³ of ammonia gas (2 volumes).

Find the volume of oxygen required and the volume of gaseous product formed when 10cm³ of methane $CH_4$ is burnt in sufficient oxygen for complete combustion.

All measurements were made at room temperature and pressure.
First write the equation.

$$CH_4(g) + 2O_2(g) \rightarrow CO_2(g) + 2H_2O(l)$$
methane + oxygen → carbon dioxide + water
    1 vol        2 vol            1 vol              0

When 10cm³ of methane gas ($CH_4$) reacts, 20cm³ of oxygen is required.
Volume of carbon dioxide produced = 10cm³
NB When gases are cooled to room temperature, the steam condenses and the volume of water produced is negligible.

### 8.8    VAPOUR DENSITY

The vapour density of a gas or vapour is given by

$$\text{vapour density} = \frac{\text{mass of a volume of the gas}}{\text{mass of an equal volume of hydrogen}}$$

at the same temperature and pressure.
Since equal volumes of all gases under the same conditions of temperature and pressure contain the same number of molecules (Avogadro's hypothesis)

$$\text{vapour density} = \frac{\text{mass of } n \text{ molecules of gas}}{\text{mass of } n \text{ molecules of hydrogen}}$$

$$= \frac{\text{mass of 1 molecule of gas}}{\text{mass of 1 molecule of hydrogen}}$$

$$\text{But relative molecular mass} = \frac{\text{mass of 1 molecule of gas}}{\text{mass of 1 atom of hydrogen}}$$

Since a molecule of hydrogen is composed of 2 atoms
Relative molecular mass = 2 × vapour density

The relative molecular mass of a gas or volatile liquid can be found by finding the vapour density of the vapour produced.

## 8.9 Relative masses of molecules by weighing known volumes of gases

1 mole of molecules of any gas can be assumed to occupy 22.4 dm$^3$ at stp. M$_r$ of a gas can be found if the volume of a known mass of the gas can be found at known temperature and pressure (see questions at the end of this Unit).

The experiment can be modified to find M$_r$ of a low boiling point (or volatile) liquid. A gas syringe is sealed with a self-sealing rubber cap placed tightly over the nozzle. The syringe is heated in a syringe oven to a steady temperature above the boiling point of the volatile liquid to be used. The volume of any expanded air in the syringe is taken. A hypodermic syringe containing the liquid is weighed and some of the liquid is injected through the cap into the syringe. The mass of the hypodermic syringe is recorded after injection to enable the mass of liquid used to be found. The liquid is vaporized and the volume of gas in the syringe is recorded. Finally the temperature of the oven and atmospheric pressure should be recorded.

Sample calculation for propanone.

Mass of hypodermic syringe containing propanone before injection = 25.246 g

Mass of hypodermic syringe after injection = 25.100 g

Mass of propanone injected = 0.146 g

Reading on gas syringe in oven before injection = 3 cm$^3$

after injection = 75 cm$^3$

0.146g of propanone on vaporization occupies 72cm$^3$ at 87°C and 101.3kPa.

Volume at stp = $72 \times \dfrac{273}{360}$cm$^3$ = 54.6 cm$^3$

0.146g of propanone would occupy 54.6cm$^3$ at stp.

Mass which would occupy 22 400 cm$^3$ at stp = $\dfrac{0.146 \times 22\ 400}{54.6}$ = 59.90g

M$_r$ (propanone) = 59.90

## 8.10 The kinetic theory of gases

In a gas, the particles of the gas are in continual rapid, random motion. The particles are very widely spaced. The pressure of a gas is the result of continuous collisions between the particles and the walls of the container.

The following assumptions are made for ideal gases:

(*i*) The particles are in a state of continuous random motion.

(*ii*) The volume of the particles is negligible compared to the total volume of the gas.

(*iii*) The attractive forces between the particles are negligible.

(*iv*) The collisions between particles of the gas are perfectly elastic – no energy is exchanged when the particles collide.

## *8.11 Real and non-ideal gases

The ideal gas does not exist in practice; two of the major assumptions of an ideal gas are that its particles are negligible in volume and the forces of attraction between the molecules are negligible.

A gas behaves more ideally at low pressure (when the molecules are widely spaced) and high temperatures (when molecules are moving rapidly and the intermolecular forces are not significant). In other words when the gas is under conditions most like a gas and least like a liquid.

The gas laws and the ideal gas equation relate only to ideal gases. The ideal gas equation is modified in van der Waals' equation.

$$(P + \frac{a}{V^2})\,(V - b) = nRT$$ where $a$ and $b$ are constants for a particular gas. The term $(P + \frac{a}{V^2})$ takes account of the

intermolecular forces in the gas and the term $(V - b)$ compensates for the volume of the particles in the gas.

The forces which exist between the molecules in a gas are mostly due to van der Waals' forces, dipole–dipole attraction or hydrogen bonding.

## 8.12 Velocities of the molecules in a gas

At a particular temperature, the molecules in a gas are moving with a wide range of velocities (and a wide range of kinetic energies). Only a relatively few molecules possess low or high velocities.

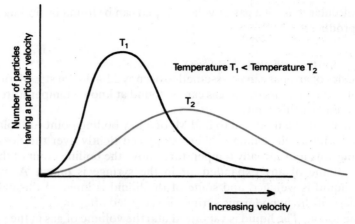

**Fig. 8.2** Distribution of velocities of molecules in a gas

Fig. 8.2 shows the distribution of velocities of the molecules in a gas. An increase in temperature causes a shift towards higher velocities and a flattening of the peak indicates that a wider range of velocities exist.

QUESTIONS

1 0.1g of a gas X has a volume of 80 cm$^3$ at 150 kPa and 27°C. The relative molecular mass of X is

A $\dfrac{150 \times 80 \times 273 \times 0.1}{300 \times 101.3}$     B $\dfrac{0.1 \times 300 \times 101.3 \times 22\,400}{150 \times 80 \times 273}$

C $\dfrac{22\,400 \times 0.1 \times 300 \times 150}{101.3 \times 80 \times 273}$     D $\dfrac{22\,400 \times 0.1 \times 27 \times 101.3}{150 \times 80 \times 273}$

Correct answer **B**

2 (*a*) By referring to the kinetic concept of the states of matter, explain concisely:

   (*i*) the melting of the solid;
   (*ii*) the evaporation of a liquid;
   (*iii*) the diffusion of a gas.

(*b*) (*i*) Assuming the air to contain 21% of oxygen by volume, calculate the partial pressure of the oxygen in air at 100 kPa pressure.
(*ii*) What is the volume composition of a helium/oxygen mixture at a total pressure of 1000 kPa which will give the same partial pressure of oxygen?
(*c*) (*i*) State the ideal gas equation.
 (*ii*) How do real gases differ from perfect gases?
(*iii*) State how van der Waals' equation accounts for these differences.
(*iv*) Calculate the value of the gas constant, *R*, given that 1 mole of an ideal gas occupies 22 400cm³ at stp
(1 atmosphere = 101.3 kPa)

(*Associated Examining Board*)

(*a*) (*i*) See Unit 10.1     (*ii*) see Unit 9.2.
(*iii*) molecules of a gas are relatively widely spaced and they are moving randomly. They move to fill uniformly all of the available space.

(*b*) (*i*) Partial pressure of oxygen $= \dfrac{\text{number of moles of oxygen}}{\text{total number of moles}} \times$ total pressure
$= 21$ kPa

(*ii*) Partial pressure of oxygen is unchanged *i.e.* 21 kPa

$$21 = \frac{\text{Percentage of oxygen by volume}}{100} \times 1000$$

$$\text{Percentage of oxygen by volume} = \frac{21 \times 100}{1000} = 2.1\%$$

Percentage of helium by volume $= 97.9\%$

(*c*) (*i*) $PV = nRT$ where *P* is pressure, *V* is volume, *n* is the number of moles of gas, *R* is the molar gas constant and *T* is the absolute temperature.
(*ii*) Real gases differ from ideal or perfect gases

   1 presence of intermolecular forces
   2 volume of molecules is not negligible.

(*iii*) For 1 mole, $(P + \frac{a}{V^2})(V - b) = RT$

(*iv*) It is important to use consistent units.

e.g. $P = 101\ 300$ Pa, $V = 0.0224\text{m}^3$, $T = 273$K, $n = 1$ mole.

Substitute in $PV = nRT$; $R = 8.31\text{J mol}^{-1}\text{K}^{-1}$

or $P = 1$ atmosphere, $V = 22.4$ litres, $T = 273$K, $n = 1$ mole

Substitute in $PV = nRT$; $R = 0.082$ litre atmospheres K$^{-1}$ mol$^{-1}$

3 Define the term partial pressure. 200cm$^3$ of hydrogen at 100 kPa (1 atm) and 20°C, and 150cm$^3$ of helium at 200 kPa (2 atm) and 20°C were mixed in a total volume of 500cm$^3$. Calculate the partial pressure at 20°C of each gas in the mixture.

(*Associated Examining Board*)

**Partial pressures see Unit 8.6**

150cm$^3$ of helium at 200 kPa and 20°C. If the pressure of this mass of gas was reduced to 100 kPa and the temperature maintained at 20°C, the volume would be 300cm$^3$ (using Boyle's law).

Mixing 300cm$^3$ of helium with 200cm$^3$ of hydrogen and using a vessel of 500cm$^3$ means that there is no change in pressure. Total pressure = 100 kPa. Now, the number of moles of gas present is proportional to the volume.

partial pressure of hydrogen $= \dfrac{200}{500} \times 100 = 40$ kPa

partial pressure of helium $= \dfrac{300}{500} \times 100 = 60$ kPa

**(Check that the sum of partial pressures equals the total pressure)**

4 Use the following information to find the relative molecular mass of a volatile liquid **X**.

0.20g of **X** on vaporization gave 40cm$^3$ of vapour measured at 373 K and a pressure of 1 atm (*i.e.* 10$^2$kPa). The molar gas constant $R$ is 0.0821 litre atm K$^{-1}$ mol$^{-1}$ or 8.31 J K$^{-1}$ mol$^{-1}$

Using $PV = nRT$ where $P = 1$ atm (10$^5$Pa), $V = 0.04$ litres (0.0004m$^3$) $T = 373$ K, $R = 0.0821$ litre atm K$^{-1}$ mol$^{-1}$ (8.31 J K$^{-1}$ mol$^{-1}$)

$n = 0.0013$ moles

Now 0.2g of **X** $\equiv 0.0013$ moles

Relative molecular mass of **X** $= \dfrac{0.2}{0.0013} = 153.8$

5 (*i*) Draw a diagram to show the distribution of velocities of molecules in a gas at certain temperature.

(*ii*) In what two ways would the graph differ at a higher temperature?

(*iii*) How is the variation in rate of a gaseous reaction with temperature related to the change in velocity distribution?

(*i*) **Fig. 8.2**

(*ii*) **Unit 8.12.**

(*iii*) **This is a frequent link between Unit 8 and Unit 14.**

The rate of a gaseous reaction depends upon the number of 'fruitful' collisions between reacting molecules. For a collision to be 'fruitful', the energy possessed by the colliding molecules must be greater than the activation energy. Increasing the temperature causes the molecules to possess greater kinetic energy and more collisions leading to reaction take place.

# **9** Liquids

## 9.1 LIQUEFYING GASES

A gas may be liquefied by cooling the gas or increasing the pressure (within limits discussed below). Andrews (1861) carried out experiments to investigate the effect of changes of temperature and pressure on the volume of carbon dioxide. The results are summarized in Fig. 9.1.

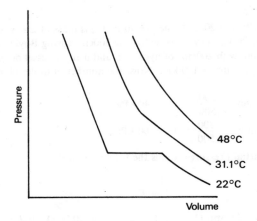

**Fig. 9.1** Andrews' isothermals

Each curve represents the variation of volume of a fixed mass of carbon dioxide with changes of pressure and at constant temperature. These curves are called **isothermals**. The isothermal at 48°C represents the usual behaviour of a gas. The isothermal at 22°C shows a horizontal portion where the volume changes rapidly for only a small change in pressure. This horizontal portion corresponds to liquefaction of carbon dioxide. There is only a slight kink in the isothermal at 31.1°C. At this temperature liquefaction barely occurs and at temperatures higher than this liquefaction cannot be achieved by simply increasing pressure. This temperature is called the **critical temperature** of carbon dioxide. The critical temperature is that temperature above which a gas cannot be liquefied by applying pressure. The minimum pressure required to liquefy a gas at its critical temperature is called the **critical pressure**.

## 9.2 THE LIQUID STATE

The particles in a liquid are much closer together than the particles in a gas. They are constantly moving and colliding. This is illustrated by **Brownian motion**. Particles of pollen on the surface of water are in a state of constant random motion. This movement of pollen grains can be seen with a microscope and can be explained by the constant bombardment of the pollen grains by water molecules.

The particles are moving with different kinetic energies. Some of the particles possess greater than average kinetic energy and, close to the surface of the liquid, may escape from the liquid into the space above. This process is called **evaporation** and, because the particles have above average kinetic energies, the temperature of the liquid falls during evaporation.

## 9.3 VAPOUR PRESSURE

Consider a liquid in a closed container with a vacuum in the space above the liquid. The system is kept at constant temperature. Molecules of the liquid escape from the liquid into the space above the liquid. Eventually, an equilibrium (Unit 15) is produced. Molecules are still leaving the liquid, but molecules are also hitting the surface of the liquid and are re-entering the liquid. In this equilibrium the pressure in the space above the liquid is unchanged and is called the **saturated vapour pressure** of the liquid at that temperature.

At a higher temperature, the molecules in the liquid possess greater kinetic energies and more escape. The gaseous molecules are moving faster and more collide with the surface and re-enter.

An equilibrium is again set up but there are more molecules in the space and, therefore, the vapour pressure is greater.

A liquid exposed to the air boils when the vapour pressure of the liquid equals atmospheric pressure.

## *9.4  PHASE DIAGRAMS FOR ONE-COMPONENT SYSTEMS

A **phase** is defined as a part of the whole system which is physically distinct from the rest of the system but its composition is the same throughout (*i.e.* it is homogeneous). A mixture of gases or a solution are examples of a single phase.

Fig. 9.2 shows a phase diagram for water. It is a diagram showing how the physical state of water changes with changes in temperature and pressure.

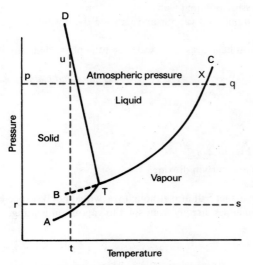

**Fig. 9.2** Phase diagram for water

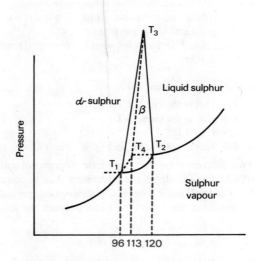

**Fig. 9.3** Phase diagram for sulphur

Referring to Fig. 9.2, TC represents the change of vapour pressure of liquid water with temperature. The line pq represents atmospheric pressure. TC meets pq at X. X is the boiling point of water.

TD represents the equilibrium between solid and liquid *i.e.* the change of melting point of ice with change in pressure. As the pressure increases, the melting point of ice decreases but big changes in pressure are required to make a significant change in melting point. For all substances apart from water, antimony and bismuth, the line DT slopes slightly forward (rather than slightly backward) and an increase in pressure produces a slight increase in the melting point.

AT represents the equilibrium between solid and vapour and TB represents the **supercooling** of liquid water.

Point T is called the **triple point** and under this set of conditions ice, water and water vapour can be in equilibrium together. For water, this unique situation exists at 0.0075°C and 530Pa pressure.

Point C is called the **critical point**. Above this pressure ($22.12 \times 10^3$kPa) and this temperature (647.4 K) a gas cannot be liquefied by decreasing only the temperature or increasing only the pressure.

If a sample of ice is heated at atmospheric pressure, it changes from solid to liquid to vapour (follow dotted line pq). If a sample of ice is heated at a low pressure, the solid ice can turn directly to a vapour. Also under these conditions a vapour can turn directly to a solid (follow dotted line rs). This process is called **sublimation**.

A sample of water vapour maintained at constant temperature undergoes changes of state with increasing pressure. As the pressure increases, the water vapour turns directly to a solid and then at a higher pressure, the solid melts (follow dotted line tu).

A simple phase diagram has to be modified for substances which can exist in different forms in the same physical state. Sulphur, for example, shows enantiotropy (Unit 26.8) and the phase diagram is shown in Fig. 9.3.

Up to 96°C, α-sulphur is the stable allotrope of solid sulphur. At the triple point $T_1$, α-sulphur, β-sulphur and sulphur vapour can exist in equilibrium. $T_2$ and $T_3$ are other triple points and $T_4$ is an unstable triple point for α-sulphur, liquid sulphur and sulphur vapour.

QUESTIONS

1 Multiple completion. Using the method of recording responses on page 3.
Which of the following statements is (are) true for the change from a gas directly to a solid?

  (*i*) The process is called sublimation.
  (*ii*) There is a decrease in entropy.
  (*iii*) The particles in the gas move closer together.
  (*iv*) The process cannot be reversed.

  (*i*), (*ii*) and (*iii*) are all correct. This corresponds to response A.

2 Draw carefully a single diagram showing:
  (*a*) how the vapour pressure of liquid water varies with temperature,
  (*b*) how the vapour pressure of ice varies with temperature,
  (*c*) the effect of pressure on the melting point of ice, and
  (*d*) the effect of an involatile solute on the vapour pressure of liquid water.
On your diagram, label the appropriate areas solid, liquid and vapour and indicate the positions of the critical point and the triple point.
  Give two differences between the phase diagram you have drawn for water and the phase diagram for carbon dioxide.

*(Oxford and Cambridge)*

The diagram required is Fig. 9.2.
  (*a*) This is the curve CT
  (*b*) This is the curve AT.
  (*c*) This is DT.
  (*d*) This is not shown on Fig. 9.2 but is shown on Fig. 12.2.

Two differences between the phase diagrams of water and carbon dioxide.
(*i*) The line DT slopes forward rather than backward.
(*ii*) The triple point T lies above atmospheric pressure (in fact 5.01 atm although you would not be expected to remember this value). At 1 atm pressure the solid changes directly from solid to vapour on heating.

# 10 Solids

## 10.1   INTRODUCTION

The particles in a solid may be atoms, ions or molecules. In a solid, the particles are regularly arranged in a lattice and this regular arrangement causes the solid to be crystalline.

One of the difficulties with answering questions on the solid state is to represent, in two dimensions, crystal lattices which exist in three dimensions. It is important to practise drawing the diagrams in this Unit.

In a lattice at 0K, the particles would be stationary but at any temperature above they are vibrating. As the temperature is raised the particles vibrate even more about their positions in the lattice. When the lattice vibrations reach a certain level, the forces between the particles in the lattice are inadequate and the lattice breaks up. This process is called **melting**. At the melting point there is a dynamic equilibrium between particles in the solid state and particles in the liquid state.

## 10.2   X-RAY DIFFRACTION

X-ray diffraction is used to determine the arrangements of particles in a solid. The wavelength of X-rays is approximately the same as the distance between the particles in the lattice ($10^{-10}$m).

X-rays are produced when cathode rays fall upon metals. A beam of X-rays (of the same wavelength) strikes a crystal and the X-rays are diffracted by the crystal and are detected by a photographic plate.
If a beam of X-rays strikes a crystal (Fig. 10.1) with the waves of the X-rays in phase, the reflected rays will only remain in phase if the difference in distance travelled is a whole number of wavelengths.

If the X-rays are to remain in phase, then
extra distance travelled $= CB + BD$ and $CB + BD = n\lambda$
where $n$ is an integer and $\lambda$ is the wave length of the X-rays.

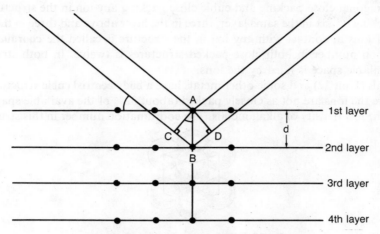

**Fig. 10.1** Beam of X-rays striking a crystal

If the angle between the crystal face and the incident X-ray is $\theta$, then $\widehat{CAB}$ and $\widehat{BAD}$ equal $\theta$.
By trigonometry, $CB = AB\sin\theta$ and $BD = AB\sin\theta$.
The distance AB is the distance between the first and second layers of particles in the crystal $= d$

$$\therefore \ 2d\sin\theta = n\lambda$$

This equation is called the **Bragg equation**.

In a particular experiment, using one face of a crystal, the wavelength ($\lambda$) and the distance between the layers ($d$) is fixed. For a given angle $\theta$, a series of spots will appear on the photographic plate corresponding to $n = 1, 2, 3$ etc. From these results $d$ can be calculated.

X-ray diffraction is used for single crystals or for a fine powder. Electron diffraction can also be used to determine the structure of a solid. Electrons, however, have a poorer penetrating power than X-rays.

## *10.3 METAL CRYSTAL STRUCTURES

A metal consists of a close-packed regular arrangement of positive ions, which are surrounded by a 'sea' of delocalized electrons that bind the ions together. The ions in a close-packed layer are arranged in a regular hexagon.

The close-packed layers of ions can be stacked in two ways – both being equally likely. Some metals have one type of stacking, some metals have the other type and some metals can exist in either depending upon the conditions. The two ways of stacking these layers are:

(*i*) ABAB or **hexagonal close packing** (Fig. 10.2 *a*). The ions in the third layer are immediately above ions in the first layer. The stacking of the layers will continue ABAB... Magnesium and zinc are examples of metals with this structure.

(*ii*) ABC or **cubic close packing** (sometimes called face centred cubic) (Fig. 10.2 *b*). The ions in the third layer are not immediately above ions in the first layer. The stacking of the layers will continue ABCABC... Aluminium and copper are examples of metals with this structure.

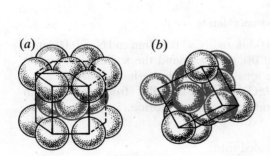

**Fig. 10.2** Hexagonal close packing (a) and cubic close packing (b)

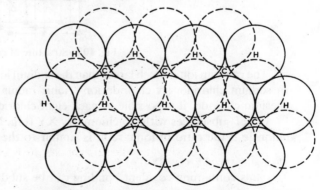

**Fig. 10.3** Difference between hexagonal close packing and cubic close packing

The difference between these two structures may be seen clearly in Fig. 10.3. The ions in the first layer are shown in continuous lines. The ions in the second layer are shown in dotted lines. There are two sets of hollows in the second layer in which ions in the third layer can rest. These are labelled H or C. Ions cannot rest in both H and C. If they rest in hollows H, hexagonal close packing is produced but if they rest in hollows C, cubic close packing results.

In both hexagonal close packing and cubic close packing any ion in the structure has twelve other ions touching it – six in the same layer, three in the layer above and three in the layer below. The number of ions in contact with any ion in the structure is called the **coordination number**. The coordination number in both close-packed structures is twelve. In both structures about 75% of the available space is filled by the ions.

Alkali metals (Unit 22) and some other metals have a **body centred cubic** structure (Fig. 10.4). In this structure the ions are not as closely packed (about 32% of the available space is unfilled). This explains the low density of alkali metals. The coordination number in this structure is eight.

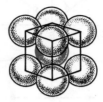

**Fig. 10.4** Body centred cubic structure

### 10.4    IONIC CRYSTALS: SODIUM CHLORIDE

An ionic crystal is a regular arrangement of positive and negative ions. The ions are held together by strong electrostatic forces. The sodium chloride lattice is shown in Fig. 10.5. The sodium chloride lattice consists of a face centred cube of sodium ions and an inter-penetrating face centred cube of chloride ions.

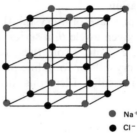

$\bullet$ Na$^+$
$\bullet$ Cl$^-$

**Fig. 10.5** Structure of sodium chloride

In simple terms, each sodium ion in the lattice is surrounded by six chloride ions and each chloride ion by six sodium ions. The coordination is called 6:6.

### *10.5    IONIC CRYSTALS: CAESIUM CHLORIDE

Although sodium chloride NaCl and caesium chloride CsCl have similar formulae they have different crystal structures. In the caesium chloride structure (Fig. 10.6), the coordination is 8:8. Eight chloride ions surround each caesium and eight caesium ions around each chloride ion.

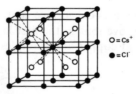

O = Cs$^+$
$\bullet$ = Cl$^-$

**Fig. 10.6** Structure of caesium chloride

The caesium ion is much larger than the sodium ion (ionic radii Cs$^+$ 0.167nm and Na$^+$ 0.098nm). In sodium chloride, six chloride ions (ionic radius 0.181 nm) surround the sodium ion but, in caesium chloride, it is possible to pack eight chloride ions around each caesium atom.

Other substances with stoichiometry XY (e.g. MgO) can have a similar structure to sodium chloride if the ratio of ionic radii is similar to the ratio in sodium chloride.

### 10.6    COVALENT CRYSTALS

Crystals containing covalent bonding can be subdivided into:

#### (i) Giant covalent structures

In a giant covalent structure, the atoms are bonded together to give a three-dimensional (or two-dimensional) structure with a high melting point.

Diamond and graphite are two allotropic forms of carbon. In diamond, all of the carbon atoms are $sp^3$ hybridized (see also Unit 5.2) and joined by four covalent bonds to four other atoms pointing towards the corners of a regular tetrahedron. All of the C–C bond lengths are the same. The bonds are very strong.

Graphite is a layer structure. The carbon atoms in each layer are $sp^2$ hybridized and the bonds between carbon atoms in the layer are strong. The forces between the layers are weak Van der Waals' forces and the layers are able to slide over one another.

### (ii) Molecular crystals

In a molecular crystal there are discrete molecules containing *strong* covalent bonds held together by *weak* Van der Waals' forces or hydrogen bonding. A substance which exists as molecular crystals melts at a low temperature because the weak forces are easily broken. Examples of molecular crystals are iodine and ice.

### 10.7    ISOTROPY AND ANISOTROPY

An isotropic substance is one whose properties are the same in whichever direction they are measured. For example the refractive index (a measure of the speed of light through a crystal) of a sodium chloride crystal is the same whichever direction light passes through the crystal. Isotropy occurs when spherical ions such as $Na^+$ and $Cl^-$ are arranged in a spherically symmetrical arrangement (Fig. 10.5). All substances that crystallize in a cubic system are isotropic. Other properties that could be considered include thermal conductivity and electrical conductivity.

An anisotropic substance is one whose properties depend upon the direction in which they are measured. For example the refractive index of calcite (a natural form of calcium carbonate) can vary between 1.49 and 1.66 depending upon the direction that light passes through the crystal. Anisotropic substances contain non-spherical ions *e.g.* $CO_3^{2-}$ or a non-spherically symmetric arrangement.

Whether a crystal is isotropic or anisotropic can be found easily by illuminating a crystal and viewing it through crossed polaroids. Isotropic substances appear dark or faintly illuminated whatever their orientation. Anisotropic substances appear dark or faintly illuminated in some orientations and brightly illuminated in others.

### QUESTIONS

1 Representations of the structures of sodium metal, diamond, caesium chloride, carbon dioxide are included in the diagrams **A–E** on page 68. The following questions relate to these substances.
   - (*a*)  (*i*) Identify the caesium chloride structure by giving the appropriate letter.
     - (*ii*) Give the coordination number of $Cs^+$ and $Cl^-$.
   - (*b*)  (*i*) Name the hardest substance.
     - (*ii*) Identify the appropriate structure.
     - (*iii*) Give reasons for its hardness.
   - (*c*)  (*i*) Name the most volatile substance.
     - (*ii*) Identify its structure.
     - (*iii*) Explain its volatility.
   - (*d*) One of the substances having a structure in which the coordination is 6:6 shatters when tapped with a hammer.
     - (*i*) Identify the structure.
     - (*ii*) Explain why the substance shatters.
   - (*e*) Only one substance has a high electrical conductance.
     - (*i*) Name it and identify its structure.
     - (*ii*) Explain this property.

                                                         *(Welsh Joint Education Committee)*

Three of these five substances are compounds and contain two different types of atom.

   **C, D** and **E** must be caesium chloride, carbon dioxide and sodium chloride. **A** and **B** are elements – diamond and sodium metal.

(*a*) (*i*) **D**    (*ii*) $Cs^+ = 8$, $Cl^- = 8$.

(*b*) (*i*) The hardest substance is diamond. In the diamond structure the four covalent bonds from each carbon atom are tetrahedral.

   (*ii*) **A**

   (*iii*) All bonds in diamond are equally strong C–C bonds. The structure cannot be broken up without breaking these bonds.

(*c*) (*i*) The most volatile (or easily vapourized) substance is carbon dioxide.

   (*ii*) **E**

   (*iii*) The bonds within each carbon dioxide molecule are strong covalent bonds but the forces between the molecules are very weak. At room temperature solid carbon dioxide turns directly to a gas.

(*d*) (*i*) **C**

   (*ii*) There are four planes within the crystal along which the crystal can be cleaved.

(*e*) (*i*) Sodium – **B**

   (*ii*) Unit 10.3

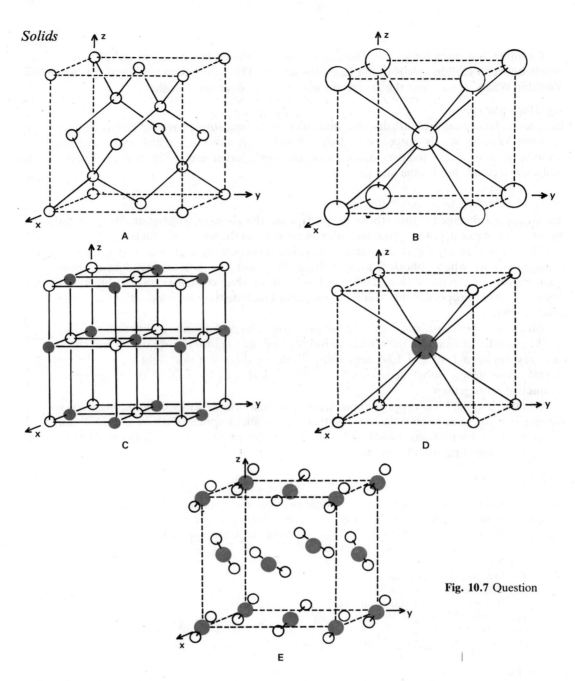

**Fig. 10.7** Question

2 Describe and explain how X-rays have been used to determine the arrangement of atoms and ions in a crystal.
   What do you understand by the terms hexagonal close-packing, cubic close-packing (face-centred cubic close-packing) and body-centred cubic packing as they apply to the structures of metals? Illustrate your answers with diagrams and give examples where possible.

*(London University)*

See Units 10.2 and 10.3.

# **11** Solutions

## 11.1   INTRODUCTION

Most of the solutions you have met on a GCSE course were solutions of solids in liquids. In this unit solutions of gases in liquids, liquids in liquids and solids in solids will be considered. This unit is closely linked with Unit 12 which deals with the colligative properties produced when a solute dissolves in a solvent.

A solution is still a single phase but there are now two or more components.

## 11.2   SOLUTIONS OF GASES IN LIQUIDS

Most gases dissolve in a solvent with the evolution of heat, *i.e.* the process is **exothermic**. In accordance with Le Chatelier's Principle (Unit 15.5), the solubility is decreased by increasing temperature.

The most important factor affecting the solubility of a gas in a solvent is expressed by **Henry's law**: the mass of gas dissolved by a given volume of solvent, to give a saturated solution, is directly proportional to the pressure of the gas, providing the temperature remains constant and there is no reaction between the gas and the solvent.

Gases such as ammonia, hydrogen chloride and carbon dioxide react when they are dissolved in water. The law does not apply to these gases but applies satisfactorily for gases such as oxygen, hydrogen, helium etc.

The solubility of a gas in a liquid is usually expressed as an **absorption coefficient**. This is defined as the number of cm³ of gas at stp which saturate 1cm³ of the liquid at a given temperature and 1 atm (101.3kPa).

If a mixture of gases is in equilibrium with a solvent (*e.g.* air with water), each gas dissolves according to its own partial pressure (Unit 8.6).

### 11.3 SOLUTIONS OF LIQUIDS IN LIQUIDS

In this section it is useful to consider the mixing of two completely miscible ideal liquids and then to consider the effects of deviations from ideality. Ideal liquids show no heat or volume change when they are mixed and obey Raoult's Law over the whole composition range.

**Raoult's Law states that the vapour pressure of a constituent of an ideal solution is equal to the vapour pressure exerted by the pure constituent at that temperature multiplied by the mole fraction of that constituent.**

**Mole fraction** is a convenient way of expressing the composition of each component in a mixture. The mole fraction of A in a mixture A and B

$$x_A = \frac{n_A}{n_A + n_B}$$

The mole fraction of B

$$x_B = \frac{n_B}{n_A + n_B}$$

The sum of the mole fractions of the components in the mixture equals one.

$$x_A + x_B = 1$$

Raoult's Law can be expressed as

$$P_A = x_A P_A^\circ$$

where $P_A$ is the vapour pressure of A over the liquid

$x_A$ is the mole fraction of A

$P_A^\circ$ is the vapour pressure of A over pure A at that temperature.

Similarly

$$P_B = x_B P_B^\circ$$

Total pressure of vapour over the liquid $P = P_A + P_B$. This is expressed in the graph in Fig. 11.1.

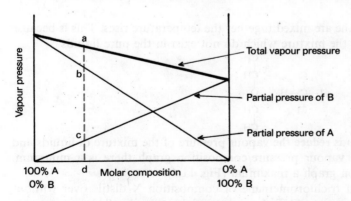

**Fig. 11.1** Pressure of vapour over a liquid for a mixture of A and B

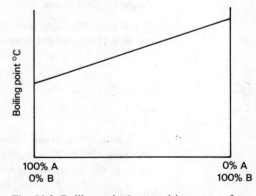

**Fig. 11.2** Boiling point/composition curves for a mixture of A and B

The vertical dotted line represents a solution of composition 20% B and 80% A. The total vapour pressure of this solution ad is equal to the sum of the partial pressure due to A (db) and the partial pressure due to B (cd).

Fig. 11.2 shows the boiling point/composition curve for the same mixture of A and B. A liquid which has a high vapour pressure will have a low boiling point and *vice versa*.

Some pairs of liquids obey Raoult's Law closely *e.g.* benzene and methylbenzene or hexane and heptane. The two components in each mixture are very similar chemically. Fig. 11.3 shows the vapour pressure and boiling point diagrams for a mixture of hexane and heptane.

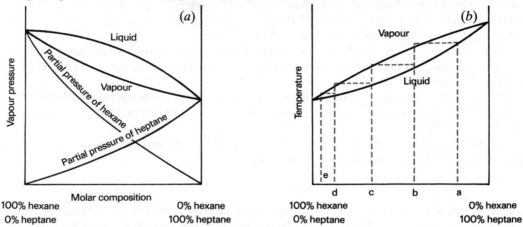

**Fig. 11.3** Vapour pressure and boiling point diagrams for a mixture of hexane and heptane

The biggest difference between the vapour pressure/composition curve in Fig. 11.3 and the ideal curve in Fig. 11.1 is the existence of separate curves for the composition of the liquid and vapour.

In the temperature/composition curve, the variation of the boiling point of the liquid with composition of the liquid is shown by the curve labelled 'liquid'. The 'vapour' curve shows the composition of the vapour in equilibrium with a liquid mixture at its boiling point. The composition of the vapour in equilibrium with a liquid of any composition is always richer in the more volatile (lower boiling point) component.

When a mixture of hexane and heptane of composition a (in Fig. 11.3 *b*) is heated, the vapour which is in equilibrium with this mixture has a composition b (richer in hexane). If this vapour is condensed the liquid still has the composition b. If this liquid is reheated the liquid of composition b is in equilibrium with the vapour of composition c. Each step produces a liquid richer in the lower boiling point component. Further repetition of this process will eventually give pure hexane.

Rather than carry this out in a series of steps, it is usual to carry it out in one stage using a **fractionating column**. A series of condensations and evaporations take place as the vapour moves up the column. Fractional distillation is used to separate the components of a pair of miscible liquids that are near ideal.

### 11.4 DEVIATIONS FROM RAOULT'S LAW

Deviations from the ideal behaviour can be of two types – positive and negative deviations.

**Negative deviations**

If trichloromethane and ethoxyethane are mixed together the temperature rises. This is because extra intermolecular forces exist in the mixture which do not exist in the pure liquids.

These extra, weak hydrogen bonds **reduce** the vapour pressure of the mixture of liquids and **increase** the boiling point. On the vapour pressure/composition graph there is a minimum, and on the temperature/composition graph a maximum (Fig. 11.4).

A mixture of ethoxyethane and trichloromethane of composition X distils over without change of composition and with boiling point T. A mixture of this type is called an **azeotropic** or **constant boiling point** mixture.

If a mixture of any other composition is heated, one of the pure components ethoxyethane or trichloromethane distils off until the composition of the mixture becomes the same as the azeotropic mixture. Then the pure azeotropic mixture distils off. Whether ethoxyethane or trichloromethane distils off first depends upon which one is in excess. Other examples of pairs of liquids behaving in this way include: hydrochloric acid and water, and sulphuric acid and water.

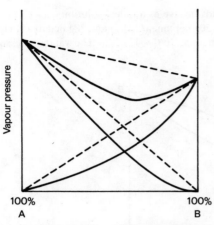
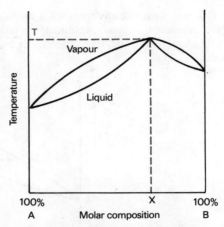

**Fig. 11.4** Vapour pressure/composition and temperature/composition graphs (minimum vapour pressure)

### Positive deviation

When trichloromethane and ethanol are mixed together the temperature falls. There are strong forces of hydrogen bonding which exist in pure ethanol (Fig. 11.5(a)).

**Fig. 11.5(a)**                                    **Fig. 11.5 (b)**

Hydrogen bonds form between the molecules on mixing the two liquids (Fig. 11.5 (b)) but are weaker than those in pure ethanol. (The important factor is how the bonding between the molecules (intermolecular bonding) in the mixture compares with the intermolecular bonding in the original separate components.) The tendency for molecules to escape, therefore, increases. This **increases** the vapour pressure and **reduces** the boiling point–there is a minimum on the temperature/composition graph (Fig. 11.6).

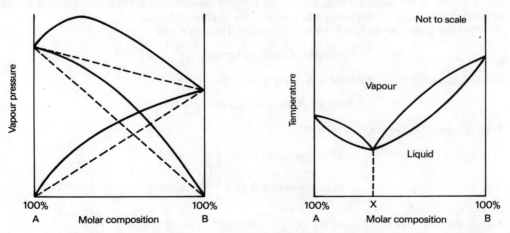

**Fig. 11.6** Vapour pressure/composition and temperature/composition graphs (maximum vapour pressure)

The mixture of composition X distils over without change of composition. Fractional distillation of a mixture will produce the azeotropic mixture X and this will distil over until one of the components is used up. The remaining component distils over pure.

A mixture of ethanol and water behaves in this way. When a mixture of 10% ethanol and 90% water is heated, the azeotropic mixture containing 95.6% ethanol and 4.4% water distils over until all the ethanol has been used up. Then pure water distils over. It is impossible to obtain pure ethanol by distilling solutions of ethanol in water.

### QUESTIONS

1 (*a*) Heptane and octane form an ideal solution.
(*i*) Give a mathematical expression for Raoult's vapour pressure law for a solution of two liquids, A and B explaining the terms used.

(*ii*) Under what circumstances will a mixture of two liquids behave as an ideal solution?

(*iii*) Calculate the vapour pressure of a solution containing 50g heptane ($C_7H_{16}$) and 38g octane ($C_8H_{18}$), at 20°C ($A_r(H) = 1$, $A_r(C) = 12$. The vapour pressure of heptane at 20°C = 473.2Pa. The vapour pressure of octane at 20°C = 139.8 Pa.)

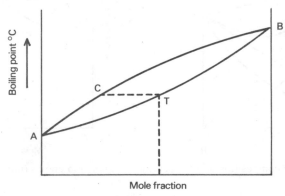

**Fig. 11.7** Question

(*b*)

The above diagram represents the boiling point composition diagram for a solution of heptane and octane.

(*i*) What do the points A and B represent?

(*ii*) An equimolar solution of heptane and octane is distilled. Mark on the diagram the temperature T at which the solution begins to boil and the point C giving the composition of the vapour given off when the solution begins to boil.

(*iii*) Explain briefly how the solution can be separated into pure heptane and pure octane.

(*Associated Examining Board*)

(*i*) Raoult's law can be expressed as follows:

$P_a = x_a P_a^\circ$ where $P_a$ represents the vapour pressure of A over the solution, $P_a^\circ$ represents the vapour pressure of A over pure A at the same temperature and $x_a$ represents the mole fraction of A in the solution.

$$x_a = \frac{n_a}{n_a + n_b}$$

$n_a$ – number of moles of A
$n_b$ – number of moles of B

Similarly $P_b = x_b P_b^\circ$ and the total pressure $P = P_a + P_b$

(*ii*) A mixture of two liquids is ideal if the forces between the molecules in the separate liquids are similar in strength and nature to the forces which exist in the mixture of liquids.

(*iii*) Relative molecular mass of heptane = $(7 \times 12) + (16 \times 1) = 100$

$$\text{Number of moles of heptane} = \frac{50}{100} = 0.5$$

Relative molecular mass of octane = $(8 \times 12) + (18 \times 1) = 114$

$$\text{Number of moles of octane} = \frac{38}{114} = 0.33$$

Total number of moles = $n_a + n_b = 0.83$

$$\text{Mole fraction of A } x_a = \frac{0.5}{0.83} = 0.602$$

$$\text{Mole fraction of B } x_b = \frac{0.33}{0.83} = 0.397$$

(Check $x_a + x_b = 1$)

Vapour pressure of A above solution = $x_a P_a^\circ = 0.602 \times 473.2 = 284.9$Pa
Vapour pressure of B above solution = $x_b P_b^\circ = 0.397 \times 139.8 = 55.5$Pa
Total pressure above the solution = $284.9 + 55.5 = 340.4$Pa.

(*b*) (*i*) A represents the boiling point of A and B the boiling point of B.

(*ii*) See the graph (Fig. 11.7). Dotted lines have been added.

(*iii*) A series of evaporations and condensations will produce pure A. This can be achieved in a fractional distillation column.

2  Ethanol (boiling point 78.5°C) and water form a constant boiling point mixture having a boiling point of 78.2°C and a composition of 95.6% ethanol.

(*a*) Define the term constant boiling point mixture.

(*b*) Sketch and label fully, the boiling point/composition diagram for ethanol and water.

(*c*) An ethanol/water mixture shows positive deviations from Raoult's Law. Explain and account for this and state the law.

(*d*) What intermolecular change takes place when ethanol is added to water.

(*e*) State qualitatively the result of distilling (*i*) a mixture containing 75% ethanol and (*ii*) a mixture containing 97.5% ethanol.

(*f*) State with reasons which one of the following pairs of substances most closely obeys Raoult's Law.
  (*i*) $C_2H_5NH_2$ and $C_6H_5NH_2$
  (*ii*) $CH_3COCH_3$ and $CH_3COC_2H_5$
  (*iii*) $C_6H_6$ and $C_6H_5CH_3$

(*a*) A constant boiling point mixture is a mixture of liquids which boils at a constant temperature and without change in composition.
(*b*) Fig. 11.5
(*c*) Unit 11.4
(*d*) Unit 11.4
(*e*) (*i*) The azeotropic mixture distils over until all the ethanol has gone. Then water distils over (*ii*) The azeotropic mixture distils over unchanged.
(*f*) (*iii*) The forces between molecules in $C_6H_6$ and between molecules in $C_6H_5CH_3$ are similar and also similar to forces between molecules of $C_6H_6$ and $C_6H_5CH_3$ in the mixture.

# 12 Colligative properties

## 12.1   INTRODUCTION

When a non-volatile solute is added to a solvent, there are changes in certain colligative properties. A **colligative property** is a property which depends upon the number of particles in a given volume of solvent and not the nature of the particles. Colligative property measurements are used to determine the relative molecular masses of dissolved solutes.

The four colligative properties are:

  (*i*) lowering of vapour pressure;
 (*ii*) elevation of boiling point;
(*iii*) depression of freezing point;
 (*iv*) osmotic pressure.

Each of these four properties will be considered separately. It is important to remember that the determination of relative molecular masses can be carried out only if **non-electrolytes** are used in **dilute** solution.

## 12.2   LOWERING OF VAPOUR PRESSURE

If a non-volatile solute, *e.g.* urea, is added to water, there is a lowering of vapour pressure of the solution.

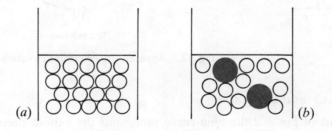

(*a*)                                     (*b*)

**Fig. 12.1** Vapour pressure of solutions

In Fig. 12.1*a*, all of the pure solvent molecules at the surface can escape into the vapour if they possess sufficient energy. In the dilute solution (*b*) the solid circles represent solute molecules which are involatile hence there are a reduced number of molecules at that temperature which can escape into the vapour and, therefore, a lowering of vapour pressure.

There is a relationship between the lowering of vapour pressure and the concentration of solution. It is given by Raoult's law (Unit 11.3).

Raoult's Law states that the relative lowering of vapour pressure is equal to the mole fraction of the solute.

$$\frac{P^\circ - P}{P^\circ} = \frac{n}{N + n}$$

where $P^\circ$ represents the vapour pressure of the pure solvent, $P$ represents the vapour pressure of the solution, $n$ is the number of moles of solute and $N$ is the number of moles of solvent.

If the masses of solvent and solute are $W$ and $w$ respectively and the molar masses of solvent and solute are $M$ and $m$, the equation becomes:

$$\frac{P^\circ - P}{P^\circ} = \frac{\dfrac{w}{m}}{\dfrac{w}{m} + \dfrac{W}{M}}$$

For very dilute solutions, the number of moles of solute is small and the above expression approximates to:

$$\frac{P^\circ - P}{P^\circ} = \frac{n}{N}$$

One problem that a candidate has is to decide whether to use the approximation or work out the answer in full. The best advice is to work out the answer without using the approximation. With a calculator it does not take much longer. Use the approximate expression if you have insufficient time or if the question asks for an approximate answer.

Raoult's law applies only to dilute solutions.

*Example*: 10g of a non-volatile solute X were dissolved in 90g of water. The vapour pressure of this solution was lowered to 742mm. The vapour pressure of water at this temperature was 750mm. Calculate the relative molecular mass of solute X.

Substitute in the above equation, knowing that the relative molecular mass of water is 18.

$$\frac{750 - 742}{750} = \frac{10/M}{10/M + 90/18} \quad \therefore \; M = 185.5$$

(If you had used the approximation, you would have obtained a value of 187.5)

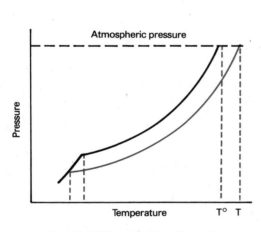

**Fig. 12.2** Elevation of boiling point

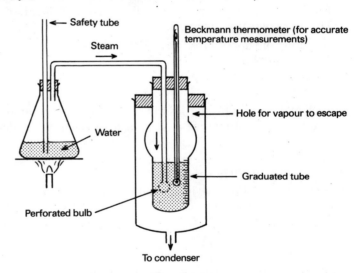

**Fig. 12.3** Apparatus to determine elevation of boiling point

## 12.3  ELEVATION OF BOILING POINT

Fig. 12.2 shows part of Fig. 9.2 the blue curve represents the vapour pressure of the dilute solution and the solid black line the vapour pressure of the pure solvent. The vapour pressure of the solution is less than the vapour pressure of the pure solvent at all temperatures. In consequence, the vapour pressure of the dilute solution does not reach atmospheric pressure until a higher temperature.

For dilute solutions, the elevation of boiling point $(T - T^0)$ is proportional to the lowering of the vapour pressure.

$$\Delta T \propto P^\circ - P$$
$$\Delta T = k(P^\circ - P) \qquad (1)$$

Raoult's law for very dilute solutions is:

$$\frac{P^\circ - P}{P^\circ} = \frac{n}{N} \qquad (P^\circ - P) = \frac{P^\circ n}{N}$$

Substituting in equation($1$)

$$\Delta T = kP^\circ \frac{n}{N}$$

The vapour pressure of a solvent is constant at a particular temperature. Therefore, $kP^\circ$ is a constant.

$$\text{Now, } n = \frac{w}{m} \text{ and } N = \frac{W}{M}$$

$$\Delta T = k' \frac{wM}{mW}$$

If a fixed mass of 1000g of solvent is taken, the ratio $\frac{M}{W}$ is a constant.

$$\Delta T = \frac{Kw}{m}$$

where $K$ is called the boiling point (or ebulliscopic) constant,
$w$ is the mass of solute
$m$ is the **relative molecular mass** of the solute.

The boiling point constant is the elevation of boiling point which would occur if one mole of a non-ionizing and non-volatile solute is dissolved in 1000g of solvent.

The experimental determination of the relative molecular mass of urea in solution in water can be carried out by elevation of boiling point measurements. Using the apparatus in Fig. 12.3, steam is passed through the water in the graduated tube until the water starts to boil. The temperature of the boiling water is recorded.

A weighed amount of solute is dissolved in the water in the graduated tube and steam is again passed until the solution boils. The temperature of the solution and the final volume of the solution are recorded. Heating the solution with steam prevents superheating.

*Example*: Calculate the relative molecular mass of a non volatile solute from the following results:

1.00g of solute dissolved in 40g of benzene causes the boiling point to be elevated by 0.20°C. The boiling point constant for benzene is 2.7°C kg⁻¹.

The equation $\Delta T = K \frac{w}{m}$ can be used where $\Delta T = 0.20°C$, $K = 2.7°C$ kg⁻¹,

$w$ = mass of solute in 1000g benzene = 25g

$$m = \frac{2.7 \times 25}{0.2} = 337.5$$

The problem with using an equation is that if you get the equation incorrect you are unlikely to pick up any marks.

An alternative method, starting from first principles, is:

1 mole (*i.e. m*g) of solute dissolved in 1000g of solvent elevates boiling point by 2.7°C.

1g of solute dissolved in 1000g of solvent elevates boiling point by $\frac{2.7°C}{m}$

(elevation of boiling point $\propto$ concentration)

1g of solute dissolved in 40g of solvent elevates boiling point by $\frac{2.7}{m} \times \frac{1000}{40}$

(This solution is more concentrated and therefore elevation of boiling point is greater).

$$\frac{2.7}{m} \times \frac{1000}{40} = 0.2$$

$$m = 337.5$$

## 12.4 DEPRESSION OF FREEZING POINT

A dilute solution freezes at a slightly lower temperature than the pure solvent. This small difference in temperature is called the **depression of freezing point**.

For dilute solutions, the depression of freezing point is proportional to the lowering of vapour pressure. The equation

$$\Delta T = K \frac{w}{m}$$

applies but in this case $K$ is called the **freezing point** (or cryoscopic) **constant**. It is the depression of freezing point which would occur if one mole of any non-ionizing and non-volatile solute were dissolved in 1000g of solvent. The depression of freezing point can be found using the apparatus in Fig. 12.4.

A known mass of the solvent is put into the centre tube and is stirred thoroughly to minimise supercooling. The temperature is recorded every half minute and a cooling curve is plotted.

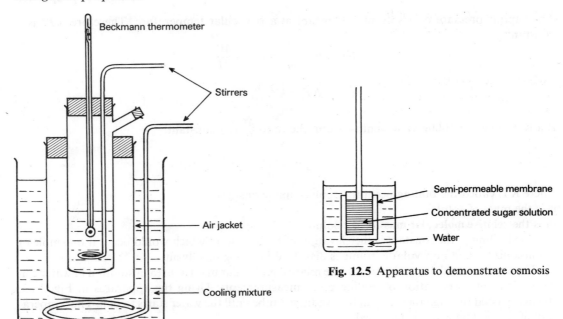

Fig. 12.4 Apparatus to determine depression
of freezing point

Fig. 12.5 Apparatus to demonstrate osmosis

The solvent is then warmed until it melts and a known mass of solute is added to the solvent. The cooling process is repeated and the new freezing point is determined.

The calculations are similar to the elevation of boiling point calculations.

### 12.5  OSMOSIS AND OSMOTIC PRESSURE

If two solutions of different concentrations are separated by a semi-permeable membrane, solvent molecules pass through the membrane and equalize the concentrations of the two solutions. This process is called **osmosis**. Solvent molecules pass through the membrane in both directions but the rate at which the solvent passes from the dilute to the concentrated solution is greater than the reverse rate.

A semi-permeable membrane is a membrane which will allow solvent molecules but not solute molecules to pass through it. The simplest semi-permeable membrane is a piece of pig's bladder but this is not very strong in practice. A good semi-permeable membrane is produced when copper(II) hexacyanoferrate(II) is precipitated in a porous pot.

The apparatus in Fig. 12.5 is used to demonstrate osmosis.

Over a period of time the water level in the tube rises as the water molecules pass through the semi-permeable membrane into the sugar solution.

When the level in the tube no longer rises, the pressure of this column of solution is equal to the **osmotic pressure**. The osmotic pressure of a solution is the pressure which has to be applied to the solution side of a semi-permeable membrane in order *just* to prevent passage of solvent molecules from pure solvent on the other side of the membrane into the solution. Solutions which have the same osmotic pressure are said to be **isotonic**.

The two laws of osmosis are:

(*i*) At a given temperature, the osmotic pressure $\Pi$ (pi) of a dilute solution is proportional to the concentration $c$.

$$\Pi \propto c \text{ (but } c \propto \frac{1}{V} \therefore \Pi \propto \frac{1}{V} \text{ )}$$

(*ii*) The osmotic pressure of a given solution is directly proportional to the absolute temperature $\Pi \propto T$.

These two laws are similar to Boyle's and Charles' laws for gases (Units 8.3, 8.4). They can be combined to form an equation.

$\Pi V = nRT$ where the value of $R$ is equal to the gas constant. The number of moles of solute $n = \frac{w}{m}$ where $w =$ mass of solute and $m$ is the relative molecular mass of the solute.

$$m = \frac{wRT}{\Pi V}$$

Osmotic pressure measurements can be used to find the relative molecular mass of a non-ionized solute. It is particularly useful for polymers which produce such small elevations of boiling point or depressions of freezing point that relative molecular masses cannot be determined accurately by these methods.

### 12.6 ABNORMAL RELATIVE MOLECULAR MASSES

Accurate relative molecular masses can be obtained for undissociated solutes in dilute solution. For partially or completely dissociated solutes the relative molecular masses obtained will be less than the true value due to the extra particles present.

For a partially dissociated electrolyte AB, the following equilibrium exists:

$$AB \rightleftharpoons A^+ + B^-$$

If one mole of AB is used and $\alpha$ moles of AB have dissociated, after dissociation

$$
\begin{array}{ccc}
(1-\alpha) & \alpha & \alpha \\
\text{moles} & \text{moles} & \text{moles}
\end{array}
$$

The ratio of the number of particles after dissociation and before dissociation is $\dfrac{1+\alpha}{1}$

This is equal to the van't Hoff factor $i$ where

$$i = \frac{\text{observed effect on colligative property}}{\text{calculated effect on colligative property}}$$

By equating, the value of $\alpha$ can be calculated. It is important to realise that an electrolyte $CaCl_2$ does not dissociate as before.

$$
\begin{array}{ccc}
CaCl_2 \rightleftharpoons & Ca^{2+} + & 2Cl^- \\
1-\alpha & \alpha & 2\alpha
\end{array}
$$

$$\frac{\text{Number of moles after dissociation}}{\text{Number of moles before}} = \frac{1+2\alpha}{1}$$

### QUESTIONS

**1** Which one of the following would freeze at the lowest temperature?

    A  pure water
    B  a solution of sodium chloride $(0.1 \text{ mol dm}^{-3})$
    C  a solution of sugar $(0.1 \text{ mol dm}^{-3})$
    D  a solution of calcium chloride $(0.1 \text{ mol dm}^{-3})$
    E  a solution of urea $(0.2 \text{ mol dm}^{-3})$

The correct answer is D because calcium chloride dissociates into $Ca^{2+}$ and $2Cl^-$ ions.

**2**

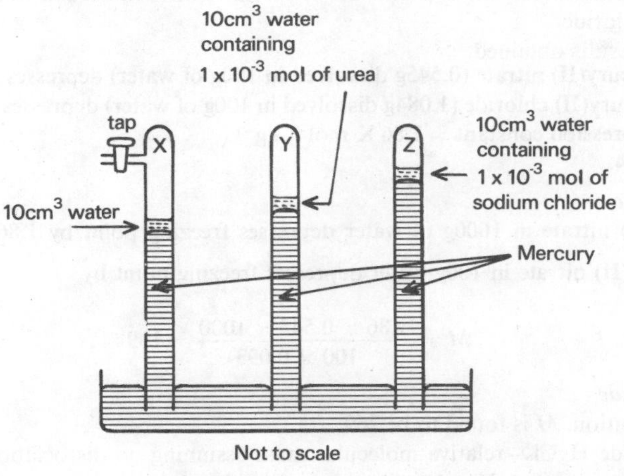

Fig. 12.6 Question

(a) (i) What is to be found in each of the spaces marked X, Y and Z?

(ii) Explain the differences in mercury levels in the three tubes in terms of the properties of dilute solutions.

(iii) What changes, if any, would occur to the mercury levels if the apparatus was placed in a thermostatic bath maintained at a higher temperature? Explain your answer.

(iv) If the pressure in X was reduced by means of a vacuum pump, explain what would happen.

(b) Calculate the boiling point of an aqueous solution of urea, $CO(NH_2)_2$, of concentration $12.0 \text{ g dm}^{-3}$ at a pressure of 101.3 kPa.

Assume that the volume of the solute is negligible compared to that of the solution, and the boiling point elevation constant for water is $0.52 \text{ mol}^{-1}\text{kg}$.

(*c*) Write an expression for the mole fraction of a solute in solution.

(*ii*) Calculate the mole fraction of sodium chloride in an aqueous solution containing 10g of sodium chloride per 100g of water [$A_r(C) = 12.0$; $A_r(H) = 1.0$; $A_r(O) = 16.0$; $A_r(Na) = 23.0$; $A_r(Cl) = 35.5$]

(*a*) (*i*) The only substance present in X, Y and Z will be water vapour. Remember urea and sodium chloride are non-volatile.

(*ii*) The different mercury levels are due to the different vapour pressures of the liquids in X, Y and Z. The vapour pressure of pure water is higher than any aqueous solution. The changes in vapour pressure are dependant upon particle numbers (colligative properties). The more particles in solution the greater the lowering of vapour pressure.

No lowering of vapour pressure for water.

Twice as many particles in sodium chloride as in urea. Sodium chloride is dissociated into ions.

$$NaCl \rightarrow Na^+ + Cl^-$$

Vapour pressure lowered more for sodium chloride than for urea.

(*iii*) Level in each tube would fall as vapour pressure increases with temperature.

(*iv*) Water would evaporate or boil. Mercury level rises. The equilibrium will be disturbed to replace water molecules removed by evaporation.

(*b*) Relative molecular mass of urea = 60.

60g of urea in 1 kg water elevates boiling point by 0.52 K.

12g of urea in 1 kg water elevates boiling point by $0.52 \times \dfrac{12}{60} = 0.104$ K

Boiling point of solution is 373.104 K (or 100.104°C)

(*c*) The mole fraction of a solute in a solution

$$= \frac{n}{n+N} \quad \text{where } n = \text{no. of moles of solute}$$
$$N = \text{no. of moles of solvent.}$$

This can be expressed as $\dfrac{w/m}{w/m + W/M}$ where $w$ is the mass of solute

$m$ is the relative molecular mass of solute

$W$ is the mass of solvent

$M$ is the relative molecular mass of solvent

Substitute in this equation

$$w = 10, m = 58.5, W = 100, M = 18$$
$$\text{Mole fraction of solute} = 0.0298$$

3  Describe briefly an experimental method for the determination of the relative molecular mass of a compound by either boiling point elevation or freezing point depression.

Elevation of boiling point and depression of freezing point are colligative properties. What are colligative properties. Give two other examples of colligative properties.

The apparent relative molecular masses of mercury(II) nitrate and mercury(II) chloride were determined. From the following results calculate the apparent relative molecular masses of mercury(II) nitrate and mercury(II) chloride.

Comment on the results obtained.

A solution of mercury(II) nitrate (0.545g dissolved in 100g of water) depresses freezing point by 0.093°C.

A solution of mercury(II) chloride (1.084g dissolved in 100g of water) depresses freezing point by 0.075°C.

Freezing point depression constant = 1.86 K mol$^{-1}$ kg$^{-1}$.

Units 12.1 and 12.4.

*Mercury(II) nitrate*

$M$g of mercury(II) nitrate in 1000g of water depresses freezing point by 1.86 K

0.545g of mercury(II) nitrate in 100g water depresses freezing point by

$$M = \frac{1.86 \times 0.545 \times 1000}{100 \times 0.093} = 109$$

*Mercury(II) chloride*

By a similar calculation, $M$ is found to be 269.

Mercury(II) chloride $HgCl_2$ – relative molecular mass assuming no dissociation is 272. Assuming slight experimental error, mercury(II) chloride is undissociated.

Mercury(II) nitrate $Hg(NO_3)_2$ – relative molecular mass assuming no dissociation is 325. The low value can be explained by the dissociation into 3 ions

$$Hg(NO_3)_2 \rightarrow Hg^{2+} + 2NO_3^-$$

# **13** Energetics

## 13.1 INTRODUCTION

At GCSE a study of chemistry can be summarized by the following triangle.

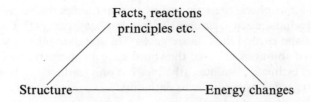

The emphasis at GCSE must be towards the learning of fact, principles, and reactions, with structure and energy changes being less important. The extent to which structure and energy changes are introduced varies from one syllabus to another.

For example, the melting of ice to give water is a simple process and a straightforward topic which GCSE candidates should know. As an A-level candidate you should appreciate that ice has an open crystal lattice with a tetrahedral structure. On heating, the water molecules in the lattice vibrate more and the structure breaks up on melting. The process requires energy:

$$i.e. \text{ ice} + \text{energy} \rightarrow \text{liquid water}$$

This is an **endothermic** process and the energy required is called the **enthalpy change of fusion**.

You should always consider changes of structure and energy changes when answering questions. Nearly all chemical reactions involve energy changes and the study of energy changes is called **energetics**.

## 13.2 FIRST LAW OF THERMODYNAMICS

This law is the equivalent of the law of conservation of energy. It states that **energy cannot be created or destroyed but may be converted from one form into another**

$$q = \Delta U + w$$

where $\Delta U$ is the change in the internal energy
$\quad q$ is the actual heat change, and
$\quad w$ is the work done by the system.

The term $w$ involves the gas produced expanding and pushing back the atmosphere. This can be estimated by using a syringe with a piston. The work done

$$w = P\Delta V \text{ where } P \text{ is external pressure and } \Delta V \text{ is the change in volume.}$$

$P\Delta V$ (and hence $w$) only becomes significant when either $P$ and/or $\Delta V$ is large. In reactions involving only solids and liquids as reactants and products $P\Delta V$ is negligible. For gases $\Delta V$ is frequently large and $w$ can be significant.

*Reactions carried out at constant volume.* In these cases $\Delta V = 0$ (*i.e.* no change in volume) and $P\Delta V = w = 0$

$$q_v = \Delta U$$

$q_v$ represents heat change at constant volume. Any heat change recorded corresponds to a change in the internal energy of the system.

*Reactions carried out at constant pressure.* Most reactions carried out in the laboratory are carried out at constant pressure *e.g.* in an open beaker. Work is done by any gas expanding against the pressure of the atmosphere.

$q_p = \Delta U + P\Delta V$ where $q_p$ represents the heat change at constant pressure

$$q_p = (U_{final} - U_{initial}) + P(V_{final} - V_{initial})$$
$$= (U_{final} + PV_{final}) - (U_{initial} + PV_{initial})$$

The term $U + PV$ represents the energy possessed by the system and is called the **enthalpy** or **heat content** $H$:

$$q_p = H_{final} - H_{initial} = \Delta H$$

Substituting,

$$\Delta H = \Delta U + P\Delta V$$

### 13.3 ENTHALPY CHANGE OF REACTION

The equation for the reaction between carbon and excess oxygen is

$$C_{(graphite)} + O_2(g) \rightarrow CO_2(g) \qquad \Delta H = -393.5 \text{ kJ mol}^{-1}$$

(Note mol$^{-1}$ means per quantity shown in the equation *not* per mole of product.)

The combustion of 12g of graphite in excess oxygen liberates 393.5 kJ. (The negative sign for $\Delta H$, by convention, tells us that energy is **liberated** or the reaction is **exothermic**).

The quantity of energy liberated or absorbed is given by the equation, $\Delta H$ is called the **enthalpy change of reaction**. Often standard enthalpy changes are quoted. These are the enthalpy changes at standard atmospheric pressure and at constant temperature, usually 25°C (or 298K). The reactants and products are in their usual or standard states at 25°C. Carbon is in the form of graphite and oxygen and carbon dioxide are gases. The standard enthalpy change of reaction is represented by $\Delta H^\ominus$. If solutions are used they must have a concentration of 1 mol dm$^{-3}$.

It is impossible to estimate absolute enthalpies but only enthalpy changes. Elements in their standard states are regarded as having zero enthalpy.

There are a number of standard enthalpy changes which are important and should be clearly defined.

### 13.4 STANDARD ENTHALPY CHANGE OF FORMATION

The standard enthalpy change of formation of a compound is the enthalpy change when one mole of the compound is formed from its constituent elements in their standard states at 25°C. Energy may be given out or taken in during the reaction (see 13.3)

For example, $\Delta H_f^\ominus = -393.5$ kJ mol$^{-1}$ for the formation of carbon dioxide.

### 13.5 STANDARD ENTHALPY CHANGE OF COMBUSTION

The standard enthalpy change of combustion of a substance $\Delta H_c^\ominus$ is the enthalpy change when one mole of the substance in its standard state undergoes complete combustion.

If the compound contains carbon, hydrogen and oxygen, the complete combustion will produce carbon dioxide and water.

$$e.g. \ C_2H_6(g) + 3\tfrac{1}{2}O_2(g) \rightarrow 2CO_2(g) + 3H_2O(l) \qquad \Delta H_c^\ominus = -1560 \text{ kJ mol}^{-1}.$$

### 13.6 STANDARD ENTHALPY CHANGE OF NEUTRALIZATION

The standard enthalpy change of neutralization is the enthalpy change when 1 mole of $H_3O^+$ ions is just neutralized by an alkali in dilute solution. The acid and base are in their standard states at 25°C and the acid and base are in solutions containing 1 mol dm$^{-3}$.

$$HCl(aq) + NaOH(aq) \rightarrow NaCl(aq) + H_2O(l) \qquad \Delta H_n^\ominus = -57.1 \text{ kJ mol}^{-1}$$

The values obtained for standard enthalpy changes of neutralization of strong acids and strong bases are constant (see question 1).

### 13.7 ENTHALPY CHANGE OF SOLUTION AND ENTHALPY CHANGE OF DILUTION

The enthalpy change of solution is the enthalpy change when 1 mole of a substance is dissolved in a specified volume of solvent. The enthalpy change will depend upon the volume of solvent used

The enthalpy change of dilution is the enthalpy change when 1 mole of a substance is dissolved in a large volume of solvent such that further dilution produces no further enthalpy change.

### 13.8 MEASURING ENTHALPY CHANGES

Enthalpy changes are determined by simple calorimetry experiments. For example to determine the enthalpy change of neutralization of sodium hydroxide and hydrochloric acid, a known volume of sodium hydroxide solution of known concentration is placed in a vacuum flask and the temperature is accurately measured. The volume of hydrochloric acid, of known concentration, required to neutralize the sodium hydroxide solution is rapidly added, the mixture stirred and the final temperature recorded.

If the initial temperature of both the acid and alkali is $t_1$°C and the final temperature is $t_2$°C, the temperature change is $t_2 - t_1$°C. If the mass of the mixture is $m$g, the specific heat capacity of the solutions is 4.2 J deg$^{-1}$ (dilute solutions can be assumed to have the same specific heat capacity as water), and the heat capacity of the vacuum flask is $W$, then

$$\text{heat evolved} = m \times 4.2 \times (t_2 - t_1) + W(t_2 - t_1) \text{ J}$$

From these results the enthalpy change of neutralization can be calculated.

## 13.9 HESS'S LAW

Hess's law can be used for calculating enthalpy changes which cannot be determined by experiment. The law states that the total enthalpy change for a reaction is independent of the route taken.

For example if X is converted to Y and Y to Z, the total enthalpy change will be the same as when X is converted directly to Z.

*Example*: Calculate the standard enthalpy change of formation of carbon monoxide from the following information:

Enthalpy changes of combustion of carbon (graphite) and carbon monoxide are $-393.5$ kJ $mol^{-1}$ and $-283$ kJ $mol^{-1}$.

The following equations can be written

Required: Enthalpy change of formation of carbon monoxide

$$C(graphite) + \tfrac{1}{2}O_2(g) \rightarrow CO(g)$$

Given: Enthalpy changes of combustion of graphite and carbon monoxide.

(*i*) $C(graphite) + O_2(g) \rightarrow CO_2(g)$ $\quad \Delta H^\circ = -393.5$ kJ

(*ii*) $CO(g) + \tfrac{1}{2}O_2(g) \rightarrow CO_2(g)$ $\quad \Delta H^\circ = -283$ kJ

A diagram can be drawn to summarize the situation

$$C(graphite) + \tfrac{1}{2}O_2(g) \xrightarrow{\Delta H^\circ_f} CO(g)$$

$+O_2(g) \qquad +\tfrac{1}{2}O_2(g)$

$\Delta H^\circ_1 = -393.5$ kJ $mol^{-1}$ $\qquad \Delta H^\circ_2 = -283.0$ kJ $mol^{-1}$

$$CO_2(g)$$

$$\Delta H^\circ_f = \Delta H^\circ_1 - \Delta H^\circ_2$$
$$= -393.5 + 283.0$$
$$= -110.5 \text{ kJ mol}^{-1}$$

Alternatively, from equations (*i*) and (*ii*) above, equation (*ii*) can be reversed when the enthalpy change is the same numerically but the sign is changed.

$$C(graphite) + O_2(g) \rightarrow CO_2(g) \quad \Delta H^\circ = -393.5 \text{ kJ}$$
$$CO_2(g) \rightarrow CO(g) + \tfrac{1}{2}O_2(g) \quad \Delta H^\circ = +283.0 \text{ kJ}$$

Adding these two equations together

$$C(graphite) + \tfrac{1}{2}O_2(g) \rightarrow CO(g) \quad \Delta H^\circ_f = -393.5 + 283.0$$
$$= -110.5 \text{ kJ mol}^{-1}$$

The **negative** sign shows, by convention, that the process is **exothermic**. A **positive** sign shows that the process is **endothermic**. Candidates lose many marks in this type of work by either getting a sign wrong or by missing out units. It is advisable when an enthalpy change is positive to put in a positive sign. This shows the examiner that you have thought about it and decided that the process is endothermic.

## 13.10 BOND ENTHALPIES

During a chemical reaction a number of bonds will be formed or broken. **Forming bonds liberates energy and breaking bonds requires energy**. It would be useful to be able to attribute specific enthalpy changes to various bond changes. The sum of these enthalpy changes would equal the total enthalpy change.

Consider the breaking of the four C —H bonds in methane:

$$\begin{array}{c} H \\ | \\ H-C-H \ (g) \rightarrow C(g) + 4H(g) \quad \Delta H^\circ = +1662 \text{ kJ mol}^{-1} \\ | \\ H \end{array}$$

It is reasonable to assume that the energy required to break each C — H bond is the same and the same as is required to break a C—H bond in any other compound, and for all practical purposes this can be taken as true at this level.

For a diatomic molecule XY, the bond enthalpy is defined as the enthalpy change for the process:

$$X-Y(g) \rightarrow X(g) + Y(g)$$

Table 13.1 lists some of the average bond enthalpies at 25°C.

(Don't try to remember these figures. They will be given if required.)

It must be remembered that these bond enthalpies can only be used as a guide. The actual value of a bond enthalpy will depend upon the nature of other atoms or groups joined to it.

The enthalpy change is positive when a bond is broken and negative on making a bond. Bond enthalpies can be used to explain a number of facts from inorganic and organic chemistry.

**Table 13.1 Some average bond enthalpies at 25°C**

| Bond | $\Delta H$ kJ mol$^{-1}$ |
|------|-------------------------|
| H — H | +436 |
| C — C | +348 |
| C = C | +612 |
| C — F | +484 |
| C — Cl | +338 |
| C — Br | +275 |
| C — I | +238 |

*E.g.* The increased reactivity of phosphorus compared with nitrogen (Unit 25) or the increased rate of substitution reactions for iodoalkanes compared with other haloalkanes.

The enthalpy change for atomization can be estimated using bond enthalpies. For example, the atomization of ethane $C_2H_6$ involves the breaking of one C—C bond and six C—H bonds. The enthalpy of atomization there is:

$$= 1(C—C) + 6(C—H)$$
$$= 348 + (6 \times 412)$$
$$= +2820 \text{ kJ mol}^{-1}$$

Bond enthalpies can also be used to estimate likely enthalpy changes in chemical reactions, *e.g.* the hydrogenation of ethene (Unit 29.6)

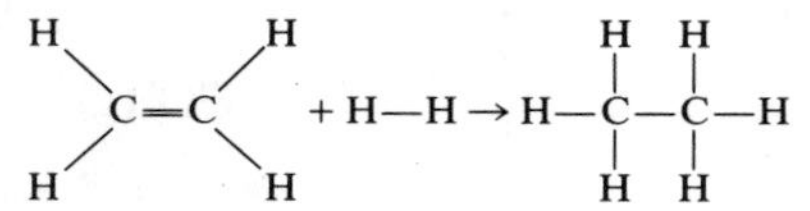

Using the information in Table 13.1

Bonds broken 1 (C=C) + 1(H—H) Enthalpy change = +612 + 436 = +1048 kJ

Bonds formed 1(C—C) + 2(C—H) Enthalpy change = −348 + 2(−412) = −1172 kJ

Net enthalpy change = 1048 − 1172 = −124 kJ

## 13.11 BORN–HABER CYCLE

Lattice enthalpies cannot be determined directly but may be found indirectly by using an **energy diagram**. These diagrams are called **Born-Haber cycles**. The most common one seen on examination papers is the one for sodium chloride.

The standard enthalpy change when one mole of sodium chloride crystal is produced from sodium metal and chlorine gas can be determined experimentally and is, in fact, the standard enthalpy change of formation of sodium chloride

$$\Delta H_f^\ominus = -411 \text{ kJ mol}^{-1}$$

We can consider the various steps which are required to bring about this change:

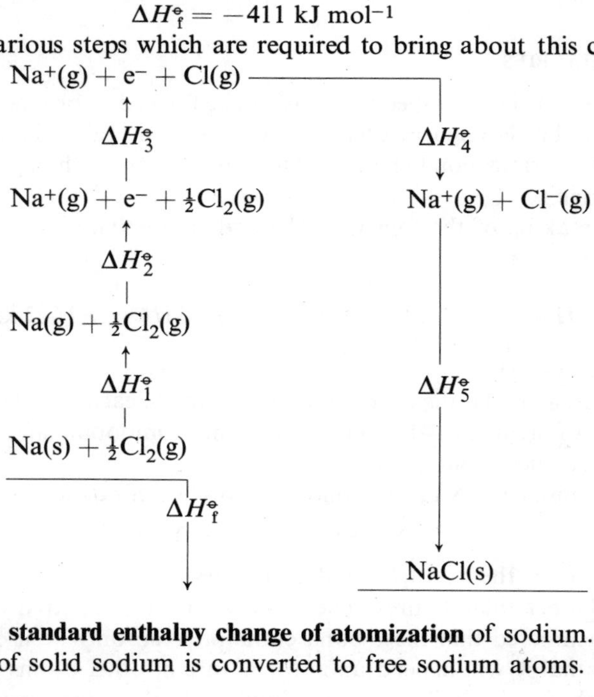

$\Delta H_1^\ominus$ corresponds to the **standard enthalpy change of atomization** of sodium. It is the enthalpy change when one mole of solid sodium is converted to free sodium atoms. $\Delta H_1^\ominus = +108.4$ kJ mol$^{-1}$

$\Delta H_2^\ominus$ corresponds to the **first ionization energy** of sodium (Unit 2.4) *i.e.* the enthalpy change when one mole of electrons in their ground state is totally removed from 1 mole of sodium atoms in the gas phase. $\Delta H_2^\ominus = +500$ kJ mol$^{-1}$.

$\Delta H_3^\ominus$ corresponds to the **standard enthalpy change of atomization** of chlorine. It is the enthalpy change when one mole of gaseous chlorine atoms is produced from chlorine molecules. $\Delta H_3^\ominus = +121$ kJ mol$^{-1}$.

$\Delta H_4^\ominus$ corresponds to the **first electron affinity** of chlorine. This is the enthalpy change when one mole of chlorine atoms each accept one electron to form one mole of chloride ions. $\Delta H_4^\ominus = -364$ kJ mol$^{-1}$.

$\Delta H_5^\ominus$ corresponds to the **lattice enthalpy** (sometimes called **lattice energy**) of sodium chloride.

By Hess's Law $\Delta H_f^\ominus = \Delta H_1^\ominus + \Delta H_2^\ominus + \Delta H_3^\ominus + \Delta H_4^\ominus + \Delta H_5^\ominus$

$\Delta H_5^\ominus = -776.4$ kJ mol$^{-1}$.

An understanding of lattice enthalpy is extremely useful in various ways at A level:

(*i*) Substances with high lattice enthalpies will have high melting points.

(*ii*) The solubility of an ionic compound will depend upon the relative values of the lattice and hydration enthalpies or energies.

(*iii*) Comparison of the calculated and experimental values for lattice enthalpy gives an indication of the degree of covalent bonding in a compound (see question 2 at the end of the Unit).

(*iv*) Lattice enthalpies can be used to explain the non-existence of some compounds.

### 13.12 THE STABILITY OF BENZENE

The enthalpy change of hydrogenation is the enthalpy change when one mole of hydrogen molecules is added to one mole of alkene. The enthalpy change of hydrogenation of cyclohexene is $-120$ kJ mol$^{-1}$.

If it is assumed that benzene contains three double bonds *i.e.* it is a triene, the enthalpy change of hydrogenation should be $3 \times (-120)$ kJ mol$^{-1}$ *i.e.* $-360$ kJ mol$^{-1}$.

The experimental value for the enthalpy change of hydrogenation of benzene is $-208$ kJ mol$^{-1}$. The difference between these two values (152kJ mol$^{-1}$) is a measure of the extra stability of benzene compared with similar non-aromatic systems.

The standard enthalpy change of formation of benzene can be obtained by using the bond enthalpies in Table 13.1 and assuming benzene is a cyclic triene.

$$\Delta H_f^\ominus = -(3 \times C{=}C + 3 \times C{-}C + 6 \times C{-}H) = -5352 \text{kJ mol}^{-1}$$

The experimental value found by a Hess's law calculation is $-5549$ kJ mol$^{-1}$. The difference between these two values (197 kJ mol$^{-1}$) again is a measure of the stability of benzene. The difference between this value and the one obtained previously from hydrogenation calculations is probably due to the use of average values for bond enthalpies.

### 13.13 SECOND LAW OF THERMODYNAMICS – ENTROPY

The majority of reactions are exothermic and it was once believed that only exothermic reactions could take place. There is another factor which governs the feasibility of a reaction.

The second factor is **entropy**, which is related to the degree of randomness or disorder in the system. If two gases in separate bulbs are allowed to come into contact, the gases mix and there is an increase in the disorder.

Given an opportunity, the disorder of a system, and hence its entropy, will spontaneously increase. The dissolving of an ionic crystal in water produces an increase in entropy. In the lattice the ions are regularly arranged. The dissolving of ammonium nitrate is an endothermic process and would not take place purely on energy considerations. There is, however, an increase in entropy which makes the process possible.

One statement of the second law of thermodynamics is that in any system a change that takes place spontaneously (*i.e.* of its own accord) will always involve an increase in entropy.

Entropy change $\Delta S$ is defined as the heat involved in a particular process divided by the temperature at which the process occurs *i.e.* $\Delta S = \dfrac{q}{T}$

Systems having perfect order *i.e.* the perfect ionic crystal lattice at 0K, possess zero entropy. The entropy of a given mass of matter increases in the order:

$$\text{solid} \rightarrow \text{liquid} \rightarrow \text{gas}$$

The entropy of substances not possessing perfect order and of all substances at temperatures above 0 K will be greater than zero, *e.g.* the standard entropy $S^\ominus$ (at 298 K and 1 atmosphere pressure) of water = 70 J K$^{-1}$ mol$^{-1}$

The entropy change in a reaction can be found.

*Example:* Find $\Delta S^\ominus$ for the reaction $H_2(g) + I2(g) \rightarrow 2HI(g)$, given that the standard entropies of hydrogen, gaseous iodine and hydrogen iodine are +131, +261 and +207 J mol$^{-1}$ K$^{-1}$ respectively.

$$\Delta S^\ominus = 2 \times S^\ominus_{HI} - S^\ominus_{H_2} - S^\ominus_{I_2} = 414 - 131 - 261 = +22 \text{ J mol}^{-1} \text{ K}^{-1}$$

There is an increase in entropy and a spontaneous reaction is feasible, although no indication is given of the possible rate of reaction.

For spontaneous processes (i.e. ones that take place of their own accord)

$$\Delta S_{TOTAL} > 0$$

$$\Delta S_{TOTAL} = \Delta S_{SYSTEM} + \Delta S_{SURROUNDINGS}$$

$$\Delta S_{SYSTEM} = \Sigma \Delta S^\ominus_{PRODUCTS} - \Sigma \Delta S^\ominus_{REACTANTS}$$

$$\Delta S_{SURROUNDINGS} = -\Delta H / T$$

$$\therefore \Delta S_{TOTAL} = \Delta S_{SYSTEM} - \Delta H / T$$

$$\therefore -T \Delta S_{TOTAL} = \Delta H - T \Delta S_{SYSTEM}$$

$$\text{But } -T \Delta S_{TOTAL} = \Delta G$$

$$\therefore \Delta G = \Delta H - T \Delta S$$

## 13.14 FREE ENERGY $G$

This is a measure of the useful work which can be obtained from a system (apart from volume changes).

A reaction can only proceed spontaneously if $\Delta G$ is negative. $\Delta G$ depends upon $\Delta H$ and $\Delta S$ and they are related by the equation $\Delta G = \Delta H - T \Delta S$.

Since for a reaction to proceed $\Delta G$ must be negative, for an endothermic reaction ($\Delta H$ positive), a reaction will take place only if $T \Delta S > \Delta H$. If $\Delta G$ is positive, no useful work can be obtained and the reaction is not feasible. If $\Delta G = 0$ the system will be in dynamic equilibrium.

### QUESTIONS

**1** State Hess's Law of Constant Heat Summation
Define (*a*) enthalpy of formation (heat of formation); (*b*) heat of neutralization.
Describe how you would determine the heat of neutralization of nitric acid by sodium hydroxide solution. Explain why approximately the same numerical result would be obtained by using hydrochloric acid, but not hydrofluoric acid, in place of nitric acid.

Using the following data collected at 25°C and standard atmospheric pressure, in which the negative sign indicates heat evolved, calculate the enthalpy of formation of rubidium sulphate $Rb_2SO_4(s)$.

|  |  | $\Delta H$ kJ mol$^{-1}$ |
|---|---|---|
| (*i*) | $H^+(aq) + OH^-(aq) \rightarrow H_2O(l)$ | − 57.3 |
| (*ii*) | $RbOH(s) \rightarrow Rb^+(aq) + OH^-(aq)$ | − 62.8 |
| (*iii*) | $Rb_2SO_4(s) \rightarrow 2Rb^+(aq) + SO_4^{2-}(aq)$ | + 24.3 |
| (*iv*) | $Rb(s) + \frac{1}{2}O_2(g) + \frac{1}{2}H_2(g) \rightarrow RbOH(s)$ | −414.0 |
| (*v*) | $H_2(g) + S(s) + 2O_2(g) \rightarrow 2H^+(aq) + SO_4^{2-}(aq)$ | −907.5 |
| (*vi*) | $H_2(g) + \frac{1}{2}O_2(g) \rightarrow H_2O(l)$ | −285.0 |

Assume that $H_2SO_4(aq)$ is fully ionized in solution.

*(Southern Universities Joint Board)*

Hess's Law (Unit 13.9)

(*a*) enthalpy of formation (or enthalpy change of formation) Unit 13.4

(*b*) heat of neutralization (or enthalpy change of neutralization) Unit 13.6.
Method used to determine the enthalpy change of neutralization will be found in Unit 13.8.
The neutralization of any strong acid with any strong base produces the same molar enthalpy change. For example, sodium hydroxide and nitric acid.

$$NaOH(aq) + HNO_3(aq) \rightarrow NaNO_3(aq) + H_2O(l)$$

Since the acid and base are fully ionised, the change is

$$OH^-(aq) + H^+(aq) \rightarrow H_2O(l) \qquad \Delta H_{neut} = -57.5 \text{ kJ mol}^{-1}$$

(The same value of $\Delta H^\ominus$ is obtained for all strong acids and bases. Effectively, it is the enthalpy change when one mole of water is formed from hydrogen and hydroxide ions in dilute solution. It may also be written $OH^-(aq) + H_3O^+(aq) \rightarrow 2H_2O(l)$.) For a weak acid (one which is not completely ionised) $\Delta H$ will be slightly less as some energy is required to complete the ionisation of the acid:
$$HF(aq) + H_2O(l) \rightleftharpoons H_3O^+(aq) + F^-(aq).$$

To find the enthalpy change for $2Rb(s) + S(s) + 2O_2(g) \rightarrow Rb_2SO_4(s)$, using the given equations:

$$\begin{array}{lr} & \Delta H \\ & \text{kJ mol}^{-1} \\ 2Rb(s) + O_2(g) + H_2(g) \rightarrow 2RbOH(s) & -828.0 \\ 2RbOH(s) \rightarrow 2Rb^+(aq) + 2OH^-(aq) & -125.6 \\ 2Rb^+(aq) + SO_4^{2-}(aq) \rightarrow Rb_2SO_4(s) & -24.3 \\ 2H^+(aq) + 2OH^-(aq) \rightarrow 2H_2O(l) & -114.6 \\ H_2(g) + S(s) + 2O_2(g) \rightarrow 2H^+(aq) + SO_4^{2-}(aq) & -907.5 \\ 2H_2O(l) \rightarrow 2H_2(g) + O_2(g) & +570.0 \end{array}$$

Adding these equations together

$$2Rb(s) + S(s) + 2O_2(g) \rightarrow Rb_2SO_4(s) \qquad \Delta H = -1430 \text{ kJ mol}^{-1}$$

**2**

$$\begin{array}{c} Cd^{2+}(g) + 2I(g) \xrightarrow{\quad (v) \quad} Cd^{2+}(g) + 2I^-(g) \\ \uparrow (iv) \qquad \qquad \qquad \big\downarrow \\ Cd^{2+}(g) + I_2(g) \\ \uparrow (iii) \\ Cd(g) + I_2(g) \\ \uparrow (ii) \\ Cd(g) + I_2(s) \\ \uparrow (i) \\ Cd(s) + I_2(s) \xrightarrow{\quad (vi) \quad} CdI_2(s) \end{array}$$

(*a*) Use the above diagram and the following data to determine the experimental value of the standard molar lattice energy of cadmium(II) iodide, $CdI_2(s) \rightarrow Cd^{2+}(g) + 2I^-(g)$

| | | | | |
|---|---|---|---|---|
| (i) $Cd(s) \rightarrow Cd(g)$ | $\Delta H^\circ = +113 \text{ kJ mol}^{-1}$ | | (iv) $I_2(g) \rightarrow 2I(g)$ | $\Delta H^\circ = +151 \text{ kJ mol}^{-1}$ |
| (ii) $I_2(s) \rightarrow I_2(g)$ | $\Delta H^\circ = +19.4 \text{ kJ mol}^{-1}$ | | (v) $2I(g) \rightarrow 2I^-(g)$ | $\Delta H^\circ = -628 \text{ kJ mol}^{-1}$ |
| (iii) $Cd(g) \rightarrow Cd^{2+}(g)$ | $\Delta H^\circ = +2490 \text{ kJ mol}^{-1}$ | | (vi) $Cd(s) + I_2(s) \rightarrow CdI_2(s)$ | $\Delta H^\circ = -201 \text{ kJ mol}^{-1}$ |

(*b*) By assuming that cadmium and iodide ions are charged spheres, a theoretical value for the molar lattice energy is calculated to be $-2050 \text{ kJ mol}^{-1}$. Give an explanation for the difference between this value and the value determined in (*a*).

*(Associated Examining Board)*

(*a*) If the standard molar lattice energy is $\Delta H^\circ$,
$$\Delta H^\circ = +113 + 19.4 + 2490 + 151 - 628 + 201 = +2346.4 \text{ kJ mol}^{-1}$$

(*b*) The difference between these two values is due to the fact that the bonding in cadmium(II) iodide is not purely ionic but there is an appreciable covalent character. To consider the lattice as an arrangement of charged spheres is incorrect.

**3**

$$CH_3OH(l) \rightleftharpoons CH_3OH(g)$$

The equation represents the equilibrium between liquid methanol and methanol vapour at 338 K. Given that $\Delta H = +35.3 \text{ kJ mol}^{-1}$, calculate the entropy change when methanol is vaporized.

The entropy change should be positive as there is an increase in disorder.
$$\Delta G = \Delta H - T\Delta S$$
The system is in equilibrium, therefore $\Delta G = 0$
$$\text{and } \Delta H = T\Delta S$$
$$\Delta S = \frac{35300}{338} = +104.4 \text{ J mol}^{-1}$$

**4** Ammonium dichromate(VI), $(NH_4)_2Cr_2O_7$, may be prepared by mixing hot concentrated solutions of an ammonium salt and a dichromate(VI) salt, cooling the mixed solutions, and collecting the crystals of ammonium dichromate(VI) which form.

In order to consider the preparation of ammonium dichromate(VI) you will need the following data:

| Salt | $M_r$ | Solubility, $m_{sat}$/mol per kg water | |
|---|---|---|---|
| | | at 15 °C | at 100 °C |
| $Na_2Cr_2O_7$ | 262 | 6.8 | 16 |
| $NH_4NO_3$ | 80 | 21 | 133 |
| $(NH_4)_2Cr_2O_7$ | 252 | 1.25 | 6.2 |
| $NaNO_3$ | 85 | 10 | 21 |

When solid ammonium dichromate(VI) is heated it decomposes in a spectacular manner that has earned the title 'the volcano reaction'.

$$(NH_4)_2Cr_2O_7(s) \rightarrow N_2(g) + 4H_2O(g) + Cr_2O_3(s)$$

In order to consider the energetics of the decomposition of ammonium dichromate(VI) you will need the following data:

$\Delta H^{\ominus}_{f,298} [(NH_4)_2Cr_2O_7(s)] = -1810$ kJ mol$^{-1}$

$\Delta H^{\ominus}_{f\,298} [H_2O(g)] \qquad = -240$ kJ mol$^{-1}$

$\Delta H^{\ominus}_{f\,298} [Cr_2O_3(s)] \qquad = -1140$ kJ mol$^{-1}$

(*a*) Consider the possibility of using sodium dichromate(VI) and ammonium nitrate to prepare ammonium dichromate(VI).

(*i*) Write a balanced equation for the production of ammonium dichromate(VI) by this process.

(*ii*) If 4 moles of sodium dichromate(VI) are available for the preparation, what is the minimum volume of boiling water that you would need to prepare a solution?

(*iii*) How many moles of ammonium nitrate in what minimum volume of boiling water would you need?

(*iv*) When your mixed solutions are cooled to 15°C what salt(s) will crystallize out?

(*v*) When your mixed solutions are cooled to 15°C what mass, in grams, of ammonium dichromate(VI) could you expect to obtain?

(*b*) (*i*) Calculate the enthalpy change at 298 K for the 'volcano reaction'.

$$(NH_4)_2Cr_2O_7(s) \rightarrow N_2(g) + 4H_2O(g) + Cr_2O_3(s)$$

(*ii*) Is the entropy change in the surroundings favourable or unfavourable for the decomposition at 298 K? Justify your answer.

(*iii*) Is the entropy change in the system likely to be favourable or unfavourable for the decomposition at 298 K? Justify your answer.

(*c*) Ammonium dichromate(VI) does not decompose when prepared in boiling water but the solid readily decomposes when heated by a Bunsen flame. Why do you think this is so?

(*Nuffield*)

(*a*) (*i*) $Na_2Cr_2O_7(aq) + 2 NH_4NO_3(aq) \rightarrow (NH_4)_2Cr_2O_7 (aq) + 2NaNO_3(aq)$

(*ii*) At 100°C, 16 mol dissolve in 1 kg of water.
Therefore, 4 mol would dissolve in 250g, i.e. 250 cm$^3$.

(*iii*) From equation in (i), 8 mol needed. 133 mol dissolve in 1 kg of water. Therefore, 8 mol in 8/133 = 0.06 kg, i.e. 60 cm$^3$.

(*iv*) $(NH_4)_2Cr_2O_7$ and $NaNO_3$.

(*v*) Mass of water = 0.25 + 0.06 = 0.31 kg.
Number of mol of $(NH_4)_2Cr_2O_7 = 4 - (1.25 \times 0.31) = 3.613$.
Mass = 3.613 × 252 = 910 g.

(*b*) (*i*) $\Delta H^{\ominus}_{298} = \Sigma \Delta H^{\ominus}_{f,298}$ (products) $- \Sigma \Delta H^{\ominus}_{f,298}$ (reactants)

$= \{O + (4x -240) + (-1140)\} - (-1810)$

$= -290$ kJ mol$^{-1}$

(*ii*) $\Delta S_{SURROUNDINGS} = -\Delta H/T$

Therefore $\Delta S_{SURROUNDINGS}$ is positive.
Therefore, the entropy change is favourable, i.e. *S* is increasing.

(*iii*) Much gas formed from a solid. Therefore, $\Delta S_{SYSTEM}$ is positive and entropy change is favourable.

(*c*) Reaction too slow at 100°C. The reaction has a high activation energy. Therefore a high temperature is needed to get the reaction started.

# 14 Reaction kinetics

## 14.1 INTRODUCTION

Most GCSE courses now contain some work on rates of chemical reactions and the factors which affect the rate of reaction. The work on reaction kinetics becomes more significant at A level, because it is closely related to structure (*e.g.* Units 5, 8, 9, 10) and to the mechanisms of reactions (Unit 38).

The factors which can affect the rate of reaction include:

(*i*) Physical state of reactants

(*ii*) Concentration (and gas pressure)

(*iii*) Temperature
(*iv*) Catalysts
 (*v*) Light

These factors will be considered in turn.

## 14.2  PHYSICAL STATE OF REACTANTS

The reaction between powdered calcium carbonate and dilute hydrochloric acid is much faster than the reaction between lumps of calcium carbonate and dilute hydrochloric acid under the same conditions. This is because the powder has a very large surface area to come in contact with the acid.

## 14.3  CONCENTRATION

This is probably the most important factor to understand at A level. During a reaction the reactants are used up and so the rate of reaction decreases.

The exact relationship between the rate of reaction and the concentration of the reactants in a particular reaction can only be determined experimentally. Most reactions take place in a series of steps and the rate of the overall reaction depends upon the rate of the slowest step, called the **rate determining step**.

It is vital for you to distinguish clearly the meaning of the terms **order of reaction** and **molecularity**. Few candidates understand the distinction and many questions test this. The order of reaction is determined experimentally. In the reaction

$$A + B \rightarrow C + D$$

experiment may show that the rate of reaction is proportional to the concentration of A to the power of $x$

$$i.e. \text{ rate} \propto [A]^x$$

and also the rate of reaction may also be proportional to the concentration of B to the power $y$

$$i.e. \text{ rate} \propto [B]^y$$

The overall equation is therefore

$$\text{rate} = k[A]^x[B]^y$$

$k$ is the **rate constant** and the overall equation is called the **rate equation**. The rate equation normally indicates what species are involved in the rate-determining step and how many of each species are involved. In this example, the order of the reaction with respect to A is $x$ and the order with respect to B is $y$ and the total order of the reaction is $x + y$.

In the reaction between peroxodisulphate and iodide ions, the equation is

$$S_2O_8^{2-}(aq) + 2I^-(aq) \rightarrow 2SO_4^{2-}(aq) + I_2(aq)$$

The rate equation determined experimentally is

$$\text{rate} = k[S_2O_8^{2-}][I^-]$$

and the total order is 2. The order is not related to the equation. The order may be integral *i.e.* 1, 2 or 3 etc, but it can also be fractional.

$$e.g. \quad H_2 + I_2 \rightarrow 2HI$$
$$\text{rate of reaction} = k[H_2][I_2]$$
$$\text{Order with respect to hydrogen} = 1$$
$$\text{Order with respect to iodine} = 1$$

Total order of reaction $= 1 + 1 = 2$

$$H_2 + Br_2 \rightarrow 2HBr$$
$$\text{rate of reaction} = k[H_2][Br_2]^{\frac{1}{2}}$$
$$\text{Order with respect to hydrogen} = 1$$
$$\text{Order with respect to bromine} = \frac{1}{2}$$

Total order of reaction $= 1 + \frac{1}{2} = 1\frac{1}{2}$

## 14.4  TEMPERATURE

Fig. 8.2 shows the distribution of velocities of molecules in a gas. The number of molecules, $n$, possessing the activation energy, $E_a$, necessary for reaction is given by

$$n = n_0 \exp(-E_a/RT)$$

Where $n_0$ is the total number of molecules, $R$ is the gas constant and $T$ is the absolute temperature.

From the simple collision theory, the rate of reaction is proportional to the number of molecules possessing energy greater than or equal to $E_a$.

$$k = A \exp(-E_a/RT)$$

Where $k$ is the rate constant and $A$ is Arrhenius constant. The equation is called the Arrhenius equation.

The activation energy can be found by plotting a graph of $\ln k$ against $1/T$. The activation energy can be found from the gradient of the straight line graph. The gradient is $\dfrac{-E_a}{R}$. (Be careful with the units of $R$.)

## 14.5  CATALYSTS

A **catalyst** is a substance which alters the rate of a chemical reaction without being used up. The catalyst may be changed physically but the mass of catalyst is unchanged at the end of the reaction.

An example of a catalyst is manganese(IV) oxide in the decomposition of hydrogen peroxide.

$$2H_2O_2(aq) \rightarrow 2H_2O(l) + O_2(g)$$

This is an example of a catalyst which speeds up a chemical reaction. A substance which slows down the rate of a reaction is called an inhibitor.

A catalyst does not alter the position of an equilibrium or increase the yield of products. It merely alters the rate at which the equilibrium is achieved or the products are obtained. A catalyst provides an alternative route with a lower activation energy barrier for the reaction. More molecules possess the lower activation energy and so the reaction is speeded up.

$$e.g. \qquad 2HI \rightarrow H_2 + I_2$$

Activation energy without catalyst    $183 \ kJ \ mol^{-1}$
with gold catalyst    $105 \ kJ \ mol^{-1}$
with platinum catalyst    $58 \ kJ \ mol^{-1}$

There are two types of catalyst–**homogeneous catalysts** and **heterogeneous catalysts**. A homogeneous catalyst is in the same phase as the reactants *e.g.* the reaction between an organic acid and an alcohol is catalysed by the presence of hydrogen ion. A heterogeneous catalyst is not in the same phase *e.g.* manganese(IV) oxide is solid and hydrogen peroxide is a liquid. There are two theories of catalysis.

### 1  Intermediate compound theory

The catalyst reacts with one of the reactants to produce an intermediate compound which eventually produces the required product and regenerates the catalyst, *e.g.* manganese(IV) oxide and hydrogen peroxide.

The intermediate compound theory explains most examples of homogeneous catalysis.

### 2  Adsorption or surface action theory

A gas is adsorbed when its particles collide with and adhere to the catalyst surface. When gas molecules are adsorbed they are brought close together, and in a state which enables them to react together. On desorption the catalyst surface is available for further reaction. An example is the reaction of an alkene with hydrogen using a nickel catalyst.

The *d*-block elements (Unit 28) and their compounds are good catalysts. The elements can exist in a variety of oxidation states and the metal surfaces are ideal for adsorption purposes.

In some reactions one of the products of the reaction acts as a catalyst for the reaction. For example, in the reaction of an ethanedioate (oxalate) with acidified potassium manganate(VII), the reaction has to be heated to about 60°C to get the reaction to start but the manganese(II) ions produced catalyse the reaction and the reaction continues even if the solution is cooled to room temperature. This is an example of what is called **autocatalysis**.

## 14.6  LIGHT

Some reactions are greatly affected by light. For example the reaction of hydrogen and chlorine is explosive in sunlight. The ultra violet light in sunlight splits some of the chlorine molecules into free chlorine atoms or chlorine **free radicals**.

$$Cl_2 \rightarrow 2Cl\cdot$$

Then a free radical chain reaction takes place producing hydrogen chloride.

$$Cl\cdot + H_2 \rightarrow HCl + H\cdot$$
$$H\cdot + Cl_2 \rightarrow HCl + Cl\cdot$$

The reaction between chlorine and methane (Unit 29.3) is another example of a reaction affected by light.

## 14.7 METHODS OF FOLLOWING THE PROGRESS OF A REACTION

There are a number of methods which can be used to follow the progress of a reaction. These include:

(*i*) Measuring the volume of gas evolved at intervals using a gas syringe, *e.g.* in the reaction between calcium carbonate and hydrochloric acid.

(*ii*) Measuring the electrical resistance of the solution.

(*iii*) Measuring the change of colour in the solution with a colorimeter.

(*iv*) Removing samples from the reacting mixture and titrating,

*e.g.* the hydrolysis of ethyl ethanoate with alkali. Portions of the reaction mixture are removed at intervals and titrated with dilute hydrochloric acid using phenolphthalein indicator.

## QUESTIONS

**1** Multiple completion question. Using the method of recording responses on page 1.
The rate equation for the reaction between X and Y is

$$rate = k[X][Y]^2$$

Which of the following statements is (are) correct?
(*i*) The order with respect to Y is 2.
(*ii*) The rate of the reaction is quadrupled if the concentration of Y is doubled.
(*iii*) The rate of reaction is halved if the concentration of X is halved.
(*iv*) The molecularity of the reaction is 3.

Answer is A

**2** Explain what you understand by the terms rate equation, rate constant and order of reaction.

Give one possible reason why the rate equation cannot be deduced from the over-all stoichiometric equation.

Using two reaction vessels of differing surface material, the reaction between A and B in the gaseous phase was investigated. The initial rate refers to the rate of removal of A (mol m$^{-3}$ s$^{-1}$) in the experimental data set out below. Deduce the rate equation in each case and assuming no errors have occurred, offer an explanation for what is found.

| | A/mol m$^{-3}$ | B/mol m$^{-3}$ | Relative rate initially |
|---|---|---|---|
| Experiment 1 | 0.15 | 0.15 | 1 |
| | 0.30 | 0.15 | 4 |
| | 0.15 | 0.30 | 2 |
| Experiment 2 | 0.15 | 0.15 | 1 |
| | 0.30 | 0.30 | 4 |
| | 0.60 | 0.30 | 16 |

*(Southern Universities Joint Board)*

The rate of reaction can be governed by an equation of the type

rate of reaction $= k[A]^x[B]^y$ where A and B are the reactants.

This is called a rate equation, $k$ is called the rate constant and the total order of the reaction $= x + y$.

The rate cannot be deduced from the overall stoichiometric equation because the reaction probably occurs in stages and the order of the reaction will be determined by the particles reacting in this stage.
Experiment 1:

$$rate = k[A]^2[B]$$

The total order is 3. In the second set of results [A] is doubled and the rate increases by a factor of 4. In the third set of results [B] is doubled compared with the first set of results and the rate is doubled.
Experiment 2:

$$rate = k'[A]^2$$

The total order is 2. In the third set of results, the rate has increased by a factor of 4 compared to the second set of results and [A] has doubled. [B] has no effect on the rate of reaction.

# **15** Equilibrium

## 15.1 INTRODUCTION

The topic of equilibrium is included in many GCSE syllabuses although many teachers leave it out or do not cover it adequately. It is an important topic at A level which needs to be understood before Units 16 and 17 are attempted.

## 15.2 WHAT IS MEANT BY CHEMICAL EQUILIBRIUM?

The reaction between iron and steam is a reversible reaction
$$3Fe(s) + 4H_2O(g) \rightleftharpoons Fe_3O_4(s) + 4H_2(g)$$
If steam is passed over heated iron in an open vessel, the forward reaction takes place and the iron oxide and hydrogen are produced. The steam displaces the hydrogen from the system, thereby limiting the reverse reaction.

When hydrogen is passed over the heated iron oxide, iron and steam are produced.

An equilibrium is established, however, if iron and steam are heated in a *closed* container so the products cannot escape. In equilibrium, the concentrations of the two products and the two reactants remain constant providing external conditions are unchanged. The reactions have not stopped however. The rate of the forward reaction is equal to the rate of the reverse reaction and so the concentrations are unchanged. Using the term **dynamic equilibrium** is probably better because it emphasises the two reactions are continuing.

## 15.3 QUANTITATIVE APPROACH TÓ EQUILIBRIUM

The **law of equilibrium** states that, at constant temperature, the rate of a reaction is proportional to the active mass of the reacting substances. For A-level purposes, active mass is the same as concentration in mol dm$^{-3}$. The law of mass action applies to an equilibrium in a homogeneous system. The sign $\rightleftharpoons$ shows the system can reach equilibrium.

In the equilibrium
$$aA + bB \rightleftharpoons cC + dD$$
the equilibrium constant $K_c = \dfrac{[C]^c[D]^d}{[A]^a[B]^b}$ at constant temperature.

By convention, the concentrations of products are divided by reactants.

Equilibrium constants can be calculated using concentrations in mol dm$^{-3}$, and the equilibrium constant is represented by $K_c$. Partial pressures can also be used to calculate equilibrium constants and the equilibrium constant is then represented by $K_p$.

*Example*:
$$CH_3COOH(l) + C_2H_5OH(l) \rightleftharpoons CH_3COOC_2H_5(l) + H_2O(l)$$
Calculate $K_c$ in terms of $a$, $b$ and $x$ if $a$ moles of ethanoic acid and $b$ moles of ethanol are mixed and left to reach equilibrium and $x$ moles of ethanoic acid are used up.

$$CH_3COOH + C_2H_5OH \rightleftharpoons CH_3COOC_5H_5 + H_2O$$

| | | | | |
|---|---|---|---|---|
| *Initially* | $a$ moles | $b$ moles | 0 | 0 |
| *At equilibrium* | $(a\text{-}x)$ moles | $(b\text{-}x)$ moles | $x$ moles | $x$ moles |

If the total volume of the mixture is $V$ dm$^{-3}$

$$[CH_3COOH]=\frac{a\text{-}x}{V} \quad [C_2H_5OH]=\frac{b\text{-}x}{V} \quad [CH_3COOC_2H_5]=\frac{x}{V} \quad [H_2O]=\frac{x}{V}$$

(All units are mol dm$^3$)

$$K_c = \frac{[CH_3COOC_2H_5][H_2O]}{[CH_3COOH][C_2H_5OH]}$$

Substituting

$$Kc = \frac{\frac{x}{V}\times\frac{x}{V}}{\frac{(a-x)}{V}\times\frac{(b-x)}{V}} = \frac{x^2}{(a-x)(b-x)}$$

In this case, the volume of the mixture has no effect on the equilibrium constant.

If 1 mole of ethanoic acid and 1 mole of ethanol come to equilibrium, two thirds of a mole of acid is used up. Calculate $K_c$.

Substitute $a = 1$ mole, $b = 1$ mole, $x = \dfrac{2}{3}$ mole

$$K_c = \frac{\frac{2}{3} \times \frac{2}{3}}{\frac{1}{3} \times \frac{1}{3}} = 4$$

## *15.4   $K_P$

In reactions involving gases, measuring the concentration of gases is difficult to do. It is more convenient to use partial pressures.

For the reaction

$$N_2(g) + 3H_2(g) \rightleftharpoons 2NH_3(g)$$

$$K_p = \frac{p_{NH_3}{}^2}{p_{N_2} \times p_{H_2}{}^3}$$

The numerical values for $K_c$ and $K_p$ for a particular reaction at the same temperature are clearly different, although mathematically related. Just be consistent.

## *15.5   EQUILIBRIUM IN A HETEROGENEOUS SYSTEM

The law of equilibrium can be modified to include heterogeneous systems. The common example is the heating of calcium carbonate in a closed system to prevent the escape of carbon dioxide

$$CaCO_3(s) \rightleftharpoons CaO(s) + CO_2(g)$$

The vapour above the solid is a homogeneous system and

$$\frac{p_{CaO}}{p_{CaCO_3}} \times p_{CO_2} = \text{constant}$$

where $p$ represents partial pressures. Thus at constant temperature the partial pressures of calcium carbonate and calcium oxide are constant. This is because the vapour pressures of solids are constant at constant temperature

$$K_p = p_{CO_2}$$

From this it follows that at any particular temperature, the partial pressure of carbon dioxide is constant.

## 15.6   QUALITATIVE APPROACH TO EQUILIBRIUM – LE CHATELIER'S PRINCIPLE

The way in which the position of an equilibrium will change if one of the conditions is changed is predicted qualitatively using **Le Chatelier's principle**.

Le Chatelier's principle states that if one of the conditions is changed, the position of equilibrium will alter in such a way as to tend to restore the original conditions.

Consider the equilibrium

$$A + B \rightleftharpoons C + D$$

At equilibrium A, B, C and D are present, and the concentrations of these substances remains constant providing conditions are unchanged. If the forward reaction is encouraged, a new equilibrium will be set up. In this equilibrium [C] and [D] will have increased and [A] and [B] will have decreased. The equilibrium is said to have 'moved to the right'.

A system 'moves to the left' if the new equilibrium established contains higher concentrations of A and B and lower concentrations of C and D.

The conditions which are commonly changed are:
 (*i*) Concentrations of reactants or products.
 (*ii*) Temperature.
(*iii*) Pressure. This is only important in reactions involving gases.
   These conditions are considered separately.

**Concentration**

If the product C is removed by liquefaction, escape into the atmosphere etc, the reverse reaction cannot occur. The position of the equilibrium moves to the right. If the concentration of A or B is increased, again the forward reaction is encouraged and again the equilibrium moves to the right. $K_c$ and $K_p$ remain constant.

**Temperature**

The change that takes place in an equilibrium when temperature is altered depends upon whether the forward reaction is exothermic or endothermic (Fig. 15.1).

If the forward reaction is exothermic, the reverse reaction will be endothermic. The equilibrium can be written

$$A + B \rightleftharpoons C + D \qquad \Delta H \text{ is negative}$$

If the temperature of the system is raised, the equilibrium moves to the left as the endothermic process will tend to reduce the temperature, *i.e.* restore the original conditions. Conversely if the temperature is decreased, the equilibrium moves to the right. $K_c$ and $K_p$ change.

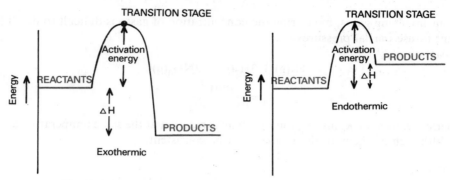

**Fig. 15.1** Energy changes in exothermic and endothermic reactions

## Pressure

For a reaction involving gases, altering the pressure may cause a change in the position of the equilibrium.

For the reaction:

$$H_2(g) + I_2(g) \rightleftharpoons 2HI(g)$$

at 700°C, the number of moles of reactants and products are the same. Pressure has no effect on the position of the equilibrium.

For a reaction where there is an increase in the number of moles from reactants to products, increasing the pressure moves the equilibrium to the left. Where there is a decrease in the number of moles from reactants to products, increasing the pressure moves the equilibrium to the right. $K_c$ and $K_p$ remain constant.

## *15.7    DISTRIBUTION OF A SOLUTION BETWEEN TWO IMMISCIBLE LIQUIDS

Ethoxyethane (diethyl ether) and water only dissolve very slightly in each other and can be considered to be **immiscible**. When mixed they form two separate layers.

If a solute is added to the mixture of ethoxyethane and water, and the mixture is shaken until an equilibrium exists, the following relationship usually applies.

$$\frac{\text{Concentration of solute in ethoxyethane}}{\text{Concentration of solute in water}} = \text{constant}$$

This constant is called the **distribution ratio** or **partition coefficient** and the relationship is called the **distribution or partition law**.

This simple relationship applies providing:

(*i*) the temperature is constant;
(*ii*) the solubility of the solute in either solvent is not exceeded;
and (*iii*) the solute is in the same molecular state in both solvents.

It is possible that either **association** or **dissociation** of the solute may occur in one or more of the solvents.

## Association

If benzenecarboxylic acid (benzoic acid) is added to a mixture of water and benzene (two immiscible solvents), association of benzenecarboxylic acid occurs in the benzene layer.

## Dissociation

In the aqueous layer, benzenecarboxylic acid is partially dissociated into ions.

The distribution law is modified where the solute is associated or dissociated in one or more of the immiscible solvents.

If solvent S is distributed between two immiscible solvents X and Y and exists in the form of single molecules in X but as $X_n$ in Y, then at constant temperature,

$$\sqrt[n]{\frac{\text{Concentration of S in X}}{\text{Concentration of S in Y}}} = \text{constant}$$

Solvent extraction (Unit 7.3) is the principle application of the distribution law.

QUESTIONS

**1** The following equilibrium is exothermic for the left to right reaction.

$$2SO_2(g) + O_2(g) \rightleftharpoons 2SO_3(g)$$

(a) Write down the equilibrium expression for $K_c$.
(b) In which units will $K_c$ be expressed?
(c) Write down the equilibrium expression for $K_p$, explaining the symbols you use.
(d) State the effect, if any, of each of the following on the equilibrium concentrations of $SO_3$ in a particular equilibrium mixture, clearly giving the reasons.
(i) decrease of pressure
(ii) decrease of temperature
(iii) addition of an inert gas to the system at equilibrium, while maintaining the pressure constant.

*(Welsh Joint Education Committee)*

Answer
(a)
$$K_c = \frac{[SO_3]^2}{[SO_2]^2 [O_2]}$$

(b) Concentrations are expressed in units of mol dm$^{-3}$.
$$K_c = \frac{(\text{mol dm}^{-3})^2}{(\text{mol dm}^{-3})^2 (\text{mol dm}^{-3})}$$
$$= \text{mol}^{-1} \text{ dm}^3$$

NB The units for an equilibrium constant will depend upon the expression for $K_c$.
For the reaction between ethanol and ethanoic acid, the $K_c$ has no units.

(c)
$$K_p = \frac{p^2{}_{SO_3}}{p^2{}_{SO_2} \times p_{O_2}}$$

The units of $K_p$ are atm$^{-1}$ or kPa$^{-1}$

(d) (i) Decreasing the pressure would move the equilibrium to the left, decreasing the concentration of $SO_3$. There are more molecules and, on decreasing the pressure, the equilibrium moves to increase the number of molecules. This is in accordance with Le Chatelier's Principle.
(ii) The forward reaction is exothermic. A decrease of temperature moves the equilibrium to the right to produce more sulphur trioxide. This is in accordance with Le Chatelier's principle. The equilibrium moves to the right to produce more heat.
(iii) $K_p$ is constant but reduction of the partial pressures of the gases moves the equilibrium to the left. The inert gas, as such, has no effect and the situation is effectively the same as in (d) (i)

**2** The diagram (Fig. 15.2) shows the general behaviour of the concentrations of reactants and products as a chemical reaction progresses.

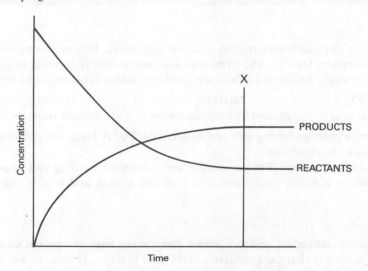

**Fig. 15.2** Question

(*a*) At what stage are the concentrations varying most rapidly?
(*b*) At what stage is the reaction at time X?
(*c*) What happens to the rate of change of concentration of the reactants as the reaction proceeds?

(*Scottish Higher*)

(*a*) The concentrations change most rapidly at the start of the reaction.
(*b*) At X the concentrations are not changing as the system is in equilibrium.
(*c*) The rate of change of concentration decreases during the reaction until it reaches zero when the system is in equilibrium.

3 The following equation represents the decomposition of hydrogen iodide into hydrogen and iodine.

$$2HI \rightleftharpoons H_2 + I_2$$

0.256g of hydrogen iodide was heated at 764K in a bulb of volume 100cm³. When an equilibrium was established, the bulb was cooled to room temperature and the iodine present determined by titration. It was found that 0.00028 moles of iodine were present.
(*a*) Why is there almost no change in the concentration of iodine when the bulb is cooled quickly to room temperature?
(*b*) Calculate
  (*i*) the number of moles of HI in 0.256g
 (*ii*) the number of moles of hydrogen formed
(*iii*) the number of moles of hydrogen iodide unreacted.
(*c*) Calculate $K_c$ for the reaction at 764K.

$$(A_r(H) = 1, A_r(I) = 127)$$

(*a*) The equilibrium mixture can be cooled quickly to room temperature or quenched without significantly changing the position of the equilibrium because the rate of this reaction is decreased dramatically by lowering the temperature.

(*b*) (*i*) Number of moles of HI $= \dfrac{0.256}{128} = 0.002$ moles

(*ii*) From the equation the number of moles of iodine present equals the number of moles of hydrogen present. Number of moles of hydrogen $= 0.00028$ moles
(*iii*) Number of moles of hydrogen iodide remaining $= 0.002 - 0.00056 = 0.00144$
NB twice as many HI molecules are used up as $I_2$ molecules produced.
(*c*)

$$K_c = \frac{[H_2][I_2]}{[HI]^2}$$

The volume is 100cm³
$[H_2] = 0.0028$ mol dm⁻³ $[I_2] = 0.0028$ mol dm⁻³ $[HI] = 0.0144$ mol dm⁻³
Substitute and calculate. $K_c = 0.0378$

# 16 Acids and bases

## 16.1 INTRODUCTION

This unit concerns various topics related to acids and bases. Before attempting this unit it is important to understand Unit 15. The terms acid and base should be familiar to you from GCSE. At GCSE an acid might be defined as a compound containing replaceable hydrogens

*e.g.*    $H_2SO_4$    →    $NaHSO_4$    →    $Na_2SO_4$
     *sulphuric acid*    *sodium hydrogensulphate*    *sodium sulphate*

An acid and a base react to form a salt and water only. For A level, the definitions of acid and base must be more clearly stated.

One important distinction for you to understand is between a strong and weak acid. A **strong acid** is an acid which is highly dissociated into ions and a **weak acid** is only slightly dissociated into ions.

## 16.2 DEFINITIONS OF ACIDS AND BASES

Brønsted and Lowry defined an **acid** as a **proton donor** and a **base** as a **proton acceptor**. A proton is the same as a positively charged hydrogen ion (H⁺). In practice, H⁺ ions do not exist in solution and are more likely to exist as $H_3O^+$.

The equation for the ionization of hydrochloric acid is:

$$HCl(aq) + H_2O(l) \rightleftharpoons H_3O^+(aq) + Cl^-(aq)$$

Hydrochloric acid is an acid because it loses an $H^+$ ion to form a $Cl^-$ ion. Water is a base because it accepts $H^+$ ions to form $H_3O^+$ ions.

Looking at the reverse reaction, $H_3O^+$ ions lose $H^+$ ions in forming $H_2O$ and, therefore $H_3O^+$ is acting as an acid. Chloride ions accept $H^+$ ions to form HCl and, therefore, act as a base. The $H_3O^+$ is called the **conjugate acid** and $Cl^-$ the **conjugate base** (*i.e.* the acid and base for the reverse process).

In the case of a general acid HA

$$HA(aq) + H_2O(l) \rightleftharpoons H_3O^+(aq) + A^-(aq)$$

The position of this equilibrium depends upon the strength of the acid. For a strong acid such as HCl the equilibrium lies well to the right, *i.e.* the acid may be considered as completely ionized. In such a case the base $Cl^-$ is a very weak base. With a weak acid the equilibrium lies well to the left. $A^-$ is a fairly strong base.

Water can also act as an acid:

$$\underset{base}{NH_3} + \underset{acid}{H_2O} \rightleftharpoons \underset{\substack{conjugate \\ acid}}{NH_4^+} + \underset{\substack{conjugate \\ base}}{OH^-}$$

Lewis widened the definition of acid and base by defining an acid as an **electron pair acceptor**. This definition includes examples of acids and bases which would not be recognized using the other definitions. For example the reaction between trimethylamine and boron trifluoride forms a solid salt.

$$CH_3-\overset{\displaystyle CH_3}{\underset{\displaystyle CH_3}{N}}\longrightarrow \overset{\displaystyle F}{\underset{\displaystyle F}{B}}-F$$

Trimethylamine has an electron pair which can be donated to the boron trifluoride which is electron deficient because it only has six electrons in the outer shell.

Trimethylamine acts as a Lewis base and boron trifluoride is a Lewis acid. Aluminium chloride acts as a Lewis acid (Unit 23.2).

## 16.3 Ostwald's dilution law

Since the ionization of weak acids and bases are equilibrium processes, the law of mass action (Unit 15.3) can be applied to them at constant temperature.

For example, consider the dissociation of 1 mole of a weak acid HA. If the volume of the solution containing one mole of acid is $V\,dm^3$ and $\alpha$ is the degree of dissociation, then

$$HA \quad \rightleftharpoons \quad H^+ \quad + \quad A^-$$

Before
ionization    1 mole        0          0

After
ionization    $1-\alpha$ mole    $\alpha$ mole    $\alpha$ mole

$$[HA] = \frac{1-\alpha}{V}\,mole\,dm^{-3} \quad \frac{\alpha}{V}\,mole\,dm^{-3} \quad \frac{\alpha}{V}\,mole\,dm^{-3}$$

Applying the law of mass action.

$$K = \frac{[H^+][A^-]}{[HA]}$$

$$K = \frac{\alpha^2}{(1-\alpha)V}\,mol\,dm^{-3}$$

This expression is called **Ostwald's dilution law**. The equilibrium constant in this case is the **dissociation constant**. For very weak electrolytes, where little dissociation takes place, $(1-\alpha)$ is approximately equal to 1 and the expression simplifies to

$$K = \frac{\alpha^2}{V} \quad or \quad K = \alpha^2 c$$

where $V$ is the volume containing one mole and $c$ is the concentration in $mol\,dm^{-3}$.

Although the degree of dissociation of the acid changes with dilution, the value of $K$ is unchanged at constant temperature.

*Example*: The degree of dissociation of 0.02M benzenecarboxylic acid, $C_6H_5COOH$, is 0.056 at 25°C. Calculate the dissociation constant of the acid at this temperature.

$$C_6H_5COOH \rightleftharpoons C_6H_5COO^- + H^+$$

Using Ostwald's dilution law

$$K = \frac{\alpha^2}{(1-\alpha)V} \qquad V = 50 \text{ dm}^{-3}$$

$$K = \frac{0.056^2}{(1-0.056)50} = 0.000063 \text{ mol dm}^3$$

### 16.4 IONIC PRODUCT OF WATER

Even if water is purified by repeated distillations, the electrical conductivity never falls to zero. This is because of **self-ionization** of water. This can be represented either by

$$2H_2O \text{ (l)} \rightleftharpoons H_3O^+(aq) + OH^-(aq)$$
or $\qquad\qquad\qquad H_2O(l) \rightleftharpoons H^+(aq) + OH^-(aq)$

Using the simpler equation

$$K_c = \frac{[H^+][OH^-]}{[H_2O]}$$

Only a few molecules of water ionise and $[H_2O]$ can be regarded as constant at constant temperature.

$$K_w = K_c \times \text{constant} = [H^+][OH^-]$$

This is called the **ionic product** of water.

At 25°C, $K_w = 1 \times 10^{-14}$ mol$^2$ dm$^{-6}$ and because, from the equation, the concentrations of $H^+$ and $OH^-$ are equal:

$$[H^+] = [OH^-] = 1 \times 10^{-7} \text{ mol dm}^{-3}$$

In any aqueous solution the product of the concentration of $H^+$ and $OH^-$ is $1 \times 10^{-14}$ mol$^2$ dm$^{-6}$. In a neutral solution both $[H^+]$ and $[OH^-]$ equal $1 \times 10^{-7}$ mol dm$^{-3}$.

The ionic product is useful in A level questions. If either the $[OH^-]$ or $[H^+]$ is known, the other one can be calculated.

*e.g.* A solution contains $[OH^-] = 10^{-1}$ mol dm$^{-3}$
Since $[H^+][OH^-] = 1 \times 10^{-14}$ mol$^2$ dm$^{-6}$ and $[OH^-] = 10^{-1}$ mol dm$^{-3}$ $[H^+] = 10^{-13}$ mol dm$^{-3}$

### 16.5 pH

A solution which is neutral contains
$$[H^+] = [OH^-] = 10^{-7} \text{ mol dm}^{-3}$$
If a solution is acidic

$$[H^+] > [OH^-]$$

and if the solution is alkaline

$$[H^+] < [OH^-]$$

The concentrations of the ions are all very small
*e.g.* a weak acid might have $[H^+] = 6.28 \times 10^{-6}$ mol dm$^{-3}$
The system of pH is designed to simplify these values by removing the awkward negative indices.
pH is defined as $- \log_{10}[H^+]$

A solution with $[H^+] = 10^{-7}$ mol dm$^{-3}$ has a pH = 7. This is the neutral solution.

Solutions that are acidic have a pH less than 7. A solution of hydrochloric acid (0.1 mol dm$^{-3}$) ionizes to produce 0.1 mol dm$^{-3}$ of $H^+$ ions ($10^{-1}$ mol dm$^{-3}$). The pH of this solution is 1(*i.e.* $-\log_{10} 10^{-1}$).

A solution with $[H^+] = 6.28 \times 10^{-6}$ mol dm$^{-3}$ has a pH of 5.20.

Solutions in which $[OH^-]$ is greater than $[H^+]$ are alkaline and have a pH greater than 7. A solution of sodium hydroxide (0.1 mol dm$^{-3}$) ionizes to produce 0.1 mol dm$^{-3}$ of $OH^-$ ($10^{-1}$ mol dm$^{-3}$)

Using the ionic product of water,

$$[H^+][OH^-] = 10^{-14} \text{ mol}^2 \text{ dm}^{-6}$$
$$[H^+] = 10^{-13} \text{ mol dm}^{-3}$$
$$pH = -\log_{10}[H^+] = 13$$

### 16.6 COMPARING THE STRENGTHS OF WEAK ACIDS AND BASES

For a weak acid HA in water, the equilibrium in the solution will be

$$HA + H_2O \rightleftharpoons H_3O^+ + A^-$$

The equilibrium constant $K_c$ is given by

$$K_c = \frac{[H_3O^+][A^-]}{[HA][H_2O]}$$

In a dilute solution $[H_2O]$ is large and effectively constant.

$$\frac{[H_3O^+][A^-]}{[HA]} = K_c \times \text{constant} = K_a$$

$K_a$ (called the **acid dissociation constant**) is the measure of the strength of an acid. For example for methanoic acid $K_a = 1.6 \times 10^{-4}$ mol dm$^{-3}$; it is obvious that the ionization of this acid is small, and many organic acids have even smaller values.

All $K_a$ values are very small and it is convenient to record the strengths of acids as $pK_a$ values. (This is similar to the use of pH for recording $[H^+]$).

$$pK_a = -\log_{10} K_a$$

The important thing to remember is that the **smaller** the value of $pK_a$, the larger is the value of $K_a$ and the **stronger** is the acid.

### *16.7 COMPARING THE STRENGTHS OF WEAK BASES

The equilibrium that exists when a base is added to water is

$$B + H_2O \rightleftharpoons BH^+ + OH^-$$

Again $[H_2O]$ is approximately constant

$$K_b = \frac{[BH^+][OH^-]}{[B]}$$

$K_b$ is called the **base dissociation constant**. It is a measure of the strength of a base and, for convenience, it is usually expressed as a $pK_b$ value. The **smaller** the value of $pK_b$ the **stronger** is the base.

A comparison of the relative strengths of acids and bases is given in Unit 39.

### 16.8 INDICATORS AND ACID–ALKALI TITRATIONS

Most indicators for acid–alkali titrations are weak acids or bases. Phenolphthalein is a weak acid and it dissociates

$$H_2O + HA \rightleftharpoons H_3O^+ + A^-$$
$$\quad \text{colourless} \qquad \text{pink}$$

HA exists in acid solution and A$^-$ in alkaline solution. If acid is added to the equilibrium mixture, the equilibrium moves to the left and the solution turns colourless. Addition of alkali removes $H_3O^+$ ions and moves equilibrium to the right turning the solution pink.

Methyl orange is a weak base which dissociates

$$BOH \rightleftharpoons B^+ + OH^-$$
$$\text{yellow} \quad \text{red}$$

BOH exists in alkaline solution and B$^+$ in acid solution. For a substance to be a good indicator, the two forms (HA and A$^-$ for phenolphthalein and BOH and B$^+$ for methyl orange) must be different colours for the indicator to function adequately.

The weak acids and bases which act as indicators have varying $K_a$ and $K_b$ values and, in consequence, they change colour at different characteristic pH values.

Table 16.1 lists the common indicators, their colour change and the pH range at which the colour change occurs.

**Table 16.1 Colour changes of some common indicators**

| Indicator | Acid solution | Approx. pH | Alkali solution |
|---|---|---|---|
| methyl orange | orange | 4 | yellow |
| methyl red | red | 5 | yellow |
| litmus | red | 7 | blue |
| phenolphthalein | colourless | 9 | red |

In choosing the indicator for an acid–alkali titration it is important to consider the pH at which the indicator changes colour.

(*i*) Strong acid – strong base
(*ii*) Weak acid – strong base
(*iii*) Strong acid – weak base
(*iv*) Weak acid – weak base.

Fig. 16.1 shows curves of pH during titrations of the four types of acid–alkali titrations.

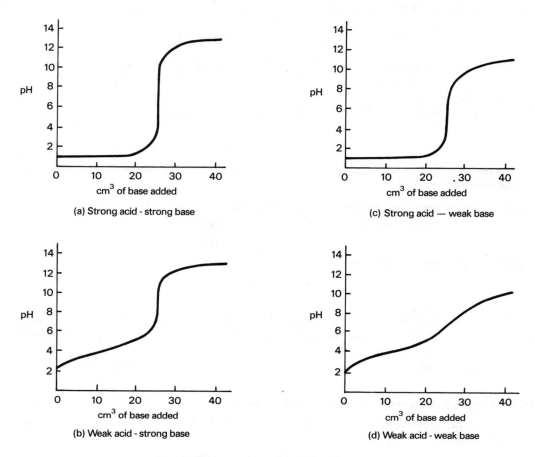

**Fig. 16.1** Four types of acid-alkali titrations

### (i) Strong acid – strong base

*E.g.* $$NaOH(aq) + HCl(aq) \rightarrow NaCl(aq) + H_2O(l)$$

0.1M sodium hydroxide solution is added to 25cm³ of 0.1M hydrochloric acid. Initially, the pH of the solution is 1 and the pH increases as alkali is added. When 25.0cm³ of sodium hydroxide solution is added the solution is exactly neutral (pH 7) and further addition of sodium hydroxide solution gives a pH greater than 7 as all the acid is used up and the alkali is in excess. A suitable indicator changes colour on the vertical portion of the graph (between 3 and 11). On the vertical portion, there is a rapid change in pH with addition of a small amount of alkali. Any of the indicators in Table 16.1 would be suitable.

### (ii) Weak acid – strong base

*E.g.* $$CH_3COOH(aq) + NaOH(aq) \rightarrow CH_3COONa(aq) + H_2O(l)$$

0.1M ethanoic acid has a pH between 2 and 3 (not 1 as with a strong acid because it is less completely ionized). On addition of 0.1M sodium hydroxide the curve in (*b*) is obtained. The vertical portion is between about 7 and 10 and phenolphthalein is the best indicator to use.

### (iii) Strong acid – weak base

*E.g.* $$HCl(aq) + NH_3(aq) \rightarrow NH_4Cl(aq)$$

This time the vertical portion of the graph is between about 3 and 7 and methyl orange or methyl red is suitable.

**(iv) Weak acid – weak base**

> *E.g.* $$CH_3COOH(aq) + NH_3(aq) \rightarrow CH_3COONH_4(aq)$$

There is no vertical portion on the graph. There is no sharp change in pH and no indicator is suitable.

It is possible to use different indicators for the same titration to detect different end points. For example, the reaction of sodium carbonate with dilute hydrochloric acid has the following overall equation.

$$Na_2CO_3(aq) + 2HCl(aq) \rightarrow 2NaCl(aq) + H_2O(l) + CO_2(g)$$

This is detected by methyl orange or methyl red.

Phenolphthalein, however, will detect the end point of the reaction.

$$Na_2CO_3(aq) + HCl(aq) \rightarrow NaHCO_3(aq) + NaCl(aq)$$

This end point requires only half of the volume of acid compared to that with methyl orange or methyl red as indicator.

## QUESTIONS

**1** For each of the following reactions, underline the substance on the left-hand side which is behaving as an acid, and the substance on the right-hand side which is its conjugate base:

> (a) $H_2O + HCl \rightarrow H_3O^+ + Cl^-$
> (b) $CH_3COOH + OH^- \rightarrow CH_3COO^- + H_2O$
> (c) $H_2SO_4 + HNO_3 \rightarrow HSO_4^- + H_2NO_3^+$
> (d) $CH_3COOH + HNO_3 \rightarrow CH_3COOH_2^+ + NO_3^-$

Underline:
> (a) HCl and $Cl^-$;
> (b) $CH_3COOH$ and $CH_3COO^-$
> (c) $H_2SO_4$ and $HSO_4^-$
> (d) $HNO_3$ and $NO_3^-$

**2** Explain why the pH of 0.1M hydrochloric acid is 1 but 0.1M ethanoic acid has a pH of approximately 3. Calculate the pH of

(a) hydrochloric acid (0.01 mol dm$^{-3}$)

(b) ammonia solution (0.1 mol dm$^{-3}$)

($K_b$ for ammonia $= 1.8 \times 10^{-5}$)
A solution of hydrochloric acid (0.1 mol dm$^{-3}$) is fully ionized.

$$HCl + H_2O \rightarrow H_3O^+ + Cl^-$$

The solution of ethanoic acid is only partially ionized.

$$CH_3COOH + H_2O \rightleftharpoons CH_3COO^- + H_3O^+$$

The number of $H_3O^+$ ions in solution is greater in hydrochloric acid than in ethanoic acid.

(a) 0.01 mol dm$^{-3}$ hydrochloric acid. The acid is completely ionized and the concentration of $H^+$ ions is 0.01 mol dm$^{-3}$ ($10^{-2}$ mol dm$^{-3}$)

$$pH = -\log_{10}[H^+] = 2$$

(b)
$$NH_3 + H_2O = NH_4^+ + OH^-$$

$$K_b = \frac{[NH_4^+][OH^-]}{[NH_3]}$$

If ammonia solution is 0.1 mol dm$^{-3}$ and assuming no ionization, $[NH_3] = 10^{-1}$ mol dm$^{-3}$. This has reduced to $(10^{-1} - x)$ mol dm$^{-3}$ because $x$ mol dm$^{-3}$ have dissociated. Concentrations of $NH_4^+$ and $OH^-$ are equal from the equation.

$$K_b = \frac{x^2}{(10^{-1} - x)} = 1.8 \times 10^{-5}$$

It is useful here to use the approximation that since $x$ is small, $(10^{-1} - x)$ is approximately equal to $10^{-1}$. This will save you having to solve an awkward quadratic equation.

$$x^2 = 1.8 \times 10^{-6}$$
$$x = 1.34 \times 10^{-3} \text{ mol dm}^{-3} = [OH^-]$$

Using the ionic product of water

$$[H^+][OH^-] = 10^{-14} \text{ mol}^2 \text{ dm}^{-6}$$
$$[H^+] = 7.46 \times 10^{-12} \text{ mol dm}^{-3}$$
$$pH = -\log_{10}[H^+]$$
$$= 12 - \log_{10} 7.46$$
$$= 11.13$$

# 17 Salt hydrolysis and buffer solutions

## 17.1 INTRODUCTION

The unit covers three important topics related to ionic equilibria. Before attempting this unit you are advised to study Units 15 and 16.

## 17.2 SALT HYDROLYSIS

The first task here is to recognize that a question is a salt hydrolysis question.

When a salt is dissolved in water, the resulting solution may be acid, alkaline or neutral. This is because the salt may have reacted to some extent with the water. This hydrolysis produces an acid and an alkali but they may not be of equal strength. There are four types of salt that should be considered.

### Salt of a strong acid and a strong base

*E.g.* sodium chloride. Sodium chloride is completely ionized in solution to form $Na^+(aq)$ and $Cl^-(aq)$. $H_3O^+(aq)$ and $OH^-(aq)$ are also present from the ionization of a few of the water molecules. There is no interaction between the sodium and chloride ions and the water and as a result the concentrations of $H_3O^+(aq)$ and $OH^-(aq)$ are the same and the solution is exactly neutral.

### Salt of a strong acid and a weak base

*E.g.* ammonium chloride or iron(III) chloride. In a solution of ammonium chloride, ammonium ions in the solution react with the solvent to produce additional $H_3O^+(aq)$ ions.

$$NH_4Cl(s) \rightarrow NH_4^+(aq) + Cl^-(aq)$$

and $$NH_4^+(aq) + H_2O(l) \rightleftharpoons NH_3(aq) + H_3O^+(aq)$$

The resulting solution is acidic because of an excess of $H_3O^+(aq)$ ions over $OH^-(aq)$ ions.
A solution of iron(III) chloride in water is appreciably acidic. It will react rapidly with magnesium to produce hydrogen. The explanation for the acidity of the solution will be found in Unit 28.9. Similarly the acidity of a solution of aluminium chloride will be found in Unit 23.5.

### Salt of a weak acid and a strong base

*E.g.* sodium ethanoate. The ionization of sodium ethanoate produces $CH_3COO^-(aq)$ and $Na^+(aq)$. The resulting solution is alkaline, however, because of reaction between the ethanoate ions and the water to produce an excess of $OH^-(aq)$ ions.

$$CH_3COONa(s) \rightarrow CH_3COO^-(aq) + Na^+(aq)$$
$$CH_3COO^-(aq) + H_2O(l) \rightleftharpoons CH_3COOH(aq) + OH^-(aq)$$

### Salt of a weak acid and a weak base

*E.g.* ammonium ethanoate. Ammonium ethanoate is ionized in solution to produce $NH_4^+(aq)$ and $CH_3COO^-(aq)$ ions. Both of these ions react with water.

$$NH_4^+(aq) + H_2O(l) \rightleftharpoons NH_3(aq) + H_3O^+(aq)$$
$$CH_3COO^-(aq) + H_2O(l) \rightleftharpoons CH_3COOH(aq) + OH^-(aq)$$

Whether the final solution is slightly acidic, slightly alkaline or neutral depends upon the position of the two equilibria. In fact this solution is almost neutral.

## 17.3 BUFFER SOLUTIONS

In this book only a qualitative account of the subject is given.

A **buffer solution** is a solution of constant pH. The pH of the solution will not change appreciably if the solution is contaminated with traces of acid or alkali.

An acidic buffer solution (*i.e.* with pH less than 7) is prepared by mixing together definite amounts of a weak acid and the sodium or potassium salt of the same acid, *e.g.* ethanoic acid and sodium ethanoate.

An alkaline buffer (*i.e.* with pH greater than 7) is prepared by mixing a weak base and a soluble salt of the base, *e.g.* ammonia solution and ammonium chloride.

The most frequent mistake at A level is to fail to appreciate that a weak acid or base is required to make a buffer solution. Statements such as 'hydrochloric acid and sodium chloride can be used to make a buffer solution' are common.

The mixture of ethanoic acid and sodium ethanoate acts as a buffer as follows:

$$CH_3COOH(aq) + H_2O(l) \rightleftharpoons CH_3COO^-(aq) + H_3O^+(aq)$$
$$CH_3COONa(s) \rightarrow CH_3COO^-(aq) + Na^+(aq)$$

The resulting mixture contains a large concentration of ethanoate ions, most of which come from the sodium ethanoate.

If $H_3O^+(aq)$ ions are added, the equilibrium in the first equation is disturbed and moves to the left in order to reduce the $H_3O^+(aq)$ concentration. This is in accordance with Le Chatelier's principle and the concentration of $H_3O^+(aq)$ (*i.e.* the acidity) remains unchanged.

If $OH^-(aq)$ ions are added, they remove the $H_3O^+(aq)$ ions from the solution.

$$H_3O^+(aq) + OH^-(aq) \rightarrow 2H_2O(l)$$

The equilibrium in the first equation is disturbed by the removal of $H_3O^+(aq)$. The equilibrium moves to the right to produce more $H_3O^+(aq)$ and restore the pH of the solution.

The mixture of ammonia solution and ammonium chloride acts as a buffer solution:

$$NH_3(aq) + H_2O(l) \rightleftharpoons NH_4^+(aq) + OH^-(aq)$$
$$NH_4Cl(s) \rightarrow NH_4^+(aq) + Cl^-(aq)$$

This buffer solution contains a large concentration of $NH_4^+(aq)$ ions. Most of these ions come from the ammonium chloride which is completely ionized.

If $OH^-(aq)$ ions are added, the equilibrium in the first equation moves to the left in accordance with Le Chatelier's principle (Unit 15.5). This reduces the concentration of the $OH^-(aq)$ and restores the original pH.

If $H_3O^+(aq)$ ions are added, they remove $OH^-(aq)$ from the solution. The equilibrium moves to the right to produce more $OH^-(aq)$.

**17.4**  QUANTITATIVE APPROACH TO BUFFER SOLUTIONS

If we consider the simple ionization of ethanoic acid

$$CH_3COOH(aq) \rightleftharpoons CH_3COO^-(aq) + H^+(aq)$$

Applying the Law of Equilibrium (Unit 15.3)

$$K_a = \frac{[CH_3COO^-][H^+]}{[CH_3COOH]} \text{ or } [H^+] = K_a \frac{[CH_3COOH]}{[CH_3COO^-]}$$

$$pH = -\log_{10}[H^+] = -\log_{10}K_a - \log_{10}\frac{[CH_3COOH]}{[CH_3COO^-]}$$

N.B.  (*i*) The ratio $\dfrac{[CH_3COOH]}{[CH_3COO^-]}$ is the ratio of concentration of acid to base.

   (*ii*) The pH of a buffer solution depends upon the ratio of concentrations of acid and base and not on the *actual* concentrations.

   (*iii*) When $[CH_3COOH]$ and $[CH_3COO^-]$ are equal, $pH = -\log_{10}K_a$.

$K_a$ can be found by measuring the pH of a solution when a solution of an acid has been half neutralized by a strong base.

*17.5**  SOLUBILITY PRODUCT

The ideas of equilibrium can be applied to the equilibrium between an almost insoluble solid and its ions in solution. This leads to the concept of **solubility product** which explains the conditions under which precipitation from a solution will occur.

For example, silver chloride AgCl is a suitable sparingly soluble salt.

$$AgCl(s) \rightleftharpoons Ag^+(aq) + Cl^-(aq)$$

An equilibrium is established when silver chloride is added to water. Using the ideas of equilibria (Unit 15), the following expression can be obtained

$$K_s = [Ag^+(aq)][Cl^-(aq)]$$

[AgCl(s)] remains constant at constant temperature, as long as any solid remains in contract with the solution.

$$K_s = [Ag^+][Cl^-] = 2 \times 10^{-10} \text{ mol}^2\text{dm}^{-6}$$

where $K_s$ is called the **solubility product**. The solubility product is the maximum value of the ionic product of $Ag^+$ and $Cl^-$ ions that can exist in solution without precipitation occurring. If the product of the ionic concentrations exceeds this maximum value precipitation will occur.

Another common example is silver chromate(VI), $Ag_2CrO_4$, whose solubility product is

$$K_s = [Ag^+]^2[CrO_4^{2-}] = 3 \times 10^{-12} \text{ mol}^3\text{ dm}^{-9}$$

For the sparingly soluble electrolyte $A_xB_y$ the equilibrium is:

$$A_xB_y(aq) \rightleftharpoons xA^{y+}(aq) + yB^{x-}(aq)$$

and the solubility product is

$$K_s = [A^{y+}]^x[B^{x-}]^y$$

The most frequent mistake with solubility product is to try to apply it to solids that are readily soluble. For example, when dry hydrogen chloride is bubbled through a saturated solution of sodium chloride, sodium chloride crystals precipitate out. Many candidates would state that the solubility product of sodium chloride had been exceeded and precipitation occurs. This statement is incorrect as the solubility product of sodium chloride is meaningless, since sodium chloride is so soluble.

The observation is explained by the **common ion effect**. The equation for the solubility of sodium chloride is

$$NaCl(s) \rightleftharpoons Na^+(aq) + Cl^-(aq)$$

Addition of a large concentration of chloride ions (from the hydrogen chloride) moves the equilibrium to the left and sodium chloride precipitates.

Solubility product can be used to calculate the solubility of a sparingly soluble solid in water. For example, the solubility product of iron(III) hydroxide $Fe(OH)_3$ at 25°C is $8 \times 10^{-40}$ mol$^4$ dm$^{-12}$. Calculate the solubility of iron(III) hydroxide in a saturated solution at 25°C. ($A_r(Fe) = 56.0$, $A_r(O) = 16.0$, $A_r(H) = 1.0$)

Now $3[Fe^{3+}] = [OH^-]$

Relative molecular mass of iron(III) hydroxide $= 56 + (3 \times (16 + 1)) = 107$

$$K_s = [Fe^{3+}][OH^-]^3 = 8 \times 10^{-40} \text{ mol}^4 \text{ dm}^{-12}$$
$$Fe(OH)_3 \rightleftharpoons Fe^{3+} + 3OH^-$$

Substitute

$$K_s = [Fe^{3+}] (3[Fe^{3+}])^3$$
$$= 27[Fe^{3+}]^4 = 8 \times 10^{-40}$$
$$[Fe^{3+}]^4 = 0.3 \times 10^{-40}$$
$$[Fe^{3+}] = 0.74 \times 10^{-10} \text{ mol dm}^{-3}$$

Mass of iron(III) hydroxide dissolving per dm$^3$ $= 0.74 \times 10^{-10} \times 107$
$$= 7.9 \times 10^{-9}\text{g}.$$

## QUESTIONS

**1** Which one of the following substances does not produce an acid solution when added to water?

A ammonium chloride    D iron(II) sulphate
B sodium carbonate     E hydrogen chloride
C sulphur dioxide

Correct answer is **B**. Sodium carbonate is a salt of a strong base and a weak acid and, therefore, produces an alkaline solution.

**2** (a) Write an expression for the solubility product of lead(II) chloride.
(b) The solubility product of lead(II) chloride at 25°C is $1.6 \times 10^{-5}$ mol$^3$ dm$^{-9}$.
(i) What is the solubility in mol dm$^{-3}$ of lead(II) chloride in water at the same temperature?
(ii) How many moles of chloride ion must be added to a solution of lead(II) nitrate containing 1 mol dm$^{-3}$ at 25°C in order to cause a precipitate of lead(II) chloride? Assume that there is no change in volume on adding the chloride ion.

(a) $K_s = [Pb^{2+}][Cl^-]^2$
(b) (i) In the saturated solution          $2[Pb^{2+}] = [Cl^-]$

Substitute in the equation          $K_s = [Pb^{2+}] (2[Pb^{2+}])^2$
$$4[Pb^{2+}]^3 = 1.6 \times 10^{-5}$$
$$[Pb^{2+}]^3 = 4 \times 10^{-6}$$
$$[Pb^{2+}] = 1.59 \times 10^{-2} \text{ mol dm}^{-3}$$

(ii) Concentration of $Pb^{2+}$ ions in solution $= 1$ mol dm$^{-3}$
Substitute          $1.6 \times 10^{-5} = 1 \times Cl^{-2}$
$$Cl^- = 4 \times 10^{-3} \text{ mol dm}^{-3}$$

**3** In what proportions must solutions of ammonia and ammonium chloride (both 0.1 mol dm$^{-3}$) be mixed to obtain a buffer solution of pH 10.0 ($K_a$ for the ammonium ion is $6 \times 10^{-10}$ mol dm$^{-3}$)?

$$NH_4^+(aq) \rightleftharpoons NH_3(aq) + H^+(aq)$$
$$\text{ACID} \qquad \text{BASE}$$

$$pH = -\log_{10}K_a - \log_{10}\frac{[NH_4^+]}{[NH_3]} \qquad 10.0 = -\log_{10}6 \times 10^{-10} - \log_{10}\frac{[NH_4^+]}{[NH_3]}$$

$$\log_{10}\frac{[NH_4^+]}{[NH_3]} = -0.8 \qquad \frac{[NH_4^+]}{[NH_3]} = 0.16$$

0.16 dm$^3$ of ammonium chloride (0.1 mol dm$^{-3}$) and 0.1 dm$^3$ of ammonia solution (0.1 mol dm$^{-3}$) need to be mixed to give a solution of pH 10.0

# **18** Further oxidation and reduction

## 18.1 INTRODUCTION

In Unit 6 an introduction to oxidation and reduction was made. Oxidation can be defined as a process where there is electron loss and reduction as one in which there is electron gain. The system of oxidation state (or oxidation number) is also explained.

In this unit the subject is developed quantitatively by the introduction of the concept of **electrode potential**.

## 18.2 ELECTROCHEMICAL CELLS

When a metal rod is dipped into a solution of metal ions an equilibrium is set up. There is a tendency for the metal to form positive metal ions and go into solution. There is also a tendency for metal ions in solution to gain electrons from the metal and be deposited. This can be summarized as:

$$M(s) \rightleftharpoons M^{n+}(aq) + ne^-$$

Whether the metal acquires an overall positive or negative charge depends upon the position of the equilibrium. This depends upon the particular metal. With reactive metals the equilibrium lies well to the right hand side.

A metal dipping into a metal salt solution is called a **half cell**. Combining two half cells together can produce an **electrochemical cell**.

For example an electrochemical cell is shown in Fig. 18.1.

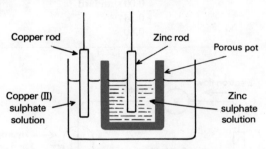

**Fig. 18.1** An electrochemical cell

A zinc rod dips into a solution of zinc sulphate and a copper rod dips into a solution of copper(II) sulphate. The porous pot allows ions to move without the solutions mixing freely. This cell is called the **Daniell cell**.

In this cell the following reactions take place

$$Zn(s) \rightarrow Zn^{2+}(aq) + 2e^-$$
$$Cu^{2+}(aq) + 2e^- \rightarrow Cu(s)$$

The overall equation is obtained by addition.

$$Zn(s) + Cu^{2+}(aq) \rightarrow Zn^{2+}(aq) + Cu(s)$$

The chemicals possess a certain amount of chemical energy that is converted into electrical energy in the cell. The cell stops working when either the zinc rod or the copper(II) sulphate solution is used up.

The zinc ions go into solution. The electrons that are released flow through the external circuit to the copper electrode and are used to deposit the copper.

The maximum potential difference between the two electrodes is known as the **electromotive force** (emf). If both solutions have concentrations of 1 mol dm$^{-3}$, the emf of the Daniell cell is 1.1 volts. The maximum potential difference is obtained when there is a large resistance in the external circuit so that a negligible current flows. The emf of a cell cannot be measured with an ordinary voltmeter as the resistance is insufficient. It is usually measured with a high resistance voltmeter or a potentiometer.

## 18.3 ELECTRODE POTENTIALS

The potential difference between a metal and a solution of its ions is dependent upon the materials used, the temperature and the concentration of the solution. The conditions are standardized with the temperature 25°C and the concentration of metal ions of 1 mol dm$^{-3}$. The **electrode potential** $E$ is the potential difference in volts between the metal and the solution of metal ions. The standard electrode potential (represented by $E^{\ominus}$) is the electrode potential that exists when the solution of metal ions has a concentration of 1 mol dm$^{-3}$ and the temperature is 25°C. The potential difference is measured relative to the standard hydrogen electrode.

The standard electrode potentials $E^{\ominus}$ for some common metals are given in Table 18.1. Do not try to remember these values, they will be given on the examination paper. Do remember that, by convention, the *reduced* form (*i.e. after* gain of electrons) is put on the *right* of the equation for which $E^{\ominus}$ is given, so that an *electropositive* metal has a *negative $E^{\ominus}$*.

It is not possible to measure the potential difference between a metal and its ions in solution directly. It is possible, however, to measure the potential difference between two electrodes. The hydrogen electrode is used as a standard reference electrode and its standard electrode potential is arbitrarily taken as zero.

The standard electrode potential of a metal M is obtained by setting up two half cells, one with a rod of M dipping into a solution (concentration 1 mol dm$^{-3}$) of a salt of M and the other a hydrogen electrode (Fig. 18.2).

**Table 18.1  Standard electrode potentials for some common metals**

| | $E^{\ominus}$/volts | |
|---|---|---|
| K$^+$ + e$^-$ → K | −2.92 | |
| Na$^+$ + e$^-$ → Na | −2.71 | |
| Mg$^{2+}$ + 2e$^-$ → Mg | −2.37 | |
| Zn$^{2+}$ + 2e$^-$ → Zn | −0.76 | |
| Fe$^{2+}$ + 2e$^-$ → Fe | −0.44 | Increasing |
| 2H$^+$ + 2e$^-$ → H$_2$ | 0.00 standard | tendency of electrode |
| Cu$^{2+}$ + 2e$^-$ → Cu | +0.34 | to release electrons |
| Ag$^+$ + e$^-$ → Ag | +0.80 | |

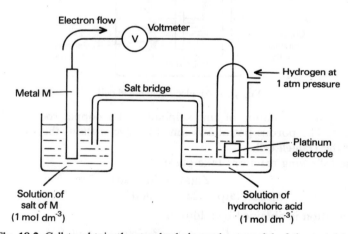

**Fig. 18.2** Cell to obtain the standard electrode potential of the metal M

Note that the arrow in the circuit from the metal M to the platinum electrode is the direction of flow of electrons when the standard electrode potential is negative. If the standard electrode potential is positive the flow of electrons is in the opposite direction. Remember that the conduction in the external circuit is electronic, but in the solution and salt bridge it is ionic.

The salt bridge replaces the porous pot in Fig. 18.1. It consists of an agar jelly with added potassium chloride (or potassium nitrate). It allows charge to pass through it to complete the circuit, but limits diffusion of the solutions. The voltage of the cell measured with a potentiometer or a high resistance voltmeter is taken to be the electrode potential of the metal.

Although a hydrogen electrode is taken as standard, it is not easy to set up. In practice a **calomel electrode** is often used as a secondary reference electrode. It consists of mercury in contact with solid mercury(I) chloride and a solution of potassium chloride saturated with the mercury(I) salt.

## 18.4 WORKING OUT THE EMF OF A CELL

In the Daniell cell (Unit 18.2), the maximum voltage is 1.1 volts. The cell can be represented by a cell diagram. In this case it is:

$$Zn(s) \mid Zn^{2+}(aq) \mid Cu^{2+}(aq) \mid Cu(s)$$

This means that a zinc rod is dipped into a solution of zinc ions and this solution is connected by a salt bridge to a solution of copper(II) ions with a copper rod dipping into it.

By convention, the more negative electrode is put on the left hand side. The emf of the cell is given by:

$$E_{cell} = E_{rhs} - E_{lhs}$$

(rhs – right hand side
lhs – left hand side)

In the case of the Daniell cell

$$E_{rhs} = 0.34 \text{ volts } (Cu^{2+}(aq) + 2e^- \rightarrow Cu(s))$$
$$E_{lhs} = -0.76 \text{ volts } (Zn^{2+}(aq) + 2e^- \rightarrow Zn(s))$$
$$E_{cell} = 0.34 -(-0.76) \text{ volts}$$
$$= 0.34 + 0.76 \text{ volts}$$
$$= +1.10 \text{ volts}$$

(NB If you get a negative value for the emf of the cell, it probably means you have written the cell diagram the wrong way round.)

Calculate the emf of the cell with the cell diagram

$$Fe(s) \mid Fe^{2+}(aq) \mid Ag^+(aq) \mid Ag(s)$$

Emf of the cell $= E_{rhs} - E_{lhs}$
Using the figures in Table 18.1

$$emf = 0.80 -(-0.44)$$
$$= +1.24 \text{ volts}$$

## 18.5 REDOX POTENTIALS

The equation

$$Zn^{2+}(aq) + 2e^- \rightarrow Zn(s)$$

is a reduction process since electrons are gained (Unit 6). The standard electrode potential ($-0.76$ volts) is a measure of the relative willingness of zinc metal to be reduced or oxidized.

For copper

$$Cu^{2+}(aq) + 2e^- \rightarrow Cu(s) \quad E^{\ominus} = 0.34 \text{ volts.}$$

This process therefore takes place more readily for copper than for zinc. The more positive the $E^{\ominus}$ value the more readily this process takes place.

There are three types of half cell
(*i*) A metal dipping into a solution of its ions *e.g.* zinc dipping into a solution of zinc sulphate.
(*ii*) A gas in contact with an inert electrode *e.g.* hydrogen in contact with a platinum electrode in the hydrogen electrode.
(*iii*) An inert metal in contact with a solution containing ions in two different oxidation states *e.g.* a platinum electrode dipping into a solution containing $Fe^{2+}(aq)$ and $Fe^{3+}(aq)$ ions.

It is possible to extend Table 18.1 by including non metals. The standard redox potentials are measured by forming a half cell of one of the three types above and measuring the voltage with reference to the hydrogen electrode. The standard redox potentials for some non metals are given in Table 18.2.

**Table 18.2 Standard redox potentials for some non metals**

|  | $E^{\ominus}$/volts |
|---|---|
| $\frac{1}{2}I_2 + e^- \rightarrow I^-$ | +0.54 |
| $\frac{1}{2}Cl_2 + e^- \rightarrow Cl^-$ | +1.36 |
| $\frac{1}{2}F_2 + e^- \rightarrow F^-$ | +2.85 |

## *18.6 THE ELECTROCHEMICAL SERIES

The arrangement of elements in order of their standard electrode potentials (Tables 18.1 and 18.2) leads to the **electrochemical series**. The element with the greatest negative $E^{\ominus}$ value is placed at the top of the list.

You may have met the electrochemical series at GCSE, or it may have been called the **reactivity series** although the two are not exactly the same. The electrochemical series is useful for a number of reasons. Usually we restrict it only to metals.

(*i*) The order of reactivity of metals decreases down the series. For non-metals the reactivity increases down the series. (Table 18.3)

**Table 18.3 Order of reactivity for metals and non-metals**

| | metals | non-metals |
|---|---|---|
| **Decreasing reactivity** | potassium<br>sodium<br>magnesium<br>zinc<br>iron<br>copper<br>silver | |
| | | iodine<br>bromine   **Increasing**<br>chlorine   **reactivity**<br>fluorine |

(*ii*) The electrochemical series can be used to predict reactions. *E.g.* if iron is added to copper(II) sulphate solution, a reaction takes place producing iron(II) sulphate and copper.

$$Fe(s) + CuSO_4(aq) \rightarrow FeSO_4(aq) + Cu(s)$$

The cell can be represented by:

$$Fe(s) \mid Fe^{2+}(aq) \mid Cu^{2+}(aq) \mid Cu(s)$$

Using the information in Table 18.1:

$$Fe \rightarrow Fe^{2+} + 2e^- \qquad E° = +0.44 \text{ V}$$
$$Cu^2 + 2e^- \rightarrow Cu \qquad E° = +0.34 \text{ V}$$

Adding the E° values, E° for the cell $= +0.78$ V. The reaction will occur if the E° value is positive

(*iii*) The electrochemical series explains why a piece of iron can be protected by galvanizing, *i.e.* coating with a layer of zinc. If the surface is scratched so that water (containing dissolved carbon dioxide to produce an electrolyte) is in contact with both metals, an electrochemical cell is set up:

$$Zn(s) \mid Zn^{2+}(aq) \mid Fe^{2+}(aq) \mid Fe(s)$$

This represents a cell with an $E^{\ominus}$ value of $+0.32$ volts; the corrosion of iron is the reverse and is therefore unfavourable.

(*iv*) The electrochemical series can predict the products of electrolysis.

(*v*) The electrochemical series is related to the methods available for extracting metals from metal ores.

### 18.7   THE FEASIBILITY OF A REACTION

If a chemical reaction is to take place spontaneously, the **standard free energy change** $\Delta G^{\ominus}$ must be negative (Unit 13). There is an equation which relates $\Delta G^{\ominus}$ to the standard redox potential $E^{\ominus}$.

$$\Delta G^{\ominus} = -nFE^{\ominus}$$

where $n$ is the number of electrons transferred and $F$ is Faraday's constant. For a reaction to be feasible, $E^{\ominus}$ should be positive.

A positive $E$ value does not give any indication of the speed of a reaction. If $E$ is greater than about $+0.4$V it can be assumed that the reaction will take place in the direction indicated, although it might take an infinitely long time.

### *18.8   NERNST EQUATION APPLIED TO THE DANIELL CELL

The variation of electrode potential with concentration is given by the Nernst equation

$$E = E^{\ominus} + \frac{2.3RT}{zF} \ln [\text{ion}]$$

where $E$ is electrode potential, $E^{\ominus}$ is standard electrode potential, $R$ is the gas constant, $T$ is absolute temperature, $z$ is the number of electrons transferred, $F$ is the Faraday constant and [ion] is the concentration of the ionic species.

In the Daniell cell

$$E_{\text{cell}} = E_{\text{Cu}} - E_{\text{Zn}} = E_{\text{Cu}}^{\ominus} - E_{\text{Zn}}^{\ominus} - \frac{2.3RT}{zF} \ln[Zn^{2+}] - \frac{2.3RT}{zF} \ln[Cu^{2+}]$$

$$= E^{\ominus}_{\text{cell}} - \frac{2.3RT}{zF} \ln\frac{[Zn^{2+}]}{[Cu^{2+}]} \qquad \text{N.B. } K_c = \frac{[Zn^{2+}]}{[Cu^{2+}]}$$

**If the cell is short-circuited by connecting the zinc and copper rods, the emf falls to zero and the cell reaction reaches equilibrium,** *i.e.* $E_{\text{cell}} \neq 0$

$$E^{\ominus}_{\text{cell}} = \frac{RT}{zF} \ln K_c \qquad\qquad K_c = 10^{37}$$

## QUESTIONS

**1** Use the equations

$$\begin{array}{ll} Fe^{3+} + e^- \rightarrow Fe^{2+} & E^{\ominus} = +0.77 \text{ volts} \\ Ag^+ + e^- \rightarrow Ag & E^{\ominus} = +0.80 \text{ volts} \end{array}$$

to predict whether iron(III) ions are reduced by silver under standard conditions.

**The equation for the reaction is**

$$Fe^{3+} + Ag \rightarrow Fe^{2+} + Ag^+$$

**If the $E^{\ominus}$ value for the cell Pt | $Fe^{2+}$, $Fe^{3+}$ ⦙ $Ag^+$ | Ag is positive the reaction is feasible.**

$$\begin{aligned} E^{\ominus}_{\text{cell}} &= E_{\text{rhs}} - E_{\text{lhs}} \\ &= +0.77 - (+0.80) \text{ volts} \\ &= -0.03 \text{ volts} \end{aligned}$$

**The reaction does not take place. For the reverse reaction the $E^{\ominus}$ value is +0.03 volts and the reaction is feasible.**

**2** The reaction between peroxodisulphate ions and iodide ions in aqueous solution is represented by the equation

$$S_2O_8^{2-}(aq) + 2I^-(aq) \rightarrow 2SO_4^{2-}(aq) + I_2(aq)$$

This reaction is catalyzed by $Fe^{3+}$ ions but not by $Cr^{3+}$. Using the following standard electrode potentials, suggest a possible mechanism for this catalytic action

$$\begin{aligned} Fe^{3+}(aq)\ Fe^{2+}(aq) &= +0.77V \\ S_2O_8^{2-}(aq)\ 2SO_4^{2-}(aq) &= +2.01V \\ I_2(aq)\ 2I^-(aq) &= +0.54V \end{aligned}$$

What can be predicted about the standard electrode potential $Cr^{3+}(aq)Cr^{2+}(aq)$ in view of the fact that the $Cr^{3+}$ does not catalyze the reaction?

**Possible mechanism in two stages:**

**(i)** $\qquad\qquad 2Fe^{3+}(aq) + 2I^-(aq) \rightarrow I_2(aq) + 2Fe^{2+}(aq)$
**(ii)** $\qquad\qquad S_2O_8^{2-}(aq) + 2Fe^{2+}(aq) \rightarrow 2Fe^{3+}(aq) + 2SO_4^{2-}(aq)$

**The $E^{\ominus}$ values for stages (i) and (ii) are +0.23 and +1.24V respectively. In both stages the $E^{\ominus}$ values are positive and therefore both stages are feasible and the mechanism is possible.**

**The fact that $Cr^{3+}$ does not catalyze the reaction suggests that in a two stage mechanism one of the $E^{\ominus}$ values is negative (and this stage is, therefore, not feasible).**

$$2I^-(aq) \rightarrow I_2(aq) + 2e^- \qquad E^{\ominus} = -0.54V$$

**The $E^{\ominus}$ value for $Cr^{3+}(aq)Cr^{2+}(aq)$ must be less negative than −0.54V if the reaction is to be feasible. It is in fact −0.41V and so no reaction takes place (you are not expected to remember this value).**

**3** The reaction between zinc powder and nickel(II) sulphate can be represented by:

$$Zn(s) + Ni^{2+}(aq) \rightarrow Zn^{2+}(aq) + Ni(s)$$

*(a)* Calculate the mass of nickel which could be formed if 1.308g of zinc was added to excess nickel(II) sulphate solution. (Ni = 59, Zn = 65.4)

*(b)* Draw a labelled diagram of the apparatus which could be used to measure the e.m.f. of the cell

$$Zn(s) \mid Zn^{2+}(aq) \vdots Ni^{2+}(aq) \mid Ni(s)$$

*(c)* Calculate $E^0$ for the cell using Table 18.1 and 8.1: and

$$Ni^{2+} + 2e^- \rightarrow Ni \qquad\qquad E^0 = -0.25 \text{ V}$$

*(d)* Calculate $\triangle G^0$ for the reaction given F = 96 500 C mol$^{-1}$

*(e)* $\triangle S_{\text{system}}$ can be calculated at 298K using $\triangle G^0$. $\triangle H^0$ and an equation on page 84:

(i) What is the equation?　　(ii) How would you attempt to find $\triangle H^0$ by experiment?

**(a)** 1 mol of Zn produces 1 mol Ni
　　1.308g of Zn = 1.308/65.4 = 0.02 mol Zn
　　0.02 mol Ni = 0.02 × 59 = 1.18g Ni

**(b)**

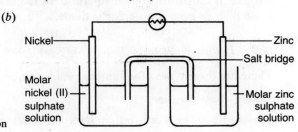

Fig. **18.3** Question

(c)

$$Ni^{2+} + 2e^- \rightarrow Ni \quad E^0 = -0.25 \text{ V}$$
$$Zn \rightarrow Zn^{2+} + 2e^- \quad E^0 = +0.76 \text{ V}$$
$$E^0 \text{ for cell} = +0.51 \text{ V}$$

(d) $\triangle G^0 = -nFE^0 \quad n = 2$ because 2 electrons transferred

$\triangle G^0 = -2 \times 96\,500 \times 0.51 = 98\,430 \text{ J mol}^{-1}$ or $98.43 \text{ kJ mol}^-$

(e) (i) $\triangle G = \triangle H - T\triangle S_{system}$

(ii) Add slight excess of zinc powder to 100 cm³ of 0.2 mol dm⁻³ nickel(II) sulphate solution (0.02 mol $NiSO_4$). Measure temperature rise, taking all precautions to minimise temperature loss. Calculate from results.

# 19 Volumetric calculations

## 19.1 INTRODUCTION

Volumetric calculations are required for theory papers and often on practical papers. These calculations involve reactions between reacting volumes of solutions. When attempting this type of question it is important to ensure that you have a balanced equation for the reaction.

*e.g.* $\quad 2NaOH(aq) + H_2SO_4(aq) \rightarrow Na_2SO_4(aq) + 2H_2O(l)$

From the equation, 2 moles of sodium hydroxide react with 1 mole of sulphuric acid. If 25cm³ of sodium hydroxide solution (0.1 mol dm⁻³, although this may still be shown as 0.1 molar or 0.1M) is neutralized by 20cm³ of sulphuric acid, we can use this equation to find the concentration of the sulphuric acid in mol dm⁻³.

*Sodium hydroxide*

1000cm³ of sodium hydroxide (0.1 mol dm⁻³) contain 0.1 moles of sodium hydroxide.

25cm³ of 0.1 mol dm⁻³ sodium hydroxide contains $\dfrac{0.1 \times 25}{1000}$ moles

$$= 0.0025 \text{ moles}$$

From the equation,

2 moles of sodium hydroxide react with 1 mole of sulphuric acid

0.0025 moles NaOH react with 0.00125 moles $H_2SO_4$

*Sulphuric acid*

0.00125 moles of sulphuric acid are present in 20cm³ of solution.

$\dfrac{0.00125 \times 1000}{20}$ moles of sulphuric acid would be present in 1000cm³.

Molar concentration of sulphuric acid = 0.0625 mol dm⁻³

All calculations can be worked out in a similar way. The method of working is far better than simply substituting in the equation

$$n_2 V_1 M_1 = n_1 V_2 M_2$$

where $n_1$ = number of moles of sulphuric acid in the equation = 1

$V_1$ = volume of sulphuric acid used = 20cm³

$M_1$ = molar concentration of sulphuric acid – unknown

$n_2$ = no. of moles of sodium hydroxide in the equation = 2

$V_2$ = volume of sodium hydroxide solution used = 25cm³

$M_2$ = molar concentration of sodium hydroxide = 0.1 mol dm⁻³

If you substitute the figures in the equation you will get the correct answer, but the problem comes when you make a mistake. The examiner has difficulty in awarding marks if you make a mistake in substituting in an equation. In the first method of calculation, marks can be awarded at each stage even if an error is made.

## 19.2 TYPES OF VOLUMETRIC CALCULATION

There are three types of volumetric analysis questions which appear on theory papers.

**1 Acid – base titration**

The one above between sodium hydroxide and sulphuric acid is an example.

## 2 Redox titrations

These can be of two types but in both cases they involve transfer of electrons.

(*i*) Reactions involving acidified potassium manganate(VII) or potassium dichromate(VI).

(*ii*) The reaction of iodine solution with sodium thiosulphate(VI). In some cases the iodine for titration is liberated from acidified potassium iodide solution by an oxidising agent such as potassium manganate(VII), potassium iodate(V) or copper(II) sulphate.

## 3 Precipitation titrations

The usual example here is the reaction of silver nitrate solution with chloride ions to precipitate silver chloride. The end point is usually detected with potassium chromate(VI) solution as indicator.

The following question section gives examples of some of these volumetric calculations.

QUESTIONS

1  (*i*) Give a half-equation for the oxidation of the ethanedioate ion, $C_2O_4^{2-}$(aq), to carbon dioxide $CO_2$(g).

   (*ii*) Give a half-equation for the reduction of the manganate(VII) ion, $MnO_4^-$(aq), to manganese(II) ions, $Mn^{2+}$ (aq), in acidic conditions.

   (*iii*) Calculate the volume of an acidified solution of 0.02 M potassium manganate(VII), $KMnO_4$, which would be needed to oxidize 100 cm$^3$ of a saturated solution of magnesium ethanedioate, $MgC_2O_4$.

Solubility of magnesium ethanedioate at 20°C is $9.3 \times 10^{-3}$ mol dm$^{-3}$

(Nuffield-part)

(i) $C_2O_4^{2-}$(aq) → $2CO_2$(g) + $2e^-$

(ii) $MnO_4^-$(aq) + $8H^+$(aq) + $5e^-$ → $Mn^{2+}$(aq) + $4H_2O$(l)

(iii) $5C_2O_4^{2-}$(aq) + $2MnO_4^-$(aq) + $16H^+$(aq) → $10CO_2$(g) + $8H_O$(l) + $2Mn^{2+}$ (aq)

100 cm$^3$ of saturated $MgC_2O_4$ solution at 20°C contains $9.3 \times 10^{-4}$ mol.

From the equation

$$5 \text{ mol of } C_2O_4^{2-} \text{ react with 2 mol } MnO_4^-$$
$$9.3 \times 10^{-4} \text{ mol} \qquad \frac{2 \times 9.3 \times 10^{-4}}{5} \text{ mol}$$

$$1000 \text{ cm}^3 \text{ of 0.02 M } MnO_4^- \text{ contains 0.02 mol } MnO_4^-$$
$$\frac{1000 \times 2 \times 9.3 \times 10^{-4}}{.02 \times 5} \text{ cm}^3 \text{ of 0.02 M } MnO_4^- \text{ contains } \frac{2 \times 9.3 \times 10^{-4}}{5} \text{ mol}$$

Answer 18.6 cm$^3$

This question would be more difficult if iron(II) ethanedioate had been used. Then both $Fe^{2+}$ and $C_2O_4^{2-}$ would be oxidized.

2 A solution of arsenic(III) oxide containing 0.248g required 50cm$^3$ of acidified potassium manganate(VII) solution (0.02 mol dm$^{-3}$) for complete oxidation. What is the oxidation state of arsenic in the product? ($A_r$(O) = 16, $A_r$(As) = 75)

$$As^{3+}(aq) \to As^{n+}(aq) + (n-3)e^-$$
$$MnO_4^-(aq) + 8H^+(aq) + 5e^- \to Mn^{2+}(aq) + 4H_2O(l)$$

$$5As^{3+}(aq) + (n-3)MnO_4^-(aq) + 8(n-3)H^+(aq) \to (n-3)Mn^{2+}(aq) + 4(n-3)H_2O(l) + 5As^{n+}(aq)$$

$$\text{No. of moles } As_2O_3 = \frac{0.248}{150 + 48} = 0.00125 \text{ moles}$$
$$\text{No. of moles } As^{3+} = 0.0025 \text{ moles}$$

50cm$^3$ of potassium permanganate(VII) (0.02 mol dm$^{-3}$) contains

$$\frac{50 \times 0.02}{1000} \text{ moles} = 0.001 \text{ moles potassium permanganate(VII)}$$

From this information,
0.0025 moles $As^{3+}$ reacts with 0.001 mole $MnO_4^-$
2.5 mole $As^{3+}$ reacts with 1 mole $MnO_4^-$

In the equation, ratio $\frac{2.5}{1} = \frac{5}{n-3}$

Oxidation state of arsenic in the product = +5.

3 When 25cm$^3$ of 0.1M potassium manganate(VIII) (permanganate) is added to an excess of acidified potassium iodide solution, the iodine liberated requires 25.0cm$^3$ of a solution of sodium thiosulphate.

$$MnO_4^- + 8H^+ + 5e^- \rightleftharpoons Mn^{2+} + 8H_2O; \qquad 2I^- \rightleftharpoons I_2 + 2e^-$$
$$2S_2O_3^{2-} \rightleftharpoons S_4O_6^{2-} + 2e^-$$

The molarity of the sodium thiosulphate solution is

A 0.1    B 0.2    C 0.3    D 0.4    E 0.5

(Northern Ireland)

Correct answer E (Write equations: 2 moles $MnO_4^- \equiv$ 10 moles $S_2O_3^{2-}$)

# 20 Radioactivity

### 20.1 INTRODUCTION

Many GCSE syllabuses contain a study of radioactivity and this would be a useful preliminary for this unit. Unfortunately some A level candidates will know nothing about the subject and, for this reason, this Unit is written to assume no previous knowledge of radioactivity.

Before studying this unit, you should have studied Unit 1.

### 20.2 RADIOACTIVITY

Radioactivity is the spontaneous decay of unstable atoms with the emission of either α (alpha), β (beta) or γ (gamma) radiation. It was first observed by Becquerel who found that a uranium salt blackened a photographic plate wrapped in paper. The radiation coming from the radioactive source passed through the paper to expose the plate.

The properties of α, β and γ radiation are as follows:

(*i*) α rays are composed of a stream of fast moving helium nuclei ($He^{2+}$). They are deflected by a magnetic field because of their positive charge. They have a very limited penetrating power being unable to penetrate even thin metal foil.

(*ii*) β rays are a stream of fast moving electrons. They are greatly deflected by a magnetic field. They will penetrate a thin metal foil but are stopped by lead.

(*iii*) γ rays are electromagnetic waves with very short wavelength. They are not deflected by a magnetic field as they are uncharged. They are more penetrating than α or β rays, being able to pass through lead. rays being able to pass through lead.

The effects of a magnetic field on α, β and γ radiation are summarised in Fig. 20.1.

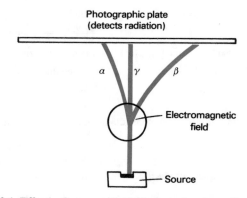

**Fig. 20.1** Effects of a magnetic field on α, β and γ radiation

### 20.3 THE CONSEQUENCES OF α, β AND γ EMISSIONS

A radioactive isotope may decay with the emission of α, β or γ rays. If an atom loses an α particle, the particle produced will contain two protons and two neutrons less than the original atom.

*E.g.*

$$^{238}_{92}U \xrightarrow{-\alpha} {}^{234}_{90}Th$$

A uranium-238 atom has a mass number of 238 and atomic number 92. On emission of an α particle, the resulting atom is an atom of thorium-234 (mass number 234 atomic number 90). Thorium is two places to the left of uranium in the Periodic Table.

The loss of a β particle from a radioactive nucleus can be explained by a change which takes place first in the nucleus. A *neutron* changes into a *proton plus an electron* and it is this electron which is lost. The resulting atom, contains one less neutron and one more proton than the original atom. The mass number is unchanged but the atomic number is increased by one.

*E.g.*

$$^{234}_{90}Th \xrightarrow{-\beta} {}^{234}_{91}Pa$$

A thorium-234 atom changes to a protactinium-234 atom. Protactinium is one place to the right of uranium in the Periodic Table.

There are no definite changes in atomic structure during γ emission.

## 20.4 HALF-LIFE OF RADIOACTIVE ISOTOPES

The **half-life** ($t_{\frac{1}{2}}$) of a radioactive isotope is the time taken for half the unstable nuclei in the sample to decay. It is independent of the original mass, unaffected by changes in temperature and cannot be catalyzed.

It can also be defined as the time required for the reactivity to drop to half its original value. Radioactive decay is a first order process.

The half-life is characteristic of a particular isotope and may vary from a fraction of a second to millions of years.

The half-life is best determined graphically by plotting radioactive count against time (Fig. 20.2). Readings of radioactive count are taken with a Geiger counter at intervals. If a convenient reading is taken (say 2000) and the lines AB and BC drawn, then the lines DE and EF are drawn at a value which is half of the original reading. The time difference between F and C is called the half-life.

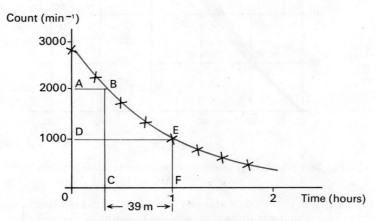

**Fig. 20.2** Radioactive decay and half life

## 20.5 ARTIFICIAL NUCLEAR REACTIONS

It is possible to carry out reactions which change one element into another. Usually one substance is bombarded by α particles, protons, neutrons or even heavier atoms such as carbon.

*E.g.* bombarding sodium atoms with neutrons

$$\,^{23}_{11}\text{Na} + \,^{1}_{0}\text{n} \rightarrow \,^{23}_{10}\text{C} + \,^{1}_{1}\text{H}$$

Neutrons from an atomic reactor are particularly useful for bombarding other atoms because they are neutral and are not repelled.

*E.g.* bombarding beryllium atoms with α particles

$$\,^{9}_{4}\text{Be} + \,^{4}_{2}\text{He} \rightarrow \,^{12}_{6}\text{C} + \,^{1}_{0}\text{n}$$

bombarding nitrogen atoms with neutrons

$$\,^{14}_{7}\text{N} + \,^{1}_{0}\text{n} \rightarrow \,^{14}_{6}\text{C} + \,^{1}_{1}\text{H}$$

Carbon-14 is formed in this process. It is present in all living material. The detection of the amount of carbon-14 remaining in a sample is used for dating objects.

Bombarding californium with boron atoms

$$\,^{250}_{98}\text{Cf} + \,^{11}_{5}\text{B} \rightarrow \,^{257}_{103}\text{Lw} + 4\,^{1}_{0}\text{n}$$

This process produces unstable radioactive **isotopes** of elements such as lawrencium which do not exist naturally on earth.

In equations of this type it is important that the total atomic numbers on both sides are equal and also the sum of the mass numbers. In the equation for the production of lawrencium-257, the mass numbers add up to 261 and the atomic numbers add up to 103.

## 20.6 STABLE ISOTOPES

Fig. 20.3 shows a graph of the numbers of protons and neutrons in stable isotopes. Each point on the graph represents a stable isotope and you will notice a certain band of stability in which all of these isotopes are located. For atoms containing less than about 20 protons, the numbers of protons and neutrons are approximately the same (*i.e.* the band is close to the line representing 1:1 neutron:proton ratio). For more complicated atoms, the isotopes have a greater neutron: proton ratio.

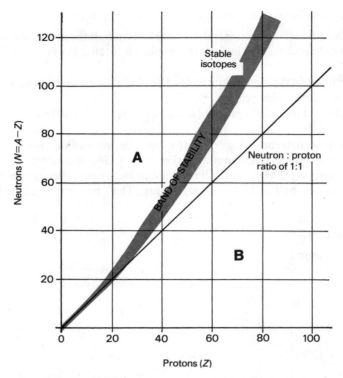

**Fig. 20.3** Numbers of protons and neutrons in stable isotopes

Isotopes not lying within the band of stability will decay until they form products which lie within this band. Nuclei which are in area A achieve stability by the emission of $\beta$ particles. Emission of a $\beta$ particle reduces the number of neutrons by one and increases the number of protons by one. $\beta$ decay has moved the point closer to the band of stability. A series of $\beta$ decay processes would produce a nucleus within this band.

Nuclei in area B are neutron-deficient and can achieve stability by converting a proton into a neutron.

$$\mathrm{{}^{1}_{1}H} \rightarrow \mathrm{{}^{1}_{0}n} + \mathrm{{}^{0}_{+1}e}$$

The positively charged electron called a **positron** is unstable and forms $\gamma$ radiation. Again the point is moving closer to the band of stability. This process can also be achieved by K-electron capture where a nucleus in region B captures an orbital electron from the first shell and converts a proton into a neutron.

$$\mathrm{{}^{1}_{1}H} + \mathrm{{}^{0}_{-1}e} \rightarrow \mathrm{{}^{1}_{0}n}$$

### 20.7   MASS DEFECT

There is always a discrepancy between the total mass of the protons and neutrons in a nucleus and the actual mass of the nucleus.

For example, for a stable $\mathrm{{}^{16}_{8}O}$ atom, which contains eight protons, eight electrons and eight neutrons.

$$\text{Mass of 8p, 8n and 8e} = (8 \times 1.0073) + (8 \times 1.0087) + (8 \times 0.0005)$$
$$= 16.132 \text{ amu (atomic mass units)}$$

Actual mass of oxygen atom (determined by mass spectroscopy) = 15.995 amu

$$\text{Difference} = 0.137 \text{ amu}$$

The difference in mass is the net loss on forming a particular isotope from its constituent particles, and is called the **mass defect**. The mass is converted into energy according to Einstein's equation where

$$E = mc^2$$

($E$, energy produced; $m$, loss of mass; $c$, velocity of light)

This energy, called the **binding energy**, is lost on forming the atom.

For comparison, the mass defect for a radioactive $\mathrm{{}^{13}_{8}O}$ isotope can be found.

Mass of 8 protons, 5 neutrons, 8 electrons = $(8 \times 1.0073) + (5 \times 1.0087) + (8 \times 0.0005)$ amu

$$\text{Actual mass} = 13.025 \text{ amu}$$
$$\text{Mass defect} = 0.081 \text{ amu}$$

The mass defect and hence the binding energy for the radioactive isotope is much less than for the stable isotope. This applies generally.

Fig. 20.4 shows a graph of the binding energy per nucleon against mass number. (A **nucleon** is a particle in the nucleus either proton or neutron). From the graph it is clear that the most stable elements are the ones with intermediate mass numbers.

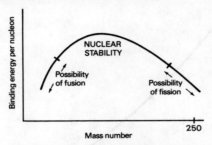

**Fig. 20.4** Graph of binding energy per nucleon against mass number

Isotopes with small mass numbers have nuclei which are relatively unstable. The **fusion** of these nuclei together to form a larger nucleus releases a vast amount of energy:

*e.g.*
$$^2_1H + ^3_1H \rightarrow ^4_2He + ^1_0n$$

This process requires a temperature of several million degrees in order to operate, but it occurs in the sun and is the source of energy in the sun. It also takes place in a hydrogen bomb.

Isotopes with a large mass number again have unstable nuclei. These nuclei can break up in a series of stages in processes called **fission** to produce smaller nuclei which are relatively stable.

*e.g.*
$$^{235}_{92}U + ^1_0n \rightarrow ^{236}_{92}U \rightarrow ^{143}_{56}Ba + ^{90}_{36}Kr + 3^1_0n$$

This process produces two smaller nuclei, when a uranium-235 nucleus is struck by a neutron. These smaller nuclei are relatively stable and a large amount of energy is released. Fission processes are the basis of the atomic bomb and (more usefully) of electrical power generation in nuclear power stations.

## QUESTIONS

**1** Multiple completion question. Use the responses on page 3.

$^{235}_{92}U$ decays by losing two $\beta$ particles and one $\alpha$ particle to form an isotope X. Which of the following statements is (are) true about X?

(*i*) The mass number of X is 235.

(*ii*) The atomic number of X is 92.

(*iii*) The product does not undergo further radioactive decay.

(*iv*) X is another isotope of uranium.

The correct answer is C corresponding to (*ii*) and (*iv*) correct.

**2** In the nuclear disintegration
$$^{238}_{92}U \rightarrow Q + \alpha \text{ particle}$$
Q is    A $^{234}_{90}Th$    B $^{238}_{92}U$    C $^{234}_{91}Pa$    D $^{234}_{92}U$    E $^{236}_{88}Ra$

*(Northern Ireland)*

**Correct answer A**

**3** The disintegration rates (measured on a Geiger counter) at 25°C of a 10g sample of radioactive sodium-24 at various times are given in Table 20.1.

**Table 20.1  Disintegration rates**

| Time (hours) | Rate of disintegration (counts sec$^{-1}$) |
|---|---|
| 0 | 670 |
| 2 | 610 |
| 5 | 530 |
| 10 | 420 |
| 20 | 270 |
| 30 | 170 |

(*a*) Plot a graph of disintegration against time. From the graph, determine the half life of sodium-24.

(*b*) On the graph, sketch the curve that would be obtained if a 5g sample of sodium-24 had been used.

(*c*) What is the effect of an increase in temperature on this process?

(*a*) The graph is shown in Fig. 20.5.

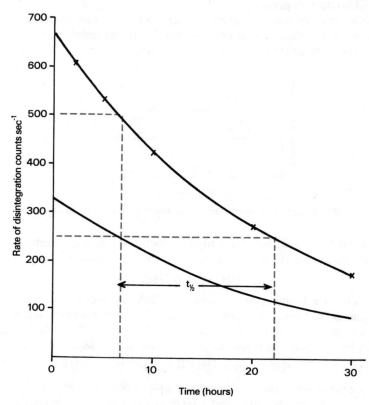

**Fig. 20.5** Question

The half-life of sodium-24 is approximately 15½ hours.

(*b*) If half the mass is used, the counts per second are halved. At time 15½ hours the reading will be 168 (½ the original vale of 335).

(*c*) Increasing temperature has no effect on the rate of disintegration of the radioactive isotope.

4 (*a*) (*i*) What is meant by the term isotope?
(*ii*) How could it be demonstrated that a sample of an element contained two or more isotopes?
(*b*) Give two ways in which the element $^{24}_{12}Mg$ could lose electrons. Give the atomic numbers and mass numbers of the species formed.
(*c*) Indicate the numbers and types of particles emitted in order to change the unstable species $^{30}_{13}Al$ and $^{57}_{25}Mn$ to the stable species $^{26}_{13}Al$ and $^{49}_{23}V$ respectively.
(*d*) If the radioactive decay of $^{63}Ni$ to $^{63}Cu$ has a half-life of 120 years, how long will it take for 15/16ths of the nickel to be changed into copper?
(*e*) Explain how isotopic labelling would enable you to determine whether the acid or the alcohol molecule is responsible for producing the oxygen of the water molecule produced during esterification.
(*f*) Complete the following equations

$$^{16}_{8}O + \quad \rightarrow ^{13}_{6}C + ^{4}_{2}He$$
$$^{27}_{13}Al + \quad \rightarrow ^{24}_{12}Mg + ^{4}_{2}He$$
$$^{14}_{7}N + ^{4}_{2}He \rightarrow \quad + ^{1}_{1}H$$

(*a*) (*i*) Isotopes are atoms of the same element containing different numbers of neutrons.
(*ii*) A mass spectrometer could be used to show that a sample of an element contained two or more isotopes. The sample is first ionized to form positive ions by removing electrons. The ions are then accelerated and passed through electric and magnetic fields where they are deflected. The amount of deflection depends upon the mass/charge ratio of the isotope. The ions are detected on a photographic plate.
(*b*) $^{24}_{12}Mg$ could, in theory, lose electrons by $\beta$ decay or ionization
$\beta$ decay produces $^{24}_{13}Al$ and ionisation produces $^{24}_{12}Mg^{2+}$.
(*c*)
$$^{30}_{13}Al \rightarrow ^{26}_{13}Al$$
This requires a decrease in mass number of 4 and the atomic number is unchanged. This could be achieved by the emission of one $\alpha$ and two $\beta$ particles.
$$^{57}_{25}Mn \rightarrow ^{49}_{23}V$$
This requires the emission of $2\alpha$ and $2\beta$ particles.
(*d*) The half life of $^{63}Ni$ is 120 years.
One sixteenth of $^{63}Ni$ remains after 4 half-lives *i.e.* 480 years

(*e*) The oxygen in the alcohol is labelled with $^{18}O$. The reaction is carried out and the ester is isolated. Using a mass spectrometer, it can be shown that the $^{18}O$ atoms are present in the ester but not in the water. The oxygen in the water, therefore, comes from the acid.

(*f*) $_0^1n, _1^1H, _8^{17}O$

5 Explain the terms mass defect and alpha emission.

For the following atoms of naturally-occurring non-radioactive isotopes

$^1H$, $^4He$, $^7Li$, $^9Be$, $^{11}B$, $^{12}C$, $^{14}N$, $^{16}O$, $^{19}F$, $^{20}Ne$, $^{23}Na$, $^{24}Mg$ and $^{27}Al$

plot the number of protons against the number of neutrons in each nucleus. Using your graph show that $^{14}C$ and $^{24}Na$ are radioactive isotopes decaying by $\beta$ emission. Write equations for the decay processes of these two isotopes.

*(Associated Examining Board Syllabus II)*

The mass deficit is the difference in mass between the combined masses of protons, neutrons and electrons in an atom and the actual mass of the atom. The difference in mass is due to the binding energy of the nucleus.

Alpha emission. Loss of helium nuclei (positively charged) from a radioactive source. Alpha particles are deflected by a magnetic field and have little penetrating power.

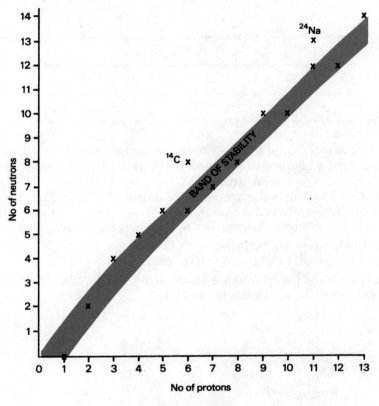

**Fig. 20.6** Question

Both $^{14}C$ and $^{24}Na$ lie outside the band of stability and decay by $\beta$ decay. $\beta$ decay reduces the number of neutrons by one and increases the number of protons by one. This puts both isotopes into the band of stability.

$$_6^{14}C \rightarrow _7^{14}N + _{-1}^0e$$
$$_{11}^{24}Na \rightarrow _{12}^{24}Mg \, _{-1}^0e$$

# 21 Chemical periodicity

### 21.1 INTRODUCTION

In Unit 2 the structure of the Periodic Table was explained. Also the variation of physical properties such as ionization energy and electronegativity were examined. It was shown that the variation was periodic *i.e.* a repeating pattern was obtained.

In this unit periodicity is shown amongst compounds of different elements. Oxides, hydrides and chlorides of the third period Na–Cl are examined.

### 21.2 OXIDES

| | | | | | | |
|---|---|---|---|---|---|---|
| $Na_2O$ | MgO | $Al_2O_3$ | $SiO_2$ | $P_4O_6$ | $SO_2$ | $Cl_2O$ |
| $Na_2O_2$ | | | | $P_4O_{10}$ | $SO_3$ | |

State at 20°C  — — — — — — — —solid — — — — — — — — — — — //— — — —gas — — — — — —

Structure  — — — — — —giant lattices — — — — — —//— — —molecules — — — — — — — —

Bonding  — — ionic — — — — //— — — — — — covalent — — — — — — — — — — — —

Type of bonding  — — basic — — — — //amphoteric //— — — — —acidic — — — — — — — — — —

In any period the oxides change gradually from ionic and basic on the left-hand side (Groups I and II) to covalent and acidic on the right-hand side (Group VII). Also a change occurs from a giant lattice structure to a molecular structure.

The ionic oxides of sodium and magnesium are discussed in Unit 22.2. These oxides are alkaline when tested with Universal indicator.

Aluminium oxide is insoluble in water but being amphoteric it reacts with acid or alkali

$$Al_2O_3(s) + 3H_2SO_4(aq) \rightarrow Al_2(SO_4)_3(aq) + 3H_2O(1)$$
$$Al_2O_3(s) + 2OH^-(aq) + 3H_2O(1) \rightarrow 2Al(OH)_4{}^-(aq)$$

The oxides of silicon and phosphorus are discussed in Units 24.7 and 25.8 respectively. The structures of $P_4O_6$ and $P_4O_{10}$ are shown in Fig 21.1.

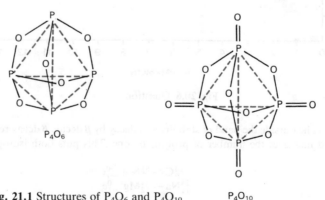

**Fig. 21.1** Structures of $P_4O_6$ and $P_4O_{10}$

There are other polymorphic forms of $P_4O_{10}$. You will notice that the structures of the two molecules are similar and based upon a tetrahedron. Each phosphorus atom in $P_4O_{10}$. is bonded to an extra oxygen atom. Because these P–O bonds are shorter, it is believed that they are double bonds.

Both oxides of sulphur are acidic, but as usual the oxide containing the greater proportion of oxygen, *i.e.* $SO_3$, is more strongly acidic.

Dichlorine oxide reacts with water to produce chloric(I) acid

$$Cl_2O(g) + H_2O(1) \rightarrow 2HOCl(aq)$$

Other oxides of chlorine are possible: chlorine dioxide $ClO_2$ (a yellow gas); chlorine hexoxide $Cl_2O_6$ (red liquid); and chlorine heptoxide $Cl_2O_7$ (colourless liquid). They are all unstable and explosive. They dissolve in water to give acidic solutions.

**21.3  HYDRIDES**

| Formula | NaH | MgH$_2$ | (AlH$_3$)$_n$ | SiH$_4$ | PH$_3$ | H$_2$S | HCl |
|---|---|---|---|---|---|---|---|
| State at 20°C | --------solid-----------------//--gas-------------------------------------- |
| Structure | ----giant lattice------------//----molecular--------------------- |
| Bonding | ----ionic---------//--------------------covalent ----------------------- |

The hydrides of sodium and magnesium are formed when hydrogen is passed over heated sodium or magnesium

$$2Na(l) + H_2(g) \rightarrow 2NaH(s)$$
$$Mg(s) + H_2(g) \rightarrow MgH_2(s)$$

These hydrides are white solids containing H$^-$ ions. A H$^-$ ion is formed when a hydrogen atom accepts an electron.

Sodium hydride and magnesium hydride react with cold water to form hydrogen

$$NaH(s) H_2O(l) \rightarrow NaOH(aq) + H_2(g)$$
$$MgH_2(s) + H_2O(l) \rightarrow MgO(s) + 2H_2(g)$$

Electrolysis of molten ionic hydrides produces hydrogen at the positive electrode (anode)

$$2H^-(l) \rightarrow H_2(g) + 2e^-$$

This is unusual because hydrogen is usually produced at the cathode.

Aluminium hydride is rather less stable and less studied, but it also reacts with water to produce hydrogen.

Silane, SiH$_4$, unlike methane, CH$_4$, dissolves in water and undergoes slow hydrolysis. This is possible because silicon can use its 3*d* orbitals to give a maximum covalency of six but carbon, with no *d* orbitals available, has a maximum covalency of 4.

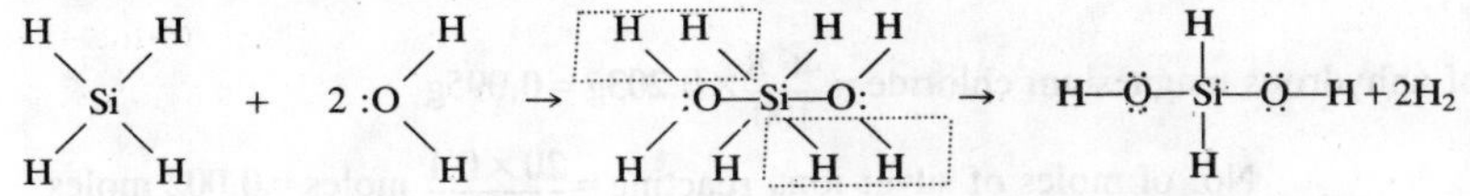

orthosilicilic acid

Phosphine, PH$_3$, is slightly soluble in water giving an almost neutral solution. It reacts with acids to form salts

$$e.g.\ PH_3(g) + HCl(aq) \rightarrow PH_4^+ Cl^-$$

Hydrogen chloride is neutral when it is absolutely dry. Under these conditions the bonding is covalent

When added to water, ionization occurs accompanied by an evolution of energy

$$HCl(g) + H_2O(l) \rightarrow H_3O^+(aq) + Cl^-(aq)$$

**21.4  CHLORIDES**

| Formula | NaCl | MgCl$_2$ | AlCl$_3$ | SiCl$_4$ | PCl$_3$ | PCl$_5$ | S$_2$Cl$_2$ | SCl$_2$ |
|---|---|---|---|---|---|---|---|---|
| State at 20°C | s | s | s | l | l | s | l | l |
| Structure | —giant lattice————//—————— molecular ———————————— |
| Bonding | — ionic ————————//—— covalent ——————————//ionic//————covalent ———— |
| Reaction with moist air | — none ——————//—————fumes producing HCl———————————— |

Anhydrous chlorides are produced by heating an element in dry chlorine

$$e.g.\ 2Al(s) + 3Cl_2(g) \rightarrow 2AlCl_3(s) \text{ (see Fig. 23.2)}$$
$$P_4(s) + 6Cl_2(g) \rightarrow 4PCl_3(l)$$

Where two chlorides are possible, the reaction with chlorine produces the higher oxidation state

$$e.g.\ 2Fe(s) + 3Cl_2(g) \rightarrow 2FeCl_3(s) \text{ iron(III) chloride}$$

Iron(II) chloride is produced by the reaction of iron with dry hydrogen chloride

$$e.g.\ Fe(s) + 2HCl(g) \rightarrow FeCl_2(s) + H_2(g)$$

Sodium chloride dissolves in water without hydrolysis. Magnesium chloride and aluminium chloride undergo some hydrolysis when dissolved in water

$$MgCl_2(s) + 2H_2O(l) \rightleftharpoons Mg(OH)_2(s) + 2HCl(aq)$$
$$AlCl_3(s) + 3H_2O(l) \rightleftharpoons Al(OH)_3(s) + 3HCl(aq)$$

The hydrolysis of silicon and phosphorus chlorides will be found in Units 24.8 and 25.9.

QUESTIONS

1 (*a*)(*i*) When 0.203g of hydrated magnesium chloride, $MgCl_m.nH_2O$, was dissolved in water and titrated with 0.1M silver nitrate ($AgNO_3$) solution, 20.0cm$^3$ of the latter were required. A sample of the hydrated chloride lost 53.2% of its mass when heated in a stream of hydrogen chloride, leaving a residue of anhydrous magnesium chloride. From these figures calculate the values of *m* and *n*.
(*ii*) Is the value of *m*, calculated in (*i*), what you would have expected? Explain your answer.
(*b*) When the hydrated chloride was heated in air instead of hydrogen chloride, the loss of mass was greater than 53.2% and both HCl and $H_2O$ were evolved.
(*i*) What reaction do you think might have occurred and what was the solid product?
(*ii*) Why was the stream of hydrogen chloride used in (*a*)?
(*c*) (*i*) What types of chemical bond exist in hydrated magnesium chloride? Indicate the particles involved in each type.
(*ii*) What is the type of bonding in sulphur dichloride, $SCl_2$?
(*iii*) Compare the physical properties of anhydrous magnesium chloride with those of sulphur dichloride, restricting your answer to two properties.
($A_r(H) = 1$. $A_r(O) = 16$, $A_r(Mg) = 24$, $A_r(Cl) = 35.5$)

*(London)*

(*a*) (*i*) Mass of anhydrous magnesium chloride $= \dfrac{46.8}{100} \times 0.203g = 0.095g$

No. of moles of silver ions reacting $= \dfrac{20 \times 0.1}{1000}$ moles $= 0.002$ moles

$$Cl^- + Ag^+ \rightarrow AgCl \text{ (reacting ratio 1:1)}$$

The sample of magnesium chloride contains 0.002 moles of chloride ions. 95g of anhydrous magnesium chloride contains 2 moles of chloride ions. The value of $m = 2$.

Percentage of water in sample $= 53.2\%$

46.8g of anhydrous magnesium chloride combines with 53.2g water

0.492 moles of anhydrous magnesium chloride combines with 2.95 moles $H_2O$

1 mole of anhydrous magnesium chloride combines with 6 moles

(*ii*) Yes. Group II. Two electrons in $s^2$ orbital to be lost forming $Mg^{2+}$ ion.
(*b*) (*i*) Hydrolysis. Magnesium oxide.
(*ii*) $MgCl_2.6H_2O \rightleftharpoons MgO + 2HCl + 5H_2O$
HCl gas ensures equilibrium moves to left and prevents hydrolysis.
(*c*) (*i*) Covalent–between atoms in water molecule.
Ionic–between $Mg^{2+}$ and $Cl^-$ ions.
Coordinate bonding between water molecules and $Mg^{2+}$ and $Cl^-$ ions.
(*ii*) Covalent.
(*iii*) High melting point of $MgCl_2$.
Solubility of $MgCl_2$ in water without extensive hydrolysis.

2 Using your knowledge of chemical reactions by reference to the appropriate data, demonstrate what is meant by periodicity of properties in relation to the oxides (where formed) of the elements lithium to argon.

*(Nuffield)*

This question refers to the second period of the Periodic Table. Similar trends can be seen. Extensive use of the book of data supplied for the examination is expected.

# 22 The *s*-block elements

## 22.1 INTRODUCTION

This unit covers the chemistry of the reactive metals in Groups I and II. These elements have one or two electrons in the outer *s* orbitals. Elements in these groups are frequently studied together and may be called *s*-block elements.

Other Units which are linked with *s*-block elements include 2, 4 and 13. In particular, questions often link the Born Haber cycle (Unit 13.11) with Groups I and II.

## 22.2 THE ELEMENTS

The elements in Groups I and II are:

| *Group I* | *Group II* |
|---|---|
| Lithium   Li | Beryllium   Be |
| Sodium   Na | Magnesium   Mg |
| Potassium   K | Calcium   Ca |
| Rubidium   Rb | Strontium   Sr |
| Caesium   Cs | Barium   Ba |
| Francium   Fr | Radium   Ra |

Francium and radium are very rare radioactive elements and they will not be included in the study which follows.

The elements in Group I (**alkali metals**) all contain a single electron in the outer *s* orbital.

*E.g.* Li  $1s^2 2s^1$
    Na $1s^2 2s^2 2p^6 3s^1$
    K  $1s^2 2s^2 2p^6 3s^2 3p^6 4s^1$

In all compounds, Group I metals show only an oxidation state of $+1$ which corresponds to the loss of a single electron from the *s* orbital.

*E.g.*         sodium atom Na $\rightarrow$ sodium ion Na$^+$ + e$^-$
                $1s^2 2s^2 2p^6 3s^1$    $1s^2 2s^2 2p^6$

The elements in Group II (**alkaline earth metals**) all contain two electrons in the outer *s* orbital.

*E.g.*    Be  $1s^2 2s^2$
        Mg  $1s^2 2s^2 2p^6 3s^2$
        Ca  $1s^2 2s^2 2p^6 3s^2 3p^6 4s^2$

With group II metals, the oxidation state $+2$ exists in all compounds. This corresponds to the loss of the two electrons in the outer *s* orbital.

*E.g.*        Magnesium atom Mg $\rightarrow$ magnesium ion Mg$^{2+}$ + 2e$^-$
                $1s^2 2s^2 2p^6 3s^2$    $1s^2 2s^2 2p^6$

In each group the ionization energies decrease and the chemical reactivities increase down the group. The ionization energies are:

| Li | 520 kJ mol$^{-1}$ | Be | 2700 kJ mol$^{-1}$ |
|---|---|---|---|
| Na | 500 kJ mol$^{-1}$ | Mg | 2240 kJ mol$^{-1}$ |
| K | 420 kJ mol$^{-1}$ | Ca | 1690 kJ mol$^{-1}$ |
| Rb | 400 kJ mol$^{-1}$ | Sr | 1650 kJ mol$^{-1}$ |
| Cs | 380 kJ mol$^{-1}$ | Ba | 1500 kJ mol$^{-1}$ |

(NB First ionization energies are given for Group I elements and the sum of first and second ionization energies for Group II)

The decrease in ionization energies down the group can be explained by the increasing atomic radius and increased shielding down the group.

The difference in reactivity can be seen in the reactions of these metals with cold water. Lithium reacts slowly with cold water to produce the alkali lithium hydroxide and hydrogen.

$$2Li(s) + 2H_2O(l) \rightarrow 2LiOH(aq) + H_2(g)$$

Potassium reacts rapidly with cold water and the hydrogen produced ignites spontaneously and burns with a pinkish-purple flame.

$$2K(s) + 2H_2O(l) \rightarrow 2KOH(aq) + H_2(g)$$

In Group II, beryllium does not react with water. Magnesium reacts very slowly with hot water but rapidly with steam.

$$Mg(s) + H_2O(g) \rightarrow MgO(s) + H_2(g)$$

(Magnesium hydroxide decomposes at high temperatures into magnesium oxide.)

Calcium, strontium and barium react with cold water.

*E.g.*     $$Ca(s) + 2H_2O(l) \rightarrow Ca(OH)_2(aq) + H_2(g)$$

The alkali metals and barium are usually stored in oil to prevent reaction with water and air.

The elements in Groups I and II react with chlorine on heating to produce chlorides.

*E.g.*     $$2Na(s) + Cl_2(g) \rightarrow 2NaCl(s)$$

A variety of oxides are produced when these metals burn in oxygen. For Group I metals, the common oxide of formula $M_2O$ is formed by all, but the more reactive metals form additional oxides. Sodium forms a peroxide and potassium, rubidium and caesium form a **peroxide** and a **superoxide**.

$$Li \rightarrow Li_2O$$
$$Na \rightarrow Na_2O \rightarrow Na_2O_2$$
$$K \rightarrow K_2O \rightarrow K_2O_2 \rightarrow KO_2$$
$$Rb \rightarrow Rb_2O \rightarrow Rb_2O_2 \rightarrow RbO_2$$
$$Cs \rightarrow Cs_2O \rightarrow Cs_2O_2 \rightarrow CsO_2$$

Group II metals all form a typical oxide of formula MO but, in addition, strontium and barium form peroxides.

$$Be \rightarrow BeO$$
$$Mg \rightarrow MgO$$
$$Ca \rightarrow CaO$$
$$Sr \rightarrow SrO \rightarrow SrO_2$$
$$Ba \rightarrow BaO \rightarrow BaO_2$$

Lithium and the Group II metals burn in nitrogen to form nitrides.

*E.g.*     $$6Li(s) + N_2(g) \rightarrow 2Li_3N(s)$$

## 22.3   OXIDES AND HYDROXIDES

The oxides of Group I react rapidly when added to water to form strongly alkaline solutions.

*E.g.*     $$Na_2O(s) + H_2O(l) \rightarrow 2NaOH(aq)$$

Lithium, however, does not form a strongly alkaline solution.

The tendency for Group II oxides to form alkaline solutions when added to water is less.

*E.g.*     $$CaO(s) + H_2O(l) \rightarrow Ca(OH)_2(aq)$$

The basic strengths of the resulting solutions from both groups increase down the group. The ionic radii of the metal ions increase down each Group (Unit 2.7). There is, therefore, less attraction between the $M^+$ or $M^{2+}$ and the $OH^-$ ion.

The hydroxides of Group II metals and lithium hydroxide decompose on heating to form the oxide.

*E.g.*     $$Ca(OH)_2(s) \rightarrow CaO(s) + H_2O(l)$$

Hydroxides of Group I metals (apart from lithium) are not decomposed by heating.

## 22.4   HYDRIDES

The hydrides produced by passing hydrogen over heated Group I or II metals are ionic hydrides (Unit 21.5).

## 22.5   CHLORIDES

The chlorides of Group I and II metals are generally ionic. They are white, crystalline solids with high melting points. They are poor conductors of electricity when solid but, when molten or dissolved in water, undergo electrolysis.

The solid chlorides have ionic crystal lattices which can be broken by melting or dissolving in water.

Beryllium chloride is different from the other chlorides of Group II. Beryllium is more electronegative (or less electropositive) than the other elements of the group. As a result it shows the greatest tendency to covalent bonding (Unit 4.4). Beryllium chloride is a solid but is a poor conductor of electricity when molten. It is soluble in organic solvents and is hydrolyzed by heating an aqueous solution of beryllium chloride.

## 22.6 CARBONATES AND HYDROGENCARBONATES

Carbonates of Group I metals dissolve in water to form an alkaline solution. The solubility of lithium carbonate is much less than the other Group I carbonates. Only lithium carbonate is decomposed by heating.

$$Li_2CO_3(s) \rightarrow Li_2O(s) + CO_2(g)$$

Group II metal carbonates are insoluble in water and are decomposed by heating into the oxide and carbon dioxide.

$$CaCO_3(s) \rightarrow CaO(s) + CO_2(g)$$

**Hydrogencarbonates** are formed by passing carbon dioxide through a solution of the carbonate or a suspension of carbonate in water.

*E.g.*
$$Na_2CO_3(aq) + H_2O(l) + CO_2(g) \rightarrow 2NaHCO_3(s)$$
$$CaCO_3(s) + H_2O(l) + CO_2(g) \rightarrow Ca(HCO_3)_2(aq)$$

The hydrogencarbonates of Group I metals can be isolated as solids which decompose on heating. On the other hand, the hydrogencarbonates of Group II metals decompose before they can be isolated.

## 22.7 SALTS OF GROUP I AND II METALS

Salts of Group II elements generally decompose more readily than salts of Group I elements, *e.g.* the decomposition of carbonates (Unit 22.6). Also nitrates of Group II elements and lithium decompose to form the oxide while the nitrates of Group I elements, apart from lithium, only partially decompose forming the **nitrite** (nitrate(III)).

$$2Ca(NO_3)_2(s) \rightarrow 2CaO(s) + 4NO_2(g) + O_2(g)$$
$$2NaNO_3(s) \rightarrow 2NaNO_2(s) + O_2(g)$$
$$4LiNO_3(s) \rightarrow 2Li_2O(s) + 4NO_2(g) + O_2(g)$$

Salts of Group I metals are generally soluble in water. Some salts of lithium, however, are insoluble, *e.g.* $Li_2CO_3$, LiF. This can be explained by the high lattice energies of these compounds caused by the small size of the lithium ion. The hydration energy produced when the substance dissolves is much less than the lattice energy.

*E.g. lithium fluoride*

lattice energy = 1022 kJ mol$^{-1}$     hydration energy = $-833$ kJ mol$^{-1}$
lattice energy > hydration energy     $\therefore$ insoluble.

*sodium fluoride*

lattice energy = 902 kJ mol$^{-1}$     hydration energy = $-912$ kJ mol$^{-1}$
lattice energy < hydration energy     $\therefore$ soluble.

Salts of Group II metals are generally less soluble, particularly if the salt contains ions with a 2– charge. This is due to the increased lattice energies. Sulphates decrease in solubility down the group. This is due to decreasing hydration energies down the group.

Many salts of alkali metals and alkaline earth metals give characteristic colours in a flame test. If one of the chlorides is mixed with concentrated hydrochloric acid, and some of the mixture introduced into a hot flame, a flame colouration is produced.

*E.g.* lithium – red; sodium – orange; potassium – pinkish purple (lilac); calcium – brick red; barium – apple green.

**1** Table 22.1 shows the atomic and ionic radii of the Group II metals.

**Table 22.1 Atomic and ionic radii of the Group II metals**

| Element | Atomic radius/nm | Ionic radius/nm |
|---|---|---|
| Beryllium | 0.112 | 0.030 |
| Magnesium | 0.160 | 0.065 |
| Calcium | 0.197 | 0.094 |
| Strontium | 0.215 | 0.110 |
| Barium | 0.221 | 0.134 |

(*a*) Explain why the atomic radius is larger than the ionic radius in each case.

(*b*) Explain why the atomic radius increase from beryllium to barium.

(*c*) The ions K$^+$ and Ca$^{2+}$ have identical electron arrangements, yet the ionic radius of K$^+$ is larger than that of Ca$^{2+}$.

(*i*) What is the electron arrangement in these two ions?

(*ii*) Explain the difference in ionic radii of the two ions.

(*d*) Give one reaction for magnesium, calcium and barium which shows the difference in reactivity between these three elements. Explain briefly why this difference exists.

(*e*) Describe clearly how to prepare good specimens of:

(*i*) magnesium sulphate $MgSO_4.7H_2O$ from magnesium carbonate;

(*ii*) barium sulphate $BaSO_4$ from barium carbonate.

(*a*) Magnesium atom $1s^22s^22p^63s^2$     magnesium ion $1s^22s^22p^6$

The ionic radius of magnesium is smaller than the atomic radius because of the removal of the $3s$ electrons. Electron orbitals in the third shell ($3s$, $3p$, $3d$) extend further than orbitals in the second shell.

In each case the number of protons in the nucleus is the same (12) and the shielding is unchanged.

(*b*) The atomic radius increases from beryllium to barium because extra electron shells are added. Although the nuclear charge increases, there is also an increase in shielding.

(*c*) The electron arrangement in $K^+$ and $Ca^{2+}$ is $1s^2 2s^2 2p^6 3s^2 3p^6$. There are the same number of filled electron orbitals and the same amount of shielding. The difference is due to the different number of protons in the ion.

$$K - 19 \text{ protons, } Ca - 20 \text{ protons.}$$

The extra nuclear charge produces extra nuclear attraction drawing the electrons slightly closer to the nucleus.

(*d*) The reactions of magnesium, calcium and barium with water (Unit 22.2).

In chemical reactions the atoms of all of these elements lose two electrons to form ions.

$$M(s) \rightarrow M^{2+}(aq) + 2e^-$$

These two electrons are lost most easily from barium because the atoms of barium are larger than those of magnesium or calcium. The ionization energies (first and second) of barium are less than calcium or magnesium.

(*e*) This part could appear also on an O level paper and would often be badly done on either O or A level scripts because of lack of detail.

In (*i*) a soluble salt is prepared but in (*ii*) an insoluble salt is prepared.

(*i*) Since a sulphate is prepared, sulphuric acid must be used.

Write and balance the equation.

$$MgCO_3(s) + H_2SO_4(aq) \rightarrow MgSO_4(aq) + H_2O(l) + CO_2(g)$$

Add powdered magnesium carbonate in small portions to warm, dilute sulphuric acid. When some magnesium carbonate remains unreacted, filter to remove excess magnesium carbonate. (Candidates often believe that when magnesium carbonate remains, the solution is saturated. This is not a dissolving process but a reacting one. The addition of magnesium carbonate uses up the dilute sulphuric acid. When all the acid is used up, the magnesium carbonate remains unreacted.)

The solution consists now of magnesium sulphate dissolved in water. The solution is evaporated until a small volume of solution remains and this is left to cool and crystallize. If the solution is heated to dryness $MgSO_4.7H_2O$ crystals will decompose to produce anhydrous magnesium sulphate.

$$MgSO_4.7H_2O \rightleftharpoons MgSO_4 + 7H_2O$$

(*ii*) Barium carbonate is insoluble in water and must be reacted with a dilute acid (but not sulphuric acid) to produce a solution containing barium ions.

$$BaCO_3(s) + 2HCl(aq) \rightarrow BaCl_2(aq) + H_2O(l) + CO_2(g)$$

Then add dilute sulphuric acid to precipitate barium sulphate.

$$BaCl_2(aq) + H_2SO_4(aq) \rightarrow BaSO_4(s) + 2HCl(aq)$$

ionic equation:     $$Ba^{2+}(aq) + SO_4^{2-}(aq) \rightarrow BaSO_4(s)$$

The precipitate is filtered off, washed with distilled water to remove soluble impurities and dried.

# 23 Group III

## 23.1 INTRODUCTION

Groups III, IV, V, VI, VII and 0 are sometimes referred to as *p*-block elements since the highest energy orbitals occupied are *p* orbitals. The *p*-block elements are discussed in Units 23–27. Questions are rarely asked about the *p*-block as a whole.

The first element in group III is boron which is a non-metallic element. The other elements, apart from boron and aluminium, are gallium, indium and thallium.

## 23.2 ALUMINIUM

The electronic arrangement of an aluminium atom is $1s^2 2s^2 2p^6 3s^2 3p^1$. The oxidation state of aluminium in compounds is +3.

Aluminium can undergo ionic or covalent bonding. For ionic bonding the two electrons in the $3s$ orbital and the one electron in the $3p$ orbital are lost to form the $Al^{3+}$ ion. The ionization energy for aluminium (*i.e.* for the loss of 3 electrons) is extremely high ($5080 \, kJ \, mol^{-1}$). The resulting $Al^{3+}$ ion is extremely small and highly polarizing (Unit 4.3). The simple $Al^{3+}$ ion is present only in the fluoride and oxide and even here the bonds will be partially covalent. The hydrated ion $[Al(H_2O)_6]^{3+}$ exists in aqueous solution since the hydration energy evolved compensates for the high ionization energy.

As an alternative to ionic bonding, an electron may be promoted from the $3s$ orbital to an empty $3p$ orbital, so there can be three covalent bonds. When aluminium forms covalent compounds, *e.g.* $AlCl_3$, there is still a vacant *p* orbital on the aluminium atom. The noble gas structure of argon is not achieved. Aluminium chloride can accept an electron pair from a suitable source, *e.g.* ammonia, to form

$$Cl-Al \leftarrow :N-H$$

with Cl and Cl on the aluminium and H and H on the nitrogen

Aluminium is acting here as a Lewis acid (Unit 16.2).

Aluminium is a dull-grey metal. It has a low density for a metal, but it is very malleable and ductile. It is not as reactive as would be expected from its position in the electrochemical series. For example, aluminium reacts with moderately concentrated hydrochloric acid.

$$2Al(s) + 6HCl(aq) \rightarrow 2AlCl_3(aq) + 3H_2(g)$$

The reaction is very slow at first due to the thin layer of aluminium oxide on the surface which prevents reaction while the film remains. When the oxide film is removed by wiping with mercury, the aluminium oxidizes rapidly in air. The oxide coating may be thickened and coloured by an electrolytic process called **anodizing**.

Aluminium shows non-metallic properties. It reacts exothermically with sodium hydroxide solution to produce hydrogen and sodium aluminate.

$$2Al(s) + 2OH^-(aq) + 6H_2O(l) \rightarrow 2[Al(OH)_4]^-(aq) + 3H_2(g)$$

DeVarda's alloy used to reduce nitrate ions to ammonia is an alloy whose main component is aluminium. It reacts with sodium hydroxide solution more slowly than aluminium.

Aluminium is a strong reducing agent because of its affinity for oxygen. For example, the **Thermit reaction** used to weld railway track

$$Fe_2O_3(s) + 2Al(s) \rightarrow 2Fe(s) + Al_2O_3(s)$$

## 23.3 EXTRACTION OF ALUMINIUM

Aluminium is extracted from the ore **bauxite**, $Al_2O_3.2H_2O$. Before extraction the ore is purified. The ore is heated with sodium hydroxide solution under pressure. The aluminium oxide reacts to form the aluminate ion which remains in solution. The solid impurities are removed by filtration.

$$Al_2O_3(s) + 2OH^-(aq) + 3H_2O(l) \rightarrow 2[Al(OH)_4]^-(aq)$$

Some freshly prepared aluminium hydroxide is added to 'seed' the solution of sodium aluminate to precipitate aluminium hydroxide.

$$[Al(OH)_4]^-(aq) \rightarrow Al(OH)_3(s) + OH^-(aq)$$

The aluminium hydroxide is heated to produce aluminium oxide.

$$2Al(OH)_3(s) \rightarrow Al_2O_3(s) + 3H_2O(l)$$

Aluminium is manufactured by the electrolysis of pure aluminium oxide dissolved in molten cryolite ($Na_3AlF_6$). The electrodes are carbon. The electrode reactions are:

Cathode $(-)$ $Al^{3+} + 3e^- \rightarrow Al$
Anode $(+)$ $\qquad\qquad 2O^{2-} \rightarrow O_2 + 4e^-$

The carbon anode is burned away and has to be frequently replaced.

### 23.4   THE HALIDES OF ALUMINIUM

Aluminium chloride is prepared by heating aluminium in dry chlorine or dry hydrogen chloride.

$$2Al(s) + 3Cl_2(g) \rightarrow 2AlCl_3(s)$$
$$2Al(s) + 6HCl(g) \rightarrow 2AlCl_3(s) + 3H_2(g)$$

Hydrated aluminium chloride $AlCl_3.6H_2O$ can be prepared by crystallizing the solution remaining from the reaction of excess aluminium with hydrochloric acid. Anhydrous aluminium chloride cannot be prepared by heating hydrated aluminium chloride because this causes hydrolysis.

$$2AlCl_3.6H_2O(s) \rightarrow Al_2O_3(s) + 6HCl(g) + 9H_2O(l)$$

Relative molecular mass determinations of aluminium chloride when dissolved in benzene or in the vapour state indicate that aluminium chloride exists as double molecules, *i.e.* $Al_2Cl_6$.

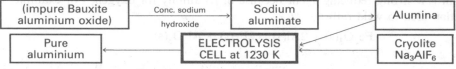

By receiving two electrons from a chlorine atom to form a dative bond, the octet of aluminium is completed. The arrangement of atoms around each aluminium atom is tetrahedral. These double molecules exist in the vapour state up to 400°C when they start to dissociate.

$$Al_2Cl_6 \rightleftharpoons 2AlCl_3$$

The anhydrous aluminium trihalides (except trifluoride) are hydrolyzed by water, and they fume in contact with moist air:

$$AlCl_3(s) + 3H_2O(l) \rightleftharpoons Al(OH)_3(s) + 3HCl(g)$$

### 23.5   SOLUTIONS OF ALUMINIUM SALTS IN WATER

Aluminium salts in aqueous solutions contain the $[Al(H_2O)_6]^{3+}$ ion. This complex cation is acidic because $Al^{3+}$ is highly polarising, which weakens the $O-H$ bonds.

$$[Al(H_2O)_6]^{3+}(aq) + H_2O(l) \rightleftharpoons [Al(H_2O)_5OH]^{2+}(aq) + H_3O^+(aq)$$
$$[Al(H_2O)_5OH]^{2+}(aq) + H_2O(l) \rightleftharpoons [Al(H_2O)_4(OH)_2]^+(aq) + H_3O^+(aq)$$
$$[Al(H_2O)_4(OH)_2]^+(aq) + H_2O(l) \rightleftharpoons [Al(H_2O)_3(OH)_3](s) + H_3O^+(aq)$$

Addition of dilute ammonia solution (a weak base) removes $H_3O^+$ ions and moves the equilibria to the right. This precipitates hydrated aluminium hydroxide. Aluminium hydroxide is **amphoteric**, *i.e.* reacts with acid and alkali.

$$[Al(H_2O)_2(OH)_4]^- \xleftarrow{\text{addition of } OH^-} [Al(H_2O)_3(OH)_3] \xrightarrow{\text{addition of } H_3O^+} [Al(H_2O)_4(OH)_2]^+$$

### QUESTIONS

The extraction and refining of aluminium can be described using a flow diagram

| (impure Bauxite aluminium oxide) | → Conc. sodium hydroxide → | Sodium aluminate | → | Alumina |
|---|---|---|---|---|
| Pure aluminium | ← | ELECTROLYSIS CELL at 1230 K | ← | Cryolite $Na_3AlF_6$ |

(a)   Give the formula for (i) alumina, (ii) sodium aluminate.

(b)   Despite being the third most abundant element in the earth's crust, aluminium was not produced on a large scale until the nineteenth century. Suggest a reason for this.

(c)   What are the positive electrodes in the electrolysis cell made of and why do they have to be replaced at regular intervals?

(d)   Write an ion-electron equation for the reaction occurring at the negative electrode in the electrolysis cell.

(e)   What mass of aluminium could theoretically be produced if a current of 289500 A flows through the electrolysis cell for two minutes?

(f)   Why is it possible to carry out the electrolysis at 1230 K if alumina normally melts at about 2300 K?

(g)   Aluminium has a standard reduction electrode potential of $-1.70V$.
    (i) What does this suggest about the 'reactivity' of aluminium?

(*ii*) In the light of your answer give a reason why aluminium is used to make window frames.
Consider the following data concerning the compounds aluminium chloride and magnesium chloride.

| | aluminium chloride | magnesium chloride |
|---|---|---|
| *Action of heat* | sublimes at 453 K | melts at 1690 K |
| *Relative molecular mass (vapour phase)* | 267 | 95 |
| *Action with water* | reacts (hydrolized) | dissolves |

Using the above data:
(*h*)  Deduce the type of bonding in (i) aluminium chloride, (ii) magnesium chloride.
(*i*)  Give the formula for aluminium chloride in the vapour phase.
(*j*)  Write a balanced equation for the reaction between aluminium chloride and water. (*Scottish Higher*)

(*a*)  (*i*) $Al_2O_3$  (*ii*) $NaAl(OH)_4$
(*b*)  Many aluminium ores contain only a small percentage of aluminium. It is very expensive to extract aluminium from its ores.
(*c*)  Carbon. They are oxidized in the oxygen which is produced at the anode.
(*d*)  $Al^{3+} + 3e^- \rightarrow Al$
(*e*)  Quantity of electricity used $= \dfrac{289500 \times 2 \times 60}{96500}$ Faradays

   $= 360$ Faradays

   27 g of aluminium is produced by 3 Faradays

   $\dfrac{27 \times 360}{3}$ g of aluminium produced by 360 Faradays            **Answer 3240 g**

(*f*)  Alumina is not melted but dissolved in molten cryolite.
(*g*)  (*i*) See Table 18.1. The very negative electrode potential suggests that aluminium is a reactive metal.
   (*ii*) Aluminium oxide forms a tough coating on the surface of the aluminium. This prevents reaction.
(*h*)  (*i*) Covalent (see 4.4); (ii) Ionic (see 4.2)
(*i*)  $Al_2Cl_6$. This corresponds to a relative molecular mass of 267.
(*j*)  $AlCl_3(s) + 3H_2O(l) \rightleftharpoons Al(OH)_3(s) + 3HCl(g)$

**2** Thallium was discovered spectroscopically by Crookes who proposed the name from the Greek word for a green twig or shoot, **thallos**. It has a melting point of 303°C, boiling point 1457°C and a density of $11.85 \, \text{g cm}^{-3}$. Two oxidation states are known, for which the following standard electrode potentials have been found:

$$Tl^{3+} + 2e^- \rightarrow Tl^+ \qquad\qquad E^0 = +1.26 \text{ V}$$
$$Tl^+ + e^- \qquad \rightarrow Tl \qquad\qquad E^0 = -0.34 \text{ V}$$

Thallium has an atomic number of 81 and a relative atomic mass of 204.4 which reflects the presence of two isotopes of mass number 203 and 205, in relative abundances of 29.5 and 70.5 per cent respectively. Its electronegativity is 1.8.

Thallium(III) fluoride decomposes at its melting point, 550°C.

Thallium iodide, $TlI_3$, has the same crystal structure as $NH_4I_3$.

Thallium and its salts are very poisonous, e.g. thallium sulphate, which is odourless and tasteless, is used as a rodenticide.

(*i*)  Suggest briefly what the phrase "discovered spectroscopically" means, and why the element was named after a "green twig".
(*ii*)  Give reasons why thallium is classified as a metal.
(*iii*)  Is the electronegativity value consistent with this classification? Explain.
(*iv*)  How is the relative atomic mass given consistent with abundances and mass numbers of the isotopes?
(*v*)  Deduce what you can about the chemistry of thallium from the standard electrode potentials given.
(*vi*)  Why is the decomposition of thallium fluoride unlikely to involve the formation of thallium(I)?
(*vii*)  What are the likely formulae of the ions in thallium iodide, $TlI_3$? Explain your answer.
(*viii*)  By examining the position of thallium in the Periodic Table, suggest why thallium's toxicity is not unexpected.            (*London 080-Sample paper*)

(*i*)  Elements emit radiation in characteristic fashion. Thallium emits in green part of spectrum.
(*ii*)  Cationic behaviour from electrode potential measurements. High density.
(*iii*)  Non-metals have high electronegativities; the value of 1.8 is modest. This is consistent.
(*iv*)  Relative atomic mass is weighted mean of relative isotopic masses.
(*v*)  $Tl^{3+}$ is quite oxidizing, so thallium would react with mild reducing agents such as iodide ions and iron (II). $Tl^+$ is quite difficult to reduce to the metal so reactions involving thallium (III) will most likely yield Tl (I) rather than the metal, though powerful reducing agents will give Tl itself.
(*vi*)  Because if Tl is reduced something must be oxidized and fluoride $\rightarrow$ fluorine is most unlikely.
(*vii*)  Since ammonium ion is 1 +, then Tl here is also 1 +. Negative ion is $I_3^-$.
(*viii*)  Bottom of group so "heavy metal". Adjacent to mercury and lead.

# **24** Group IV

## 24.1   INTRODUCTION

Group IV chemistry frequently appears in A-level questions. Perhaps this is due to the fact that there are four elements in Group IV that the candidate should have met in practical sessions. This contrasts with Group III, for example, where only one element, aluminium, will be familiar to the student.

Also within Group IV the transition down the Group from non-metal to metal is clearer than in other Groups.

The elements in group IV are:

| | | |
|---|---|---|
| Carbon | C | $1s^2 2s^2 2p^2$ |
| Silicon | Si | $1s^2 2s^2 2p^6 3s^2 3p^2$ |
| Germanium | Ge | $1s^2 2s^2 2p^6 3s^2 3p^6 3d^{10} 4s^2 4p^2$ |
| Tin | Sn | $1s^2 2s^2 2p^6 3s^2 3p^6 3d^{10} 4s^2 4p^6 4d^{10} 5s^2 5p^2$ |
| Lead | Pb | $1s^2 2s^2 2p^6 3s^2 3p^6 3d^{10} 4s^2 4p^6 4d^{10} 5s^2 5p^6 5d^{10} 6s^2 6p^2$ |

All of these elements contain electronic arrangements with $s^2 p^2$ in the outer energy level. They could achieve noble gas electronic arrangements by losing these four electrons or by gaining four electrons. The ionization energies for losing four electrons are:

| | |
|---|---|
| C | 6200 kJ mol$^{-1}$ |
| Si | 4400 kJ mol$^{-1}$ |
| Ge | 4400 kJ mol$^{-1}$ |
| Sn | 3900 kJ mol$^{-1}$ |
| Pb | 4100 kJ mol$^{-1}$ |

There is no possibility of gaining sufficient energy to produce $M^{4+}$ ions.

Carbon and silicon are non-metals while tin and lead are metals. Germanium has properties intermediate between metal and non-metal and is called a **metalloid**.

## 24.2   OXIDATION STATES

There are two oxidation states seen within the group. The oxidation state of +4 decreases in stability down the group. It corresponds to the formation of four covalent bonds

*E.g.*

$$H-\underset{\underset{H}{|}}{\overset{\overset{H}{|}}{C}}-H$$

For this to occur, one electron is promoted from the $s$ orbital to the empty $p$ orbital, *i.e.* $s^2 p^2 \rightarrow sp^3$. Four identical bonds arranged tetrahedrally by this $sp^3$ hybridization.

The other oxidation state is +2. This increases in stability down the group and becomes noticeable from germanium downwards. For lead, the +2 oxidation state is more stable than the +4. In the +2 oxidation state only the two electrons in the $p$ orbital are lost. The remaining ion has a 2+ charge and retains the pair of s electrons. The reluctance of these two electrons to engage in bond formation is called the **inert pair effect**.

With germanium and tin, the oxidation state +4 is more stable. Germanium(II) and tin(II) compounds are strong reducing agents.

*E.g.*          $SnCl_2(aq) + 2HgCl_2(aq) \rightarrow SnCl_4(aq) + Hg_2Cl_2(aq)$
          $SnCl_2(aq) + Hg_2Cl_2(aq) \rightarrow 2Hg(l) + SnCl_4(aq)$

With lead, however, the +2 oxidation state is more stable than the +4. Lead(IV) compounds, therefore, act as oxidizing agents. For example, when concentrated hydrochloric acid is warmed with lead(IV) oxide.

$$PbO_2(s) + 4HCl(aq) \rightarrow PbCl_2(s) + 2H_2O(l) + Cl_2(g)$$

## 24.3   THE ALLOTROPY (POLYMORPHISM) OF GROUP IV ELEMENTS

**Allotropy** (or **polymorphism**) is the existence of two or more forms of the same element in the same physical state. The two allotropes of carbon are diamond and graphite (Unit 10.5). Silicon and germanium do not show allotropy but both have structures similar to diamond.

There are three allotropes of tin. Each allotrope is stable over a definite temperature range.

| **grey tin** | **white tin** | **rhombic tin** |
|---|---|---|
| or α-tin | or β-tin | or γ-tin |
| stable below 13°C | stable between 13°C and 161°C | stable between 161°C and melting point |
| *diamond structure* | *metal structure* | *metal structure* |

grey tin $\xrightarrow{\longleftarrow}$ white tin $\xrightarrow{\longleftarrow}$ rhombic tin

Lead does not show allotropy but has a metal structure.

Carbon and tin are examples of the two types of allotropy. Carbon is said to show **monotropy** because although more than one allotrope can exist under particular conditions one is more stable than the others at all temperatures. Diamond and graphite both exist at room temperature and pressure, but graphite is more stable than diamond.

Tin is an example of **enantiotropy** because each allotrope exists over a definite temperature range.

## 24.4  REACTIONS OF THE ELEMENTS WITH ACIDS

Carbon reacts with hot oxidizing acids such as concentrated nitric and sulphuric acids.

*E.g.* $$C(s) + 2H_2SO_4(l) \rightarrow CO_2(g) + 2H_2O(l) + 2SO_2(g)$$

Silicon is not attacked by any acid except hydrofluoric acid.

$$Si(s) + 6HF(l) \rightarrow H_2SiF_6(l) + 2H_2(g)$$
*hexafluorosilicic acid*

Germanium, tin and lead react with nitric acid.

$$3Ge(s) + 4HNO_3(l) \xrightarrow{conc} 3GeO_2(s) + 4NO(g) + 2H_2O(l)$$
$$3Sn(s) + 4HNO_3(l) \xrightarrow{conc} 3SnO_2(s) + 4NO(g) + 2H_2O(l)$$
$$Pb(s) + 4HNO_3(l) \xrightarrow{conc} Pb(NO_3)_2(aq) + 2NO_2(g) + 2H_2O(l)$$

## 24.5  REACTIONS OF THE ELEMENTS WITH ALKALIS

Carbon does not react with alkali. Silicon reacts with dilute sodium hydroxide solution to produce sodium silicate(IV).

$$Si(s) + 2OH^-(aq) + H_2O(l) \rightarrow SiO_3^{2-}(aq) + 2H_2(g)$$

Germanium and tin react with hot concentrated sodium hydroxide solution to form a germanate(IV) and stannate(IV) respectively.

$$Ge(s) + 2OH^-(aq) + H_2O(l) \rightarrow GeO_3^{2-}(aq) + 2H_2(g)$$
$$Sn(s) + 2OH^-(aq) + H_2O(l) \rightarrow SnO_3^{2-}(aq) + 2H_2(g)$$

Lead reacts with hot concentrated sodium hydroxide solution to produce the plumbate(II) (plumbite).

$$Pb(s) + 2OH^-(aq) \rightarrow PbO_2^{2-}(aq) + H_2(g)$$

This illustrates the greater stability of the +2 oxidation state down the group.

## 24.6  CATENATION

**Catenation** is the ability of an element to form bonds between its own atoms to form chains. Carbon can form long chains, *e.g.* alkanes (Unit 29). The tendency to catenation decreases markedly down the group (Unit 24.9).

The strength of the $C-C$ bond is due to the small size of the carbon atom and the closeness of the bonding electrons to the two nuclei.

Silicon–oxygen bonds are much stronger than silicon–silicon bonds, and these bonds are the basis of the structure of silicates and silicones.

*E.g.*

$$\left[ \begin{array}{c} CH_3 \\ | \\ -Si-O- \\ | \\ CH_3 \end{array} \right]_n$$

## 24.7  OXIDES

All of the elements form dioxides. The stability of these dioxides decreases down the group.

Carbon dioxide can be prepared by burning carbon in excess oxygen. Alternatively, it can be produced by treating a metal carbonate with a dilute acid, or by heating most carbonates and hydrogencarbonates.

*E.g.* $$CaCO_3(s) + 2HCl(aq) \rightarrow CaCl_2(aq) + H_2O(l) + CO_2(g)$$

Carbon dioxide is an acidic oxide, dissolving in water to form weak carbonic acid.

$$H_2O(l) + CO_2(g) \rightleftharpoons H_2CO_3(aq)$$

Carbon dioxide is composed of discrete molecules. These molecules have a structure:

$$O = C = O$$

The carbon–oxygen bonds are however shorter than expected. This suggests that two resonance structures are

$$^+O \equiv C - O^- \leftrightarrow O^- - C \equiv O^+$$

Silicon(IV) oxide (silicon dioxide) is produced by the hydrolysis of silicon(IV) chloride or silicon(IV) fluoride. It occurs, with various degrees of impurity, as sand.

$$SiCl_4(l) + 2H_2O(l) \rightarrow SiO_2(s) + 4HCl(g)$$

Silicon(IV) oxide is an acidic oxide. It reacts with an alkali to form a silicate(IV).

$$SiO_2(s) + 2OH^-(aq) \rightarrow SiO_3^{2-}(aq) + H_2O(l)$$

The structure of silicon(IV) oxide is a giant structure (see Fig. 4.2).

Germanium(IV) oxide, tin(IV) oxide and lead(IV) oxide are **amphoteric oxides**. For example, tin(IV) oxide reacts with acids and alkalis.

Lead(IV) oxide cannot be prepared by heating lead in oxygen. It is prepared by heating dilead(II) lead(IV) oxide (red lead) with dilute nitric acid.

$$Pb_3O_4(s) + 4HNO_3(aq) \rightarrow 2Pb(NO_3)_2(aq) + PbO_2(s) + 2H_2O(l)$$

The elements also form monoxides. The monoxides are more important with elements lower in the group. Carbon monoxide is produced when carbon is burned in a limited supply of oxygen.

$$2C(s) + O_2(g) \rightarrow 2CO(g)$$

It can also be prepared by dehydration of methanoic acid (formic acid) or its salts with warm concentrated sulphuric acid.

$$HCOOH(l) \rightarrow H_2O(l) + CO(g)$$

Carbon monoxide is only very slightly soluble in water and the solution does not show acidic properties. Carbon monoxide does react with sodium hydroxide solution at 150°C and under high pressures to form sodium methanoate.

$$NaOH(aq) + CO(g) \rightarrow HCOONa(aq)$$

Carbon monoxide is a good reducing agent. It is produced in the blast furnace (Unit 28.9) and reduces iron(III) oxide to iron.

$$Fe_2O_3(s) + 3CO(g) \rightarrow 2Fe(l) + 3CO_2(g)$$

Carbon monoxide combines with chlorine in the presence of ultra violet light to form carbonyl chloride (phosgene).

$$CO(g) + Cl_2(g) \rightarrow COCl_2(g)$$

The structure of carbon monoxide is a resonance hybrid of two structures.

$$C = O \leftrightarrow C \equiv O \qquad \left[ \cdot \! \cdot C \! \begin{smallmatrix} x \\ x \\ \cdot \\ x \end{smallmatrix} \! O \begin{smallmatrix} x \\ x \\ x \end{smallmatrix} \right]$$

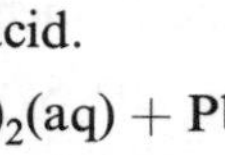

The resulting structure contains a lone pair of non-bonding electrons on the carbon atom. This can form coordinate bonds with *d*-block elements (Unit 28.6).

Silicon(II) oxide is an unstable and unimportant oxide. Germanium(II) oxide is also unstable and tends to disproportionate on heating.

$$2GeO(s) \rightarrow GeO_2(s) + Ge(s)$$

Both tin(II) oxide and lead(II) oxide are amphoteric. They react with acids to form tin(II) and lead(II) salts. With alkalis they form stannate(II) (stannite) and plumbate(II) (plumbite).

Dilead(II) lead(IV) oxide (red lead) is a mixed oxide behaving as if it is a mixture of two parts of lead(II) oxide and one part of lead(IV) oxide.

*E.g.* with concentrated hydrochloric acid

$$2PbO(s) + 4HCl(aq) \rightarrow 2PbCl_2(s) + 2H_2O(l)$$
$$PbO_2(s) + 4HCl(aq) \rightarrow PbCl_2(s) + 2H_2O(l) + Cl_2(g)$$
$$\overline{Pb_3O_4(s) + 8HCl(aq) \rightarrow 3PbCl_2(s) + 4H_2O(l) + Cl_2(g)}$$

## 24.8   HALIDES

Germanium, tin and lead form dihalides which are solid and show some ionic character in the bonding. Tin(II) chloride can be prepared by reacting tin with concentrated hydrochloric acid.

$$Sn(s) + 2HCl(aq) \rightarrow SnCl_2(aq) + H_2(g)$$

Lead(II) chloride is prepared by adding dilute hydrochloric acid to lead(II) nitrate solution.

$$Pb(NO_3)_2(aq) + 2HCl(aq) \rightarrow PbCl_2(s) + 2HNO_3(aq)$$

NB lead(II) chloride precipitates as a white solid. It is soluble in hot water but insoluble in cold water.

Tin(II) chloride and lead(II) chloride form complex ions with concentrated hydrochloric acid.

*E.g.* $$PbCl_2 + 2Cl^- \rightarrow [PbCl_4]^{2-}$$

Germanium(II) and tin(II) halides are reducing agents because the oxidation state $+4$ is more stable.

All of the elements form tetrahalides. The tetrahalides are more important higher up the group. The tetrabromide and tetraiodide of lead do not exist. All of the tetrahalides apart from lead(IV) fluoride are volatile and show covalent bonding.

Tetrachloromethane (carbon tetrachloride) is prepared by passing chlorine through boiling carbon disulphide.

$$CS_2(l) + 3Cl_2(g) \rightarrow CCl_4(l) + S_2Cl_2(l)$$

Alternatively, it can be prepared by the reaction of excess chlorine with methane in the presence of ultra violet light (Unit 29.3).

Tetrachlorides, bromides and iodides of silicon, germanium and tin are prepared by the reaction of chlorine, bromine or iodine with the heated elements.

*E.g.* $$Si(s) + 2Cl_2(g) \rightarrow SiCl_4(l)$$

Lead(IV) chloride cannot be prepared by a similar method as this leads to lead(II) chloride. Lead(IV) chloride is prepared by the reaction of cold concentrated hydrochloric acid with lead(IV) oxide.

$$PbO_2(s) + 4HCl(aq) \rightarrow PbCl_4(l) + 2H_2O(l)$$

If the temperature rises above $0°C$, the lead(IV) oxide oxidizes the hydrochloric acid to chlorine.

Complex ions such as $[SnCl_6]^{2-}$ and $[PbCl_6]^{2-}$ can be prepared by adding concentrated hydrochloric acid to tin(IV) chloride or lead(II) chloride.

All of the tetrachlorides are covalent liquids. All (except tetrachloromethane) fume in moist air because they are hydrolyzed by water.

$$SiCl_4(l) + 2H_2O(l) \rightarrow SiO_2(s) + 4HCl(g)$$

During the hydrolysis of silicon(IV) chloride, there is an intermediate stage when the water molecules are forming bonds with the silicon as the chloride ions are leaving.

At this stage the silicon atom is using $3s$, $3p$ and $3d$ orbitals.

For tetrachloromethane, no reaction takes place because this kind of intermediate is not possible. The carbon atom does not have the opportunity of hybridization of $s$, $p$ and $d$ orbitals because $2d$ orbitals do not exist. The non-availability of $d$ orbitals is a major reason for the fact that the first elements in each Group behave oddly.

### 24.9 HYDRIDES

Carbon forms a wide range of hydrides and these are collectively called hydrocarbons (Unit 29). The existence of so many hydrides is due to the strong bonds between carbon atoms (Unit 24.6).

Carbon–carbon double and triple bonds are also possible. Other elements in Group IV do not form multiple bonds.

Silicon forms a limited number of hydrides similar to alkanes. They range from $SiH_4$ to $Si_6H_{14}$ and are called **silanes**. **Germanes** range from $GeH_4$ to $Ge_3H_8$ and tin hydrides are only $SnH_4$ and $Sn_3H_8$. The hydride $PbH_4$ is extremely unstable. The stability of hydrides decreases down the group.

### QUESTIONS

1 The elements in Group IV of the Periodic Table are carbon (6), silicon (14), germanium (32), tin (50) and lead (85). (The numbers in brackets are the atomic numbers).

(*a*) Give the electronic arrangements of silicon and germanium.

(*b*) Explain why the total energy for the process shown is greatest for the element carbon.

$$M(g) \rightarrow M^{2+}(g) + 2e^-$$

(*c*) Why are compounds containing the $M^{4+}$ or $M^{4-}$ ions comparatively rare in this group? Which of these elements is most likely to form these compounds?

(*d*) What is 'catenation'? Which of these elements has the greatest tendency to catenation?

(*a*)    Si      $1s^2 2s^2 2p^6 3s^2 3p^2$
        Ge    $1s^2 2s^2 2p^6 3s^2 3d^{10} 4s^2 4p^2$

(b) The process is the sum of the first and second ionizations of M. Carbon has the greatest total ionization energy for this process and the value decreases down the group.

The ionization energies are greatest for carbon because the electrons being removed are much closer to the nucleus and, therefore, there are greater forces of attraction between these electrons and the nucleus. This explains the fact that $C^{2+}$ ions are never formed and the stability of ions with a 2+ charge increases down the group.

(c) Compounds containing $M^{4+}$ ions are rare because of the large amount of energy which would be required to remove four electrons (sum of 1st, 2nd, 3rd and 4th ionization energies).

For $M^{4-}$ ions four electrons would have to be accepted. The steps

$$M^- + e^- \rightarrow M^{2-}$$
$$M^{2-} + e^- \rightarrow M^{3-}$$
$$M^{3-} + e^- \rightarrow M^{4-}$$

are all very unfavourable since considerable electrostatic repulsions would have to be overcome.
Lead(Pb) is most likely to form $M^{4+}$ ions. (c.f. ionization energies, p.126). Similarly lead will also be most likely to gain electrons to form a $M^{4-}$ ion as the repulsion of electrons in the larger ion will be less than for the other elements in the group.
NB Neither $Pb^{4+}$ nor $Pb^{4-}$ is actually formed.

(d) Catenation is the formation of chains of atoms (Unit 24.6). Carbon is better at catenation because the C–C bond is much stronger than Si–Si etc.

2 Give an account of the chemistry of Group IV elements (carbon–lead) using the following headings as a guide to answering.

(a) The variety of oxidation states possible and the relative stabilities of these oxidation states.

(b) The properties of carbon and carbon compounds which are different from those of silicon and its compounds.

(c) The changes in ionic and covalent character of the compounds of these elements.

In this question the examiner is asking for a comparison of the Group IV elements. Too many candidates reproduce a poorly written set of Group IV notes without concentrating on the specific points mentioned. The content of Unit 24 is sufficient. The following is a brief guide to the content but a good essay style is important in this question.

(a) Oxidation states +2 and +4. +2 increases in stability and +4 decreases in stability down the group. Give examples.

(b) Differences between the elements and compounds of the elements include:

allotropy of carbon
reaction of carbon with hot, concentrated oxidizing acids
silicon reacts with alkalis
carbon dioxide is a gas, silicon(IV) oxide is a solid
hydrolysis of silicon tetrachloride.

(c) Compounds in oxidation state +4 are covalent and +2 are ionic. Give examples and relate to physical properties, *e.g.* solubility in water.

# 25 Group V

## 25.1 INTRODUCTION

Included in Group V there are three elements which may figure in A-level questions. These are nitrogen, phosphorus and, rarely, bismuth. Of these three, nitrogen is the most important. As with Group IV there is an obvious trend down the group from non-metal (nitrogen and phosphorus) to metal (bismuth).

The elements in Group V all contain electron arrangements $s^2 p^3$.

## 25.2 NITROGEN

Nitrogen is the first member of Group V and has an electron arrangement of $1s^2 2s^2 2p^3$. A nitrogen atom contains a single electron in each $2p$ orbital.

Nitrogen can form bonds in three ways.
(i) It can form three covalent bonds by overlap with orbitals on other atoms – each bond contains one pair of shared electrons.
(ii) Nitrogen can accept three electrons to form nitride $N^{3-}$ ions.

(*iii*) Molecules containing a nitrogen atom often contain a pair of non-bonding electrons. These electrons can be donated to another particle to form a coordinate bond, *e.g.*

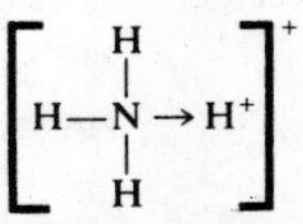

$$\left[\begin{array}{c} H \\ | \\ H-N \rightarrow H^+ \\ | \\ H \end{array}\right]^+$$

ammonium ion

Nitrogen can show a range of oxidation states. These are shown in Table 25.1 with examples.

**Table 25.1  Oxidation states of nitrogen**

| Oxidation state | Example |
|---|---|
| $-3$ | $NH_3$ ammonia |
| $-2$ | $N_2H_4$ hydrazine |
| $-1$ | $NH_2OH$ hydroxylamine |
| 0 | $N_2$ nitrogen |
| $+1$ | $N_2O$ dinitrogenmonoxide |
| $+2$ | $NO$ nitrogen monoxide |
| $+3$ | $NO_2^-$ nitrite (nitrate(III)) |
| $+4$ | $N_2O_4$ dinitrogentetraoxide |
| $+5$ | $NO_3^-$ nitrate (nitrate(V)) |

Nitrogen is a colourless, odourless and tasteless gas, and is particularly unreactive at room temperature. It reacts with reactive metals to form nitrides (Unit 22) and with hydrogen to form ammonia (Unit 15). The unreactivity of nitrogen can be explained by its high dissociation energy ($945$ kJ mol$^{-1}$) caused by the strong triple bond between the two nitrogen atoms.

Nitrogen can be produced in the laboratory by the action of heat on ammonium nitrite (ammonium nitrate(III)). Ammonium nitrite is prepared by mixing ammonium chloride and sodium nitrite (sodium nitrate(III)):

$$NH_4Cl(s) + NaNO_2(s) \rightarrow NH_4NO_2 + NaCl$$
$$NH_4NO_2(s) \rightarrow N_2(g) + 2H_2O(g)$$

Industrially nitrogen is obtained by the fractional distillation of liquid air. Nitrogen (boiling point $-196°C$) boils off first, followed by oxygen (boiling point $-183°C$). Much of the nitrogen produced is used for making ammonia (Unit 15).

## *25.3  OXIDES OF NITROGEN

There are three common oxides of nitrogen and the preparation and properties of these three are summarized in Table 25.2.

## 25.4  NITROGEN TRICHLORIDE

Nitrogen forms only a single chloride (compare with phosphorus Unit 25.9). Nitrogen trichloride is formed when excess chlorine reacts with ammonia:

$$4NH_3(g) + 3Cl_2(g) \rightarrow NCl_3(l) + 3NH_4Cl(s)$$

It is a covalent liquid composed of $NCl_3$ molecules. It is liable to explode.

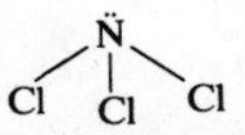

$$\begin{array}{c} \ddot{N} \\ / | \backslash \\ Cl \quad Cl \quad Cl \end{array}$$

Nitrogen trichloride is readily hydrolyzed to ammonia and chloric(I) acid (hypochlorous acid):

$$NCl_3(l) + 3H_2O(l) \rightarrow NH_3(g) + 3HOCl(aq)$$

## 25.5  HYDRIDES OF NITROGEN

The most important hydride of nitrogen is ammonia $NH_3$. Ammonia is prepared in the laboratory by heating a mixture of an ammonium compound and an alkali. (Revise Chemistry GCSE Unit 27).

*E.g.* $\quad NH_4Cl(s) + NaOH(s) \rightarrow NaCl(s) + H_2O(g) + NH_3(g)$

or $\quad NH_4^+ + OH^- \rightarrow NH_3 + H_2O$

**Table 25.2  The oxides of nitrogen**

| Name | dinitrogen oxide | nitrogen monoxide | nitrogen dioxide (or dinitrogen tetraoxide) |
|---|---|---|---|
| *Alternative name* | nitrous oxide | nitric oxide | — — |
| *Formula* | $N_2O$ | $NO$ | $NO_2$ (or $N_2O_4$) |
| *Oxidation state of nitrogen* | $+1$ | $+2$ | $+4$ |
| *Laboratory preparation* | Action of heat on ammonium nitrate. $$NH_4NO_3(s) \rightarrow N_2O(g) + 2H_2O(g)$$ Ammonium nitrate is prepared from ammonium chloride and sodium nitrate. | Action of dilute nitric acid on copper. $$3Cu(s) + 8HNO_3(aq) \rightarrow 3Cu(NO_3)_2(aq) + 4H_2O(l) + 2NO(g)$$ | Heating lead(II) nitrate crystals $$2Pb(NO_3)_2(s) \rightarrow 2PbO(s) + 4NO_2(g) + O_2(g)$$ On cooling, $NO_2$ dimerizes to form dinitrogen tetraoxide. This is a pale yellow liquid. $2NO_2 \rightleftharpoons N_2O_4$ |
| *Structure* | Resonance hybrid of two resonance structures $$^-\!:\!\ddot{N}=\overset{+}{N}=\ddot{O}: \leftrightarrow :N\equiv\overset{+}{N}-\ddot{O}:^-$$ (Isoelectronic with carbon dioxide) | Contains an unpaired electron (paramagnetic). Resonance hybrid of $$\cdot\dot{N}=\ddot{O}: \leftrightarrow :\ddot{N}=\overset{+}{N}-\ddot{O}:$$ This odd electron can be lost or gained $NO - 1e \rightarrow NO^+$   *e.g.* nitrosyl tetrafluoroborate $(NO)^+(BF_4)^-$ $NO + 1e \rightarrow NO^-$   *e.g.* $Na^+NO^-$ sodium nitrosyl | Contains an unpaired electron (paramagnetic). Resonance hybrid of 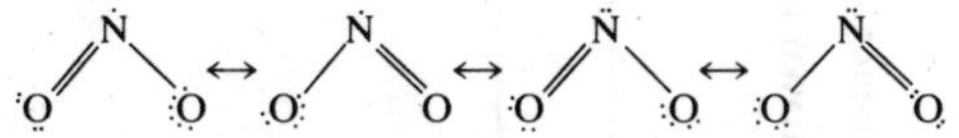 When it dimerizes, no unpaired electron 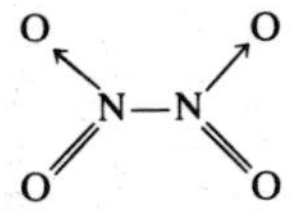 |
| *Properties* | Slightly soluble in water producing a neutral solution (neutral oxide) | Slightly soluble in water producing a neutral solution (neutral oxide). Turns brown in contact with air $$2NO(g) + O_2(g) \rightarrow 2NO_2(g)$$ Forms a brown complex with iron(II) sulphate solution. $$[Fe(H_2O)_6]^{2+} + NO \rightarrow [Fe(H_2O)_5NO]^{2+} + H_2O$$ (brown ring test) | Soluble in water forming an acidic solution (acidic oxide) $$2NO_2(g) + H_2O(l) \rightarrow HNO_2 + HNO_3$$ nitric(III)  nitric(V) acid  acid (or nitrous acid) Decomposes on strong heating $$2NO_2(g) \rightleftharpoons 2NO(g) + O_2(g)$$ |
| *Test* | Relights a glowing splint but sweetish smell and more soluble in water than oxygen. | Turns brown in contact with air. | Brown gas which turns blue litmus red. No bleaching properties (Unlike bromine). |

Ammonia is also produced by the action of water on metal nitrides.

*E.g.*  $$N^{3-}(s) + 3H_2O(l) \rightarrow 3OH^-(aq) + NH_3(g)$$

Ammonia is produced industrially by the Haber process (Unit 15).

The structure of the ammonia molecule is shown in Unit 5. The lone pair of electrons on the nitrogen atom has a profound effect on the properties of ammonia. Because nitrogen is a highly electronegative element, hydrogen bonds can be formed with other ammonia molecules and water molecules.

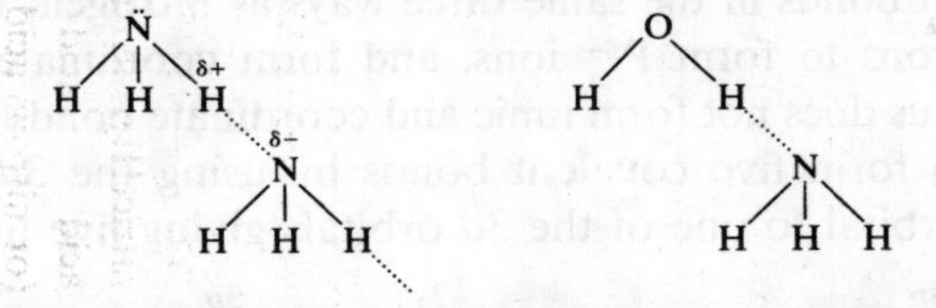

The hydrogen bonds formed with water molecules explain the high solubility of ammonia in water.

Ammonia can also use its lone pair of electrons to remove a proton ($H^+$ ion) from a water molecule. In doing this, ammonia is acting as a Lewis base.

$$NH_3 + H_2O \rightleftharpoons NH_4^+ + OH^-$$

Ammonia solution (sometimes called ammonium hydroxide although this is not strictly correct) is a weak base and will precipitate certain metal hydroxides:

*e.g.*  $$CuSO_4(aq) + 2OH^-(aq) \rightarrow Cu(OH)_2(s) + SO_4^{2-}(aq)$$
$$\text{copper(II) hydroxide – blue ppt.}$$

Some metal hydroxide precipitates redissolve in excess ammonia solution forming soluble complexes:

*e.g.*  $$Cu(OH)_2(s) + 4NH_3(aq) + 2H_2O(l) \rightarrow [Cu(NH_3)_4(H_2O)_2]^{2+}(aq) + 2OH^-(aq)$$
$$\text{deep blue solution}$$

Hydrazine $NH_2NH_2$ is another hydride of nitrogen. It is much less stable than ammonia. Hydrazine is to $NH_3$ as hydrogen peroxide is to $H_2O$.

## 25.6  NITRIC ACID

Nitric acid (nitric(V) acid), $HNO_3$, is prepared industrially from ammonia. A mixture of ammonia and air is passed over a heated platinum catalyst at 900°C. The reactions are

$$4NH_3(g) + 5O_2(g) \rightarrow 4NO(g) + 6H_2O(g)$$

On cooling  $$2NO(g) + O_2(g) \rightarrow 2NO_2(g)$$

The gases are passed through water

$$4NO_2(g) + O_2(g) + 2H_2O(l) \rightarrow 4HNO_3(l)$$

In the laboratory, it is prepared by the action of concentrated sulphuric acid on potassium nitrate.

$$KNO_3(s) + H_2SO_4(l) \rightarrow HNO_3(g) + KHSO_4(s)$$

With cold, very dilute nitric acid, magnesium reacts to form magnesium nitrate and hydrogen.

$$Mg(s) + 2HNO_3(aq) \rightarrow Mg(NO_3)_2(aq) + H_2(g)$$

In other reactions with metals, nitric acid acts as an oxidizing agent and oxides of nitrogen are formed. The products depend upon the conditions.
*E.g.* copper and dilute nitric acid:

$$3Cu(s) + 8HNO_3(aq) \rightarrow 3Cu(NO_3)_2(aq) + 4H_2O(l) + 2NO(g)$$

or copper and concentrated nitric acid:

$$Cu(s) + 4HNO_3(l) \rightarrow Cu(NO_3)_2(aq) + 2H_2O(l) + 2NO_2(g)$$

Nitric acid oxidizes iron(II) compounds to iron(III):

$$2HNO_3(aq) + 6H_3O^+(aq) + 6Fe^{2+}(aq) \rightarrow 6Fe^{3+}(aq) + 10H_2O(l) + 2NO(g)$$

The following ionic half-equations represent the oxidizing agent properties of nitric acid:

$$4HNO_3 + 2e^- \rightarrow 2NO_3^- + 2H_2O + 2NO_2$$
$$8HNO_3 + 6e^- \rightarrow 6NO_3^- + 4H_2O + 2NO$$
$$2HNO_3 + 6H_3O^+ + 6e^- \rightarrow 10H_2O + 2NO$$
$$10HNO_3 + 8e^- \rightarrow 9NO_3^- + 3H_2O + NH_4^+$$

The equations in Unit 25.6 should be examined in the light of these half-equations to see how electrons are transferred in the redox processes.

### 25.7 Phosphorus

Phosphorus has an electron arrangement of $1s^2\,2s^2\,2p^6\,3s^2\,3p^3$ and each of the $3p$ orbitals contains a single electron.

Phosphorus can form bonds in the same three ways as nitrogen. It can form three covalent bonds, gain three electrons to form $P^{3-}$ ions, and form coordinate bonds with electron deficient species. Phosphorus does not form ionic and coordinate bonds as readily as nitrogen.

Phosphorus can also form five covalent bonds by using the $3d$ orbitals. An electron is promoted from the $3s$ orbital to one of the $3d$ orbitals giving five unpaired electrons.

$dsp^3$ hybridization produces five orbitals which can form five covalent bonds by overlap with five other atoms. Nitrogen is unable to form five covalent bonds by overlap with five other atoms because there are no available $2d$ orbitals. This explains why nitrogen forms one chloride $NCl_3$ but phosphorus forms two, $PCl_3$ and $PCl_5$.

Phosphorus can show allotropy (polymorphism) – compare with nitrogen (Unit 25.2). The properties of the two common allotropes are summarized in Table 25.3.

**Table 25.3 Two allotropes of phosphorus**

| White phosphorus | Red phosphorus |
|---|---|
| Whitish-yellow waxy solid | Red powder |
| Stored under water | Stored dry |
| Composed of $P_4$ molecules | Macromolecular structure |
| Low melting point | High melting point |
| Soluble in organic solvents | Insoluble in organic solvents |
| Ignites at 30°C in air | Ignites above 300°C in air |
| Reacts with chlorine at room temperature | Reacts with chlorine on heating |
| Reacts with sodium hydroxide solution to form phosphine | No reaction with sodium hydroxide solution |

The $P_4$ tetrahedra in white phosphorus also exist in the phosphorus vapour.

### 25.8 Oxides of Phosphorus

Two oxides of phosphorus exist corresponding to the oxidation states $+3$ and $+5$.

Phosphorus(III) oxide is made by burning phosphorus in a limited amount of oxygen:

$$P_4(s) + 3O_2(g) \rightarrow P_4O_6(s)$$

Phosphorus(V) oxide is the product of the reaction between phosphorus and excess oxygen.

$$P_4(s) + 5O_2(g) \rightarrow P_4O_{10}(s)$$

(It was originally believed that these two oxides had formulae of $P_2O_3$ and $P_2O_5$ but relative molecular mass studies in the solution and vapour states confirmed formulae of $P_4O_{10}$ and $P_4O_6$.)

Phosphorus(III) and phosphorus(V) oxides are acidic oxides reacting with water to form acids.

$$P_4O_6(s) + 6H_2O(l) \rightarrow 4H_3PO_3(aq) \quad \text{phosphonic acid}$$
$$\text{(phosphorous acid)}$$
$$P_4O_{10}(s) + 6H_2O(l) \rightarrow 4H_3PO_4(aq) \quad \text{phosphoric(V) acid}$$
$$\text{(orthophosphoric acid)}$$

### 25.9 Chlorides of Phosphorus

There are two chlorides of phosphorus – $PCl_3$ and $PCl_5$. Phosphorus trichloride is prepared when phosphorus reacts with a limited amount of chlorine and phosphorus pentachloride when phosphorus reacts with excess chlorine.

$$P_4(s) + 6Cl_2(g) \rightarrow 4PCl_3(l)$$
$$P_4(s) + 10Cl_2(g) \rightarrow 4PCl_5(s)$$

In the solid state phosphorus pentachloride exists in the form of ions.

$$[PCl_4^+] \; [PCl_6^-]$$

The structures of phosphorus trichloride and phosphorus pentachloride are given in Unit 5.

On heating, phosphorus pentachloride dissociates:

$$PCl_5 \rightleftharpoons PCl_3 + Cl_2$$

Both chlorides of phosphorus are hydrolyzed by water and fume in moist air:

$$PCl_3(l) + 3H_2O(l) \rightarrow H_3PO_3(aq) + 3HCl(g)$$
$$\text{phosphonic acid (phosphorous acid)}$$

Phosphorus pentachloride reacts in two stages.

$$PCl_5(s) + H_2O(l) \rightarrow POCl_3(l) + 2HCl(g)$$
$$\text{phosphorus trichloride oxide (phosphorus}$$
$$\text{oxychloride)}$$

$$POCl_3(l) + 3H_2O(l) \rightarrow H_3PO_4(aq) + 3HCl(g)$$
$$\text{phosphoric(V) acid (orthophosphoric acid)}$$

### 25.10 HYDRIDES OF PHOSPHORUS

Phosphine $PH_3$ can be prepared by the action of sodium hydroxide solution on phosphonium iodide:

$$PH_4I(s) + NaOH(s) \rightarrow NaI(s) + H_2O(g) + PH_3(g)$$

(NB This is similar to the preparation of ammonia Unit 25.5)

A more common method is to boil white phosphorus with a concentrated aqueous solution of sodium hydroxide.

$$P_4(s) + 3NaOH(aq) + 3H_2O(l) \rightarrow 3NaH_2PO_2(aq) + PH_3(g)$$
$$\text{sodium phosphinate}$$

NB disproportionation $4P(0) \rightarrow P(-III) + 3P(+I)$

The reaction does not take place with red phosphorus. An impurity in the phosphine is the unstable hydride diphosphine $P_2H_4$ which is spontaneously inflammable.

Phosphine is less basic than ammonia. It is only slightly soluble in water and the following equilibrium lies well over to the left:

$$PH_3 + H_2O \rightleftharpoons PH_4^+ + OH^-$$

The phosphorus atom is larger than a nitrogen atom. Ammonia is better at donating its pair of non bonding electrons.

The boiling points of $NH_3$ and $PH_3$ are $-33°C$ and $-90°C$. Because phosphine molecules have a higher relative molecular mass than ammonia molecules, the boiling point of phosphine should be higher than that of ammonia. This assumption neglects the hydrogen bonding (see Unit 4.8) which is present in ammonia but not in phosphine. Phosphorus is not electronegative enough for hydrogen bonding to exist. The lack of hydrogen bonding also explains the low solubility of phosphine in water.

### 25.11 PHOSPHORIC(V) ACID $H_3PO_4$

This acid is a very deliquescent crystalline solid and exists in a viscous solution owing to a large amount of hydrogen bonding. It is tribasic and three end points should be detected (Unit 16.7)

$$NaOH(aq) + H_3PO_4(aq) \rightarrow NaH_2PO_4(aq) + H_2o(1) \text{ (indicator–methyl orange)}$$
$$NaOH(aq) + NaH_2PO_4(aq) \rightarrow Na_2HPO_4(aq) + H_2O(1) \text{ (indicator–phenolphthalein)}$$
$$NaOH(aq) + Na_2HPO_4(aq) \rightarrow Na_3PO_4(aq) + H_2O(1) \text{ The last end-point cannot be detected}$$

as $PO_4^{3-}$ is a strong base and hydrolyzes in solution (Unit 17.2)

### 25.12 BISMUTH

Bismuth is more metallic than nitrogen or phosphorus. It forms $Bi^{3+}$ ions by the loss of the three electrons in the $p$ orbitals. It shows the **inert pair effect** (Unit 24.2). Bismuth reacts with dilute nitric acid to produce bismuth(III) nitrate.

$$Bi(s) + 6HNO_3(aq) \rightarrow Bi(NO_3)_3(aq) + 3H_2O(l) + 3NO_2(g)$$

The hydride of bismuth $BiH_3$ called bismuthine is much less stable and more difficult to form than ammonia or phosphine.

Bismuth in the $+5$ oxidation state is a very powerful oxidizing agent.

QUESTIONS

1 (a) State how each of the following compounds could be prepared.
   (i) dinitrogen oxide $N_2O$
   (ii) nitrogen oxide NO
   (iii) dinitrogen tetraoxide $N_2O_4$
   (iv) phosphorus(V) oxide $P_4O_{10}$
In each case, give the reactants, conditions and a suitable equation.
   (b) How, if at all, do the above compounds react with water?

(a) (i) Prepared by mixing sodium nitrate and ammonium chloride and heating the mixture. Ammonium nitrate is prepared in the mixture.

$$NH_4Cl(s) + NaNO_3(s) \rightarrow NH_4NO_3(s) + NaCl(s)$$
$$NH_4NO_3(s) \rightarrow N_2O(g) + 2H_2O(g)$$

(ii) Action of dilute nitric acid on copper.

$$3Cu(s) + 8HNO_3(aq) \rightarrow 3Cu(NO_3)_2(aq) + 4H_2O(l) + 2NO(g)$$

No heat is required.
(iii) Dinitrogen tetraoxide is prepared by heating lead(II) nitrate crystals.

$$Pb(NO_3)_2(s) \rightarrow 2PbO(s) + 4NO_2(g) + O_2(g)$$
$$\text{lead(II) oxide}$$

The nitrogen dioxide is cooled by passing it through a U-tube surrounded by ice to condense dinitrogen tetraoxide.

$$2NO_2 \rightleftharpoons N_2O_4$$

(iv) Prepared by burning red or white phosphorus in excess air or oxygen. White phosphorus is spontaneously inflammable but red phosphorus requires heating.

$$P_4(s) + 5O_2(g) \rightarrow P_4O_{10}(s)$$

(b) $N_2O$ – slightly soluble in cold water. Collected over hot water.
   NO – virtually insoluble in cold water.
   $N_2O_4$ – reacts with cold water:

$$N_2O_4(l) + H_2O(l) \rightarrow HNO_2(aq) + HNO_3(aq)$$
$$\text{nitrous acid} \quad \text{nitric acid}$$
$$\text{(nitric(III) acid)}$$

$P_4O_{10}$ – reacts with water on heating:

$$P_4O_{10}(s) + 6H_2O(l) \rightarrow 4H_3PO_4(aq)$$
$$\text{phosphoric(V) acid}$$

2 (a) Compare the bonding in the elements nitrogen, phosphorus and bismuth.
   (b) Describe the bonding in the three chlorides nitrogen trichloride, phosphorus trichloride and phosphorus pentachloride.
      Why can phosphorus form a pentachloride but nitrogen cannot?
   (c) Write equations and name the products for the reactions of the chlorides in (b) with water.

*(Associated Examining Board)*

(a) Bismuth is a metal and has a metallic lattice (Unit 10.3). Nitrogen consists of diatomic molecules $N_2$. The bonds between nitrogen atoms are triple covalent bonds. White phosphorus consists of a regular arrangement of $P_4$ tetrahedra, and red phosphorus has a covalent macromolecular structure.
(b) In nitrogen trichloride and phosphorus trichloride the bonding is covalent. In phosphorus pentachloride solid the bonding is ionic (Unit 25.9). In the gas phase the bonding is covalent.
The fact that a pentachloride of nitrogen is impossible is explained in Unit 25.7.
(c) Units 25.4 and 25.9.

3 (a) Explain in outline how compounds of nitrogen in oxidation states +2, −3 and +4 could be prepared from nitric acid.
   (b) The overall equation for the preparation of iodine monochloride is

$$N_2H_6O(aq) + IO_3^-(aq) + 2H_3O^+(aq) + Cl^-(aq) \rightarrow N_2(g) + ICl(g) + 6H_2O(l)$$

During this reaction the oxidation states of all elements except nitrogen and iodine are unchanged. Using the equation, give the oxidation states of nitrogen in reactants and products.

(a) Three nitrogen compounds with oxidation states +2, −3 and +4 are nitrogen monoxide (NO), ammonia ($NH_3$) and nitrogen dioxide ($NO_2$).
   Nitrogen monoxide is prepared by the reaction of copper with dilute nitric acid (Table 25.2).
   Ammonia can be prepared by adding sodium hydroxide solution and aluminium powder (or De Varda's alloy Unit 23.2). Nitrogen is reduced from oxidation state +5 to −3.

$$HNO_3 + 8[H] \rightarrow NH_3 + 3H_2O$$

Nitrogen dioxide is prepared by the action of copper on concentrated nitric acid (Table 25.2).
(b) Oxidation state of iodine changes from +5 to +1. In the products nitrogen is the oxidation state zero (element). In the compound $N_2H_6O$, nitrogen must be in oxidation state −2.

# 26 Group VI

## 26.1 INTRODUCTION

Oxygen and sulphur are the two common elements in Group VI of the Periodic Table.

Both of these elements have electron arrangements $s^2p^4$ in the highest energy level. Atoms of both elements can form ionic or covalent bonds.

## 26.2 OXYGEN

Oxygen (electron arrangement $1s^2 2s^2 2p^4$) can form two covalent bonds or gain two electrons to form $O^{2-}$ ions.

Oxygen is the product in several common reactions:

(i) catalytic decomposition of hydrogen peroxide (laboratory preparation)

$$2H_2O_2(aq) \rightarrow 2H_2O(l) + O_2(g)$$

(ii) thermal decomposition of oxygen-rich compounds:

e.g. potassium manganate(VII)

$$2KMnO_4(s) \rightarrow K_2MnO_4(s) + MnO_2(s) + O_2(g)$$

potassium chlorate(VII)

$$2KClO_3(s) \rightarrow 2KCl(s) + 3O_2(g)$$

potassium nitrate

$$2KNO_3(s) \rightarrow 2KNO_2(s) + O_2(g)$$

(iii) Electrolysis of aqueous solutions

$$4OH^-(aq) \rightarrow 2H_2O(l) + O_2(g) + 4e^-$$

(iv) Action of a peroxide on water:

e.g. $$2Na_2O_2(s) + 2H_2O(l) \rightarrow 4NaOH(aq) + O_2(g)$$

Industrially, oxygen is prepared by the fractional distillation of liquid air.

Oxygen is an extremely electronegative element. Most elements, apart from the noble gases, combine with oxygen to form oxides. Elements which combine with oxygen (or with fluorine in Group VII) tend to show their maximum oxidation state:

e.g. $Cl_2O_7$ dichlorine heptoxide – chlorine shows a maximum oxidation state of $+7$. The stability of high oxidation states is due to the high electronegativity and small size of the oxygen atom.

## 26.3 CLASSIFICATION OF OXIDES

Oxides can be divided into six groups, though the boundaries between some of these groups are rather arbitrary.

### (i) Basic oxides

These are oxides of metals containing the $O^{2-}$ ion. The oxidation state of a metal in a basic oxide is low. A basic oxide reacts with an acid to form a salt and water only.

E.g. $$CuO(s) + H_2SO_4(aq) \rightarrow CuSO_4(aq) + H_2O(l)$$
ionic equation: $O^{2-}(s) + 2H_3O^+(aq) \rightarrow 3H_2O(l)$

Basic oxides of metals high in the electrochemical series react with water to form soluble hydroxides or alkalis.

E.g. $$CaO(s) + H_2O(l) \rightarrow Ca(OH)_2(s)$$
ionic equation: $O^{2-}(s) + H_2O(l) \rightarrow 2OH^-(s)$

### (ii) Acidic oxides

Acidic oxides are oxides of non-metals or d-block elements in high oxidation states.

E.g. $SO_3$ sulphur(VI) oxide dissolves in water to form an acid.

$$SO_3(s) + H_2O(l) \rightarrow H_2SO_4(aq)$$

Chromium(VI) oxide dissolves in water to form chromic(VI) acid.

$$CrO_3(s) + H_2O(l) \rightarrow H_2CrO_4(aq)$$

These acidic oxides are usually simple molecules with mainly covalent bonding. Oxides dissolving in water to form an acid are called **acid anhydrides**.

Where an element forms a number of different oxides with different oxidation states, the higher the oxidation state the more acidic the oxide is.

*E.g.*   $SO_3$   oxidation state of sulphur is $+6$
$SO_2$   oxidation state of sulphur is $+4$
$SO_3$ is more strongly acidic than $SO_2$

Acidic oxides react with basic oxides to form **salts**.

### (iii)  Amphoteric oxides

These are oxides which can behave as acidic or basic oxides depending upon conditions. They are usually oxides of the less electropositive metals, *e.g.* zinc, aluminium, lead.

*(i)*                            $ZnO(s) + 2HCl(aq) \rightarrow ZnCl_2(aq) + H_2O(l)$
*(ii)*            $ZnO(s) + 2OH^-(aq) + H_2O(l) \rightarrow Zn(OH)_4^{2-}(aq)$

In equation *(i)* zinc oxide is acting as a basic oxide and in *(ii)* as an acidic oxide.

### (iv)  Neutral oxides

A neutral oxide does not react either with an acid or an alkali to form a salt, examples are nitrogen oxide, NO, and dinitrogen oxide, $N_2O$.

### (v)  Mixed oxides

They behave as if they were composed of mixtures of simple oxides, *e.g.* dilead(II) lead(IV) oxide, $Pb_3O_4$, behaves as if it is a mixture of lead(II) oxide, PbO (2 parts), and lead(IV) oxide, $PbO_2$ (1 part).

### (vi)  Peroxides

Peroxides are higher oxides of electropositive metals containing $O_2^{2-}$ ions. They are strong oxidizing agents. When acidified with cold, dilute acid, hydrogen peroxide is produced:

$$Na_2O_2(s) + H_2SO_4(aq) \rightarrow Na_2SO_4(aq) + H_2O_2(aq)$$

ionic equation:    $O_2^-(s) + 2H_3O^+(aq) \rightarrow H_2O_2(aq) + 2H_2O(l)$

### 26.4   TRIOXYGEN(OZONE)

Oxygen gas is composed of diatomic molecules $O_2$. Trioxygen is an unstable allotrope of oxygen where the molecules contain three oxygen atoms *i.e.* $O_3$.

Trioxygen is a non-linear molecule which is a resonance hybrid of two resonance structures.

In trioxygen the two oxygen–oxygen bonds are of equal length.

Trioxygen occurs naturally in the upper atmosphere. The trioxygen in the so called '**ozone layer**' absorbs most of the ultra violet radiation approaching the earth. There is concern about the possible destruction of the trioxygen by fluorocarbons from aerosol propellants.

Trioxygen can be produced by passing a high electrical discharge through dry oxygen.

$$3O_2(g) \rightleftharpoons 2O_3(g)$$

Only about 10% of the oxygen is converted to trioxygen, which is very unstable.

The strong oxidizing agent properties of trioxygen in acid solution can be represented by:

$$O_3(g) + 2H_3O^+(aq) + 2e^- \rightarrow O_2(g) + 3H_2O(l)$$

The two important reactions of trioxygen for examination purposes are:
*(i)* The reaction with alkenes to split carbon–carbon double bonds (called **ozonolysis** Unit 29.6).
*(ii)* Trioxygen will oxidize a sulphide to a sulphate

$$PbS(s) + 4O_3(g) \rightarrow PbSO_4(s) + 4O_2(g)$$
lead(II) sulphide    lead(II) sulphate
black                white

ionic equation:        $S^{2-}(s) + 4O_3(g) \rightarrow SO_4^{2-}(s) + 4O_2(g)$

### 26.5   WATER

Water is a hydride of oxygen and its properties should be compared with the properties of the hydride of sulphur – hydrogen sulphide (Unit 26.8).

Many of the physical properties of water differ from those expected, including:
  (*i*) maximum density at 4°C,
 (*ii*) high boiling point,
(*iii*) high enthalpies of fusion and vaporization,
(*iv*) high dielectric constant (relative permittivity).
   The main reasons for these differences are:
(*i*) Water is highly polarized. Because of the high electronegativity of oxygen there is a partial separation of charges in the molecule.

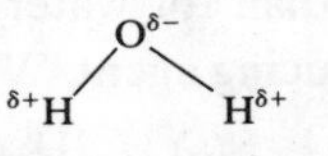

$$\overset{\delta-}{O} \diagup \diagdown$$
$${}^{\delta+}H \qquad H^{\delta+}$$

This can cause solvation of ions in solution (Fig. 4.1).
(*ii*) As a result of the presence of these partial charges, hydrogen bonding can exist between water molecules, and between water molecules and polar molecules such as alcohols (Unit 4).
   Water is composed of $H_2O$ molecules but a very few molecules are dissociated into ions.
$$2H_2O(l) \rightleftharpoons H_3O^+(aq) + OH^-(aq)$$
   Many substances are split up by water and this is called **hydrolysis** (e.g. Units 24.8, 34.2).

### 26.6  HYDROGEN PEROXIDE $H_2O_2$

Hydrogen peroxide is an unstable hydride of oxygen containing $-O-O-$ grouping. The bond between the two oxygen atoms is weak and this accounts for its instability.
   It is prepared in the laboratory by the action of dilute sulphuric acid on barium peroxide. The temperature must be maintained around 0°C to prevent decomposition of the hydrogen peroxide.
$$BaO_2(s) + H_2SO_4(aq) \rightarrow BaSO_4(s) + H_2O_2(aq)$$

The barium sulphate is removed by filtration.
   Industrially, 2-butylanthraquinone is reduced to 2-butylanthraquinol using a catalyst and hydrogen. The 2-butylanthraquinol is oxidised by oxygen to regenerate the 2-butylanthraquinone and produce hydrogen peroxide.

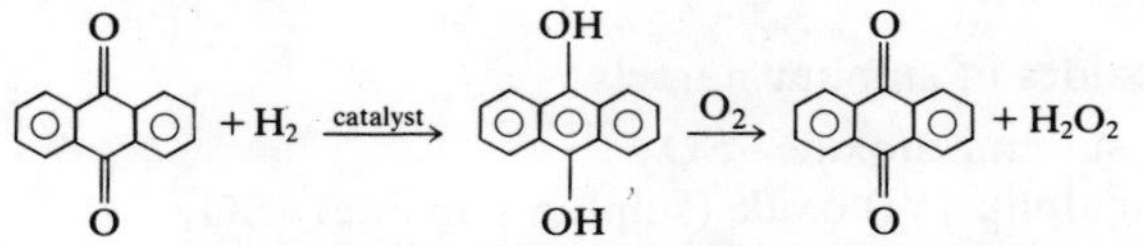

Hydrogen peroxide can act as an oxidizing agent or as a reducing agent depending upon the conditions (Unit 6).

### 26.7  SULPHUR

Sulphur can exist in two solid allotropes (polymorphs). These allotropes are α-sulphur (**rhombic sulphur**) and β-sulphur (**monoclinic sulphur**).
   In addition to forming two covalent bonds and ions with a 2− charge, sulphur can also form compounds where it forms four or six covalent bonds, *i.e.* where the oxidation states of sulphur are +4 and +6. It achieves this by promotion of electrons from 3s or 3p orbitals into 3d orbitals.

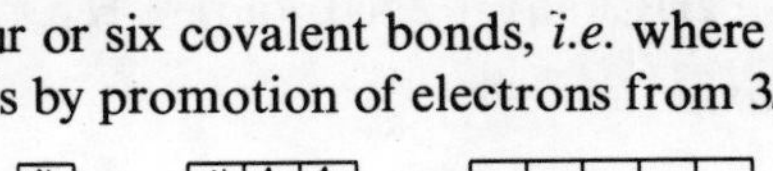

Two unpaired electrons – can form two covalent bonds.

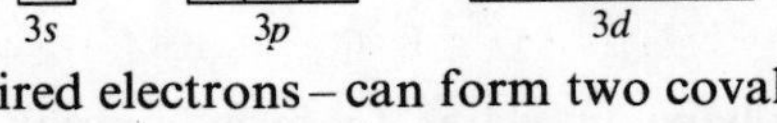

Four unpaired electrons – can form four covalent bonds.

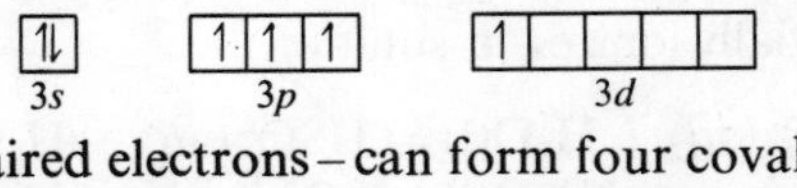

Six unpaired electrons – can form six covalent bonds.
Examples of compounds of sulphur with oxidation states +4 and +6 are:
$$SO_3 \text{ sulphur(VI) oxide } +6$$
$$SO_2 \text{ sulphur dioxide } \quad +4$$

The oxidation state of +6 is only achieved with fluorine and oxygen in fluorides and oxides.
   Sulphur reacts with most metals to form sulphides which are much less ionic than the corresponding oxides.
   *E.g.*  $$Zn(s) + S(s) \rightarrow ZnS(s)$$

Sulphur does not react with water or dilute acids. With hot, concentrated sulphuric acid, sulphur is oxidized to sulphur dioxide:
$$S(s) + 2H_2SO_4(l) \rightarrow 3SO_2(g) + 2H_2O(l)$$

## 26.8    HYDROGEN SULPHIDE

Hydrogen sulphide, $H_2S$, is the hydride of sulphur. Its properties differ greatly from those of water. It is a poisonous gas with an unpleasant smell of rotten eggs.

Hydrogen sulphide has a slight tendency to ionize producing $H_3O^+$ ions:

$$H_2S(aq) + H_2O(l) \rightleftharpoons HS^-(aq) + H_3O^+(aq)$$
$$H_2S(aq) + 2H_2O(l) \rightleftharpoons S^{2-}(aq) + 2H_3O^+(aq)$$

This tendency for ionization is greater than for water.

Hydrogen sulphide is a powerful reducing agent. When it acts in this way, sulphur is produced.

$$H_2S(aq) + 2H_2O(l) \rightarrow S(s) + 2H_3O^+(aq) + 2e^-$$

An example of hydrogen sulphide as a reducing agent is the reaction which occurs when hydrogen sulphide is mixed with moist chlorine:

$$Cl_2(g) + 2e^- \rightarrow 2Cl^-(aq)$$
$$H_2S(aq) + 2H_2O(l) \rightarrow S(s) + 2H_3O^+(aq) + 2e^-$$

$$\overline{Cl_2(g) + H_2S(aq) + 2H_2O(l) \rightarrow 2Cl^-(aq) + 2H_3O^+(aq) + S(s)}$$

Most metallic sulphides are insoluble in water. When hydrogen sulphide is passed through a metal salt solution, a precipitate of a metallic sulphide is often formed.

*E.g.*          $Pb^{2+}(aq) + S^{2-}(aq) \rightarrow PbS(s)$
                                  lead(II) sulphide
                                  black precipitate

This reaction is the basis of the test for hydrogen sulphide. A piece of filter paper soaked in a solution of lead(II) ions (usually the nitrate or ethanoate) turns black in contact with hydrogen sulphide gas.

## 26.9    OXIDES OF SULPHUR

There are two simple oxides of sulphur namely:

         sulphur dioxide $- SO_2$
         sulphur(VI) oxide (sulphur trioxide) $- SO_3$

Sulphur dioxide is a pungent smelling gas produced by the action of hot, concentrated sulphuric acid on copper:

$$Cu(s) + 2H_2SO_4(l) \rightarrow CuSO_4(aq) + 2H_2O(l) + SO_2(g)$$

It can also be produced by burning sulphur or warming a sulphite (sulphate(IV)) with a dilute acid:

$$Na_2SO_3(s) + 2HCl(aq) \rightarrow 2NaCl(aq) + H_2O(l) + SO_2(g)$$

The structure of sulphur dioxide is

It is an acidic oxide dissolving in water to form a weak acid called sulphurous acid (sulphuric(IV) acid). This partially ionizes in solution:

$$H_2SO_3(aq) + H_2O(l) \rightleftharpoons HSO_3^-(aq) + H_3O^+(aq)$$
$$H_2SO_3(aq) + 2H_2O(l) \rightleftharpoons SO_3^{2-}(aq) + 2H_3O^+(aq)$$

Sulphurous acid and moist sulphur dioxide act as a reducing agent. As a result of this process sulphur is oxidized from oxidization state $+4$ to $+6$.

$$SO_3^{2-}(aq) + 3H_2O(l) \rightarrow SO_4^{2-}(aq) + 2H_3O^+(aq) + 2e^-$$

*e.g.* when sulphur dioxide is bubbled through a purple solution of potassium manganate(VII), the purple solution is decolourised.

Sulphur(VI) oxide (sulphur trioxide) cannot be produced directly by burning sulphur in oxygen. It can be produced by passing a mixture of sulphur dioxide and oxygen over a heated platinum or vanadium(V) oxide catalyst.

$$2SO_2(g) + O_2(g) \rightleftharpoons 2SO_3(g)$$

Sulphur(VI) oxide collects in a cooled receiver as white needle-shaped crystals. This reaction is also the basis of the **Contact process** for manufacturing sulphuric acid. (See page 142).

The structure of sulphur(VI) oxide in the gaseous state is:

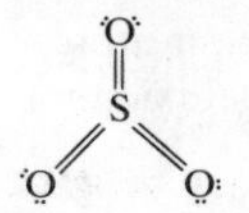

The molecule is planar.

Sulphur(VI) oxide reacts exothermically with water to produce sulphuric acid (sulphuric(VI) acid).

$$SO_3(s) + H_2O(l) \rightarrow H_2SO_4(aq)$$

## 26.10 SULPHURIC ACID (SULPHURIC(VI) ACID)

Sulphuric acid is produced when sulphur(VI) oxide reacts with water. The reactions of sulphuric acid include:

### (i) Acidic properties

Sulphuric acid in aqueous solution ionizes as follows:

$$H_2SO_4(aq) + H_2O(l) \rightleftharpoons H_3O^+(aq) + HSO_4^-(aq)$$
$$H_2SO_4(aq) + 2H_2O(l) \rightleftharpoons 2H_3O^+(aq) + SO_4^{2-}(aq)$$

Neutralizing sulphuric acid with alkali can produce sulphates and hydrogensulphates.

### (ii) Dehydration reactions

Concentrated sulphuric acid acts as a dehydrating agent. It removes water or the elements of water from carbohydrates.

*E.g.* $\qquad C_6H_{12}O_6(s) \xrightarrow{\text{conc } H_2SO_4} 6C(s) + 6H_2O(g)$

### (iii) As an oxidizing agent

Hot concentrated sulphuric acid acts as an oxidizing agent usually producing sulphur dioxide:

$$2H_3O^+(aq) + H_2SO_4(l) + 2e^- \rightarrow 4H_2O(l) + SO_2(g)$$

Hydrogen fluoride and hydrogen chloride are not oxidized by sulphuric acid. Hydrogen bromide is oxidized by concentrated sulphuric acid to bromine:

$$2H_2O(l) + 2HBr(aq) \rightarrow Br_2(g) + 2H_3O^+(aq) + 2e^-$$
$$\underline{2H_3O^+(aq) + H_2SO_4(l) + 2e^- \rightarrow 4H_2O(l) + SO_2(g)}$$
$$2HBr(aq) + H_2SO_4(l) \rightarrow Br_2(g) + 2H_2O(l) + SO_2(g)$$

Hydrogen iodide is oxidized even more thoroughly by concentrated sulphuric acid. The sulphuric acid is reduced, in this case, partly to hydrogen sulphide because hydrogen iodide is an even stronger reducing agent than hydrogen bromide.

$$H_2SO_4(l) + 8H_3O^+(aq) + 8e^- \rightarrow H_2S(g) + 12H_2O(l)$$
$$\underline{8I^-(aq) \rightarrow 4I_2(aq) + 8e^-}$$
$$H_2SO_4(l) + 8H_3O^+(aq) + 8I^-(aq) \rightarrow H_2S(g) + 12H_2O(l) + 4I_2(aq)$$

### (iv) Sulphonation

Sulphuric acid is used as a sulphonating agent (Unit 29.12).

The structure of sulphuric acid is

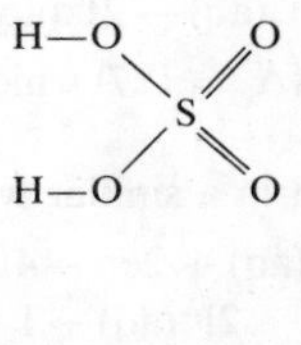

## 26.11 HALIDES OF SULPHUR

When chlorine is passed over heated molten sulphur, disulphur dichloride is formed.

$$2S(l) + Cl_2(g) \rightarrow S_2Cl_2(l)$$

Disulphur dichloride is an amber liquid which combines with more chlorine at 0°C to form sulphur dichloride.

$$S_2Cl_2(l) + Cl_2(g) \rightarrow 2SCl_2(l)$$

Fluorine reacts with sulphur to form sulphur hexafluoride.

$$S(s) + 3F_2(g) \rightarrow SF_6(s)$$

QUESTIONS

**1** (*a*) Give the electron arrangement for a sulphur atom.

(*b*) For each of the following species give the oxidation state, the shape and distribution of electrons in the molecule or ion.

(*i*) $SO_2$

(*ii*) $H_2S$

(*iii*) $SO_4^{2-}$

(*c*) Describe in outline how substances containing sulphur in oxidation states $+6$ and $0$ respectively could be obtained from sulphur dioxide.

(*a*) $1s^2 2s^2 2p^6 3s^2 3p^4$

(*b*) (*i*) Oxidation state $+4$. The shape and distribution of electrons is shown in Unit 26.9.

(*ii*) Oxidation state $-2$. (iii) Oxidation state $+6$.

(*c*) A compound with sulphur in oxidation state $+6$ is sulphur(VI) oxide (sulphur trioxide).

Pass a mixture of sulphur dioxide and oxygen over a heated platinum catalyst. Collect the sulphur(VI) oxide in a receiver cooled in ice.

$$2SO_2(g) + O_2(g) \rightleftharpoons 2SO_3(s)$$

Sulphur is in oxidation state zero in elemental sulphur. Plunge burning magnesium into a gas jar of sulphur dioxide:

$$2Mg(s) + SO_2(g) \rightarrow 2MgO(s) + S(s)$$

or mix with damp $H_2S$:

$$2H_2S(g) + SO_2(g) \rightarrow 2H_2O(l) + 3S(g)$$

**2** (*a*) Describe in outline how oxygen (dioxygen) may be converted into

(*i*) a sample of oxygen containing trioxygen (ozone) and

(*ii*) an aqueous solution of hydrogen peroxide.

(*b*) Suggest how the concentrations of trioxygen in sample (*i*) and of hydrogen peroxide in sample (*ii*) could be determined.

(*a*) (*i*) Pass an electrical discharge through dry oxygen (Unit 26.4)

$$3O_2(g) \rightleftharpoons 2O_3(g)$$

(*ii*) Burn a reactive metal, *e.g.* sodium, barium, in excess oxygen.

$$Ba(s) + O_2(g) \rightarrow BaO_2(s) \text{ barium peroxide.}$$

Add this solid residue to cold, dilute sulphuric acid, stirring the mixture after each addition.

$$BaO_2(s) + H_2SO_4(aq) \rightarrow BaSO_4(s) + H_2O_2(aq)$$

Filter off the barium sulphate and retain the solution.

(*b*) Bubble a known volume of oxygen containing trioxygen through 25cm³ of acidified potassium iodide solution (0.1 mol dm⁻³). The trioxygen oxidizes some of the iodide ions to iodine.

$$O_3(g) + 2H_3O^+(aq) + 2e^- \rightarrow O_2(g) + 3H_2O(l)$$
$$2I^-(aq) \rightarrow I_2(aq) + 2e^-$$
$$\overline{O_3(g) + 2H_3O^+(aq) + 2I^-(aq) \rightarrow O_2(g) + 3H_2O(l) + I_2(aq)}$$

Then titrate the liberated iodine in the solution with a standard solution of sodium thiosulphate.

$$I_2(aq) + 2e^- \rightarrow 2I^-(aq)$$
$$2S_2O_3^{2-}(aq) \rightarrow S_4O_6^{2-}(aq) + 2e^-$$
$$\overline{2S_2O_3^{2-}(aq) + I_2(aq) \rightarrow 2I^-(aq) + S_4O_6^{2-}(aq)}$$

1 mole of trioxygen liberates 254g of iodine ($A_r = 127$) which in turn would react with 2000cm³ of sodium thiosulphate solution (1 mol dm⁻³).

Hydrogen peroxide could be determined in a similar way.

$$H_2O_2(aq) + 2H_3O^+(aq) + 2e^- \rightarrow 4H_2O(l)$$
$$2I^-(aq) \rightarrow I_2(aq) + 2e^-$$
$$\overline{H_2O_2 + 2H_3O^+(aq) + 2I^-(aq) \rightarrow 4H_2O(l) + I_2(aq)}$$

**3** (*a*) The important step in the Contact process is the following reversible reaction:

$$2SO_2(g) + O_2(g) \rightleftharpoons 2SO_3(g) \quad \triangle H = -98 \text{ kJ mol}^{-1}$$

Describe the manufacture of sulphuric acid, explaining carefully and making use of the information given, the reasons for the chemical and physical conditions used.

(*b*) Why is it important to recycle the gases after the absorption of sulphur trioxide?

(*a*) There are three stages in the manufacture of sulphuric acid.

(*i*) Oxidation of sulphur or sulphur containing minerals

$$e.g. \quad S(s) + O_2(g) \rightarrow SO_2(g)$$
$$4FeS_2(s) + 11O_2(g) \rightarrow 2Fe_2O_3(s) + 8SO_2(g)$$

The sulphur dioxide is purified to remove impurities which would poison the catalyst.

(*ii*) The oxidation of sulphur dioxide to produce sulphur trioxide (equation given in the question). The conditions are low temperature to increase the yield of sulphur trioxide (Le Chatalier's Principle). However, low temperatures would cause reaction to be very slow. Temperatures about 450°C and vanadium (V) oxide catalyst. High pressure increases yield of sulphur trioxide as there are fewer molecules on the right hand side of equation.

(*iii*) Absorption of sulphur trioxide in concentrated sulphuric acid to produce oleum which, on dilution with the correct volume of water, produces concentrated sulphuric acid.

$$SO_3(g) + H_2SO_4(l) \rightarrow H_2S_2O_7(l)$$
$$H_2S_2O_7(l) + H_2O(l) \rightarrow 2H_2SO_4(l)$$

(*b*) Sulphur is the most expensive raw material. It is important not to lose any expensive sulphur dioxide. Sulphur dioxide also causes acid rain problems if it escapes into the atmosphere.

# 27 Group VII

## 27.1 INTRODUCTION

The elements in Group VII are called the **halogens**. This group frequently appears in A-level questions, probably because there are three elements in the group familiar to students, and the elements show a good gradation of properties.

## 27.2 THE ELEMENTS

The elements in Group VII are shown in Table 27.1.

**Table 27.1 The elements of Group VII**

| Element | symbol | appearance | electron arrangement |
|---|---|---|---|
| Fluorine | F | pale greenish-yellow gas | [He] $2s^2 2p^5$ |
| Chlorine | Cl | greenish yellow gas | [Ne] $3s^2 3p^5$ |
| Bromine | Br | dark red liquid | [Ar] $3d^{10} 4s^2 4p^5$ |
| Iodine | I | grey-black solid | [Kr] $4d^{10} 5s^2 5p^5$ |

Astatine is a further member of the halogen family. It is very rare because it is intensely radioactive. It is estimated that there is only 0.029g of astatine in the earth's crust.

All of the halogen atoms are one electron short of a noble gas arrangement.

*E.g.*     fluorine $1s^2 2s^2 2p^5$     neon $1s^2 2s^2 2p^6$

Halogen atoms can combine by ionic and covalent bonding. The halogen atoms can gain one electron to form negatively charged halide ions.

*E.g.*     $Cl + e^- \rightarrow Cl^-$

A halogen atom can also form a covalent bond by overlap of the *p* orbital containing one electron with a partially filled orbital on another atom.

Ionic halides are formed with electropositive metals, *e.g.* sodium chloride. Ionic halides are high melting point solids which are soluble in water. Covalent halides are formed with a halogen and a non-metal or a less electropositive metal.

Fluorine exhibits a maximum covalency of one. The other elements can exhibit covalencies of three, five or seven corresponding to the promotion of electrons into the available *d*-orbitals.

*E.g.* iodine can show higher covalencies

| 5s | 5p | 5d |
|----|----|----|
| ↑↓ | ↑↓ ↑↓ ↑ | ☐ ☐ ☐ ☐ ☐ |

One unpaired electron – forms one covalent bond, *e.g.* ICl.
One electron promoted into a 5d orbital

| 5s | 5p | 5d |
|----|----|----|
| ↑↓ | ↑↓ ↑ ↑ | ↑ ☐ ☐ ☐ ☐ |

Three unpaired electrons – forms three covalent bonds, *e.g.* $ICl_3$.
Two electrons promoted into 5d orbitals

| 5s | 5p | 5d |
|----|----|----|
| ↑↓ | ↑ ↑ ↑ | ↑ ↑ ☐ ☐ ☐ |

Five unpaired electrons – forms five covalent bonds, *e.g.* $ICl_5$.
Three electrons promoted into 5d orbitals

| 5s | 5p | 5d |
|----|----|----|
| ↑ | ↑ ↑ ↑ | ↑ ↑ ↑ ☐ ☐ |

Seven unpaired electrons – forms seven covalent bonds, *e.g.* $ICl_7$

All halogens are composed of diatomic molecules. Within the molecule, the two halogen atoms are joined by a single covalent bond. The enthalpies of atomization of these elements are:

| | |
|---|---|
| fluorine | 79.1 kJ mol⁻¹ |
| chlorine | 121.1 kJ mol⁻¹ |
| bromine | 112.0 kJ mol⁻¹ |
| iodine | 106.6 kJ mol⁻¹ |

From chlorine to iodine, the strength of the bond between halogen atoms decreases. Fluorine has a much weaker bond than might be expected due to the extra repulsion caused by the close proximity of the two fluorine nuclei and repulsion between non-bonding electrons on both atoms.

The halogen elements have the highest electron affinities of all the elements. They are:

| | |
|---|---|
| fluorine | −332 kJ mol⁻¹ |
| chlorine | −364 kJ mol⁻¹ |
| bromine | −342 kJ mol⁻¹ |
| iodine | −295 kJ mol⁻¹ |

Again, the electron affinity of fluorine is less than might be expected.

The electronegativities of these elements decrease down the group. The high electro-negativity of fluorine is the reason for the existence of hydrogen bonding in hydrogen fluoride. It also explains why fluorine (like oxygen in Group VI, Unit 26) causes elements to show their maximum oxidation state.

Since iodine is the least electronegative, it is more inclined to show metallic properties. It is possible to form compounds of iodine containing the $I^+$ ion, *e.g.* $I^+CNO^-$.

All of the halogens are oxidizing agents. Fluorine is the strongest of all chemical oxidizing agents.

*E.g.*  $\frac{1}{2}F_2 + e^- \rightarrow F^-$

Despite the fact that the electron affinity of fluorine is less than that of chlorine, the dissociation energy of the fluorine molecule is also less. The difference in oxidizing agent power of the halogens is shown by the standard electrode potentials $E^{\ominus}$

| | |
|---|---|
| fluorine/fluoride | +2.87 V |
| chlorine/chloride | +1.36 V |
| bromine/bromide | +1.09 V |
| iodine/iodide | +0.54 V |

The difference in oxidizing agent power of the halogens explains the displacement reactions of halogens. If chlorine is bubbled through a colourless solution of potassium iodide, the solution turns brown due to the liberation of iodine.

$$Cl_2(g) + 2I^-(aq) \rightarrow 2Cl^-(aq) + I_2(aq)$$

Chlorine is a stronger oxidizing agent than iodine.

## 27.3  PROPERTIES OF THE HALOGENS

### (i) Reactions with water

Fluorine reacts vigorously with cold water to form hydrogen fluoride and oxygen:

$$2F_2(g) + 2H_2O(l) \rightarrow 4HF(g) + O_2(g)$$

In addition to these products, traces of trioxygen, hydrogen peroxide and oxygen difluoride $F_2O$ are detected.

Chlorine reacts less readily with water. A solution of chlorine in water (chlorine water) is acidic due to the formation of hydrochloric acid and chloric(I) acid (hypochlorous acid):

$$Cl_2(g) + H_2O(l) \rightleftharpoons HCl(aq) + HOCl(aq)$$

When this solution is exposed to strong sunlight the chloric(I) acid decomposes to produce oxygen:

$$2HOCl(aq) \rightarrow 2HCl(aq) + O_2(g)$$

It is this available oxygen which accounts for the bleaching properties of chlorine.

Bromine is much less soluble in water than chlorine. The reaction of bromine with water is similar to that of chlorine but to a lesser extent.

Iodine is only very sparingly soluble in water. Iodine dissolves better in other solvents. The colour of an iodine solution depends upon the solvent. With solvents containing oxygen (*e.g.* ethers, alcohols) the solution is brown. Hydrocarbons and tetrachloromethane solutions are violet in colour. Iodine dissolves in an aqueous solution of potassium iodide to form a brown solution due to the formation of a complex ion:

$$I_2(aq) + I^-(aq) \rightleftharpoons I_3^-(aq)$$

### (ii) Reactions with alkalis

Fluorine reacts with an aqueous solution of potassium hydroxide. With a cold, dilute potassium hydroxide solution, oxygen difluoride (fluorine monoxide) is produced:

$$2KOH(aq) + 2F_2(g) \rightarrow 2KF(aq) + F_2O(g) + H_2O(l)$$

With hot, concentrated potassium hydroxide solution, a different reaction takes place producing oxygen:

$$4KOH(aq) + 2F_2(g) \rightarrow 4KF(aq) + O_2(g) + 2H_2O(l)$$

Chlorine and bromine also react with aqueous potassium hydroxide solution with the products again depending upon the conditions. With cold, dilute potassium hydroxide solution, potassium chloride and potassium chlorate(I) (potassium hypochlorite) are formed.

$$Cl_2(g) + 2OH^-(aq) \rightarrow Cl^-(aq) + ClO^-(aq) + H_2O(l)$$

With hot, concentrated potassium hydroxide, the chlorate(I) or bromate(I) disproportionate.

$$6OH^-(aq) + 3Cl_2(g) \rightarrow 5Cl^-(aq) + ClO_3^-(aq) + 3H_2O(l)$$
$$\text{chlorate(V)}$$

The reaction of iodine with potassium hydroxide solution is reversible.

$$2KOH(aq) + I_2(s) \rightleftharpoons KI(aq) + KIO(aq) + H_2O(l)$$

### (iii) Reactions of halogens with hydrogen

See Unit 21.4.

### 27.4 HYDROGEN HALIDES

The hydrogen halides can all be prepared by direct combination (Unit 21.4). The laboratory preparations of the halogen halides differ. Hydrogen chloride and hydrogen fluoride are prepared by the action of concentrated sulphuric acid on sodium chloride or calcium fluoride.

$$NaCl(s) + H_2SO_4(l) \rightarrow NaHSO_4(s) + HCl(g)$$
$$CaF_2(s) + H_2SO_4(l) \rightarrow CaSO_4(s) + 2HF(g)$$

The reaction between a bromide or iodide and concentrated sulphuric acid does not produce hydrogen bromide or hydrogen iodide since the hydrogen bromide or hydrogen iodide reduce the sulphuric acid (Unit 26.10).

Hydrogen bromide and hydrogen iodide are prepared by the hydrolysis of phosphorus tribromide or phosphorus triiodide.

$$PBr_3(l) + 3H_2O(l) \rightarrow H_3PO_3(aq) + 3HBr(aq)$$
$$PI_3(s) + 3H_2O(l) \rightarrow H_3PO_3(aq) + 3HI(aq)$$

Hydrogen fluoride is a liquid at room temperature and pressure while the others are gases. The boiling points are:

| | |
|---|---|
| HF | 20°C |
| HCl | −85°C |
| HBr | −67°C |
| HI | −35°C |

The anomalous boiling point of hydrogen fluoride is due to hydrogen bonding (Unit 4.8) between hydrogen fluoride molecules.

Hydrogen bonding exists in the liquid and in the vapour up to about 90°C.

Solutions of hydrogen halides in water produce acidic solutions. All except hydrogen fluoride produce strong acids.

*E.g.*  $$HCl(aq) + H_2O(l) \rightarrow H_3O^+(aq) + Cl^-(aq)$$

Hydrogen fluoride is a weak acid due to the dissociation:

$$2HF(aq) + H_2O(l) \rightleftharpoons H_3O^+(aq) + HF_2^-(aq)$$

(NB 2 moles of hydrogen fluoride produce 1 mole of $H_3O^+(aq)$ ions.)

## 27.5  HALIDES

The halides formed by electropositive metals are ionic while halides of non-metals and less electropositive metals are covalent. Where an element forms two chlorides, the one with the other element in the higher oxidation state is more covalent *e.g.* $SnCl_4$ (oxidation state of tin +4) is more covalent than $SnCl_2$ (oxidation state +2).

In the series lithium fluoride, lithium chloride, lithium bromide, lithium iodide the bonding is predominately ionic for lithium fluoride but there is a tendency towards covalency across the series. This is due to the increasing size of the halogen ion.

Anhydrous halides are prepared by passing dry halogen vapour over the heated element.

*E.g.*  $$2Fe(s) + 3Cl_2(g) \rightarrow 2FeCl_3 \text{ iron(III) chloride}$$

(NB The reaction of iron with dry hydrogen chloride produces iron(II) chloride.)

$$Fe(s) + 2HCl(g) \rightarrow FeCl_2(s) + H_2(g)$$

Most ionic halides are not hydrolyzed by water and can be produced by reacting a metal, metal oxide, metal hydroxide or metal carbonate with a hydrogen halide, followed by evaporation and crystallization.

*E.g.*  $$NaOH(aq) + HCl(aq) \rightarrow NaCl(aq) + H_2O(l)$$

Metal halides containing water of crystallization are often hydrolyzed on heating. It is not possible to produce an anhydrous metal halide by heating the hydrated metal halide. This process can be carried out by heating the hydrated halide in an atmosphere of hydrogen halide.

Insoluble metal halides can be produced by mixing two suitable solutions, *e.g.* lead(II) iodide is precipitated when solutions of lead(II) nitrate and potassium iodide are mixed.

$$Pb(NO_3)_2(aq) + 2KI(aq) \rightarrow PbI_2(s) + 2KNO_3(aq)$$
ionic equation:  $$Pb^{2+}(aq) + 2I^-(aq) \rightarrow PbI_2(s)$$

The yellow lead(II) iodide is removed by filtration and washed and dried.

The silver halides are produced by adding silver nitrate solution to a solution of halide ions.

$$Ag^+(aq) + X^-(aq) \rightarrow AgX(s) \quad \text{where X represents a halogen.}$$

Silver fluoride is soluble in water because the fluoride ion has a particularly high hydration energy due to its small size. The other halides precipitate as follows:

| | | |
|---|---|---|
| silver chloride | AgCl | white precipitate |
| silver bromide | AgBr | cream precipitate |
| silver iodide | AgI | pale yellow precipitate |

The precipitates can also be distinguished by their solubilities in ammonia solution. Silver chloride is readily soluble forming the diammine silver(I) ion.

$$AgCl(s) + 2NH_3(aq) \rightarrow [Ag(NH_3)_2]^+(aq) + Cl^-(aq)$$

Silver bromide is partially soluble and silver iodide virtually insoluble.

Calcium chloride, bromide and iodide are soluble in water but calcium fluoride is insoluble due to the high lattice energy of calcium fluoride. This is caused by the small sizes of calcium and fluoride ions and the double charge on the calcium ion.

QUESTIONS

**1** (*a*) Complete the electron configuration of an iodine atom.

[Kr]

(*b*) What is the highest oxidation state of iodine?
(*c*) Name a substance which oxidizes elemental iodine into this state.
(*d*) What is the lowest oxidation state of iodine?
(*e*) Give one example of a redox reaction involving iodine in its lowest oxidation state.
(*f*) Which of the hydrogen halides has the lowest bond dissociation energy?
(*g*) In dilute aqueous solution, which of the hydrogen halides is the strongest acid? Explain your answer.

(*a*) $[Kr] 4d^{10}5s^25p^5$
(*b*) Highest oxidation state of iodine $+7$
(*c*) Fluorine

$$I_2(s) + 7F_2(g) \rightarrow 2IF_7(s)$$

(*d*) $-1$
(*e*)
$$2I^-(aq) + Cl_2(g) \rightarrow I_2(aq) + 2Cl^-(aq)$$

Pass chlorine into a solution of potassium iodide. The iodide ions are oxidized to iodine (oxidation state $-1 \rightarrow 0$)
(*f*) Hydrogen iodide has the lowest bond dissociation energy.
(*g*) Hydrogen iodide is the strongest acid. The HI bond is weakest and the reaction

$$HI(aq) + H_2O(l) \rightarrow H_3O^+(aq) + I^-(aq)$$

occurs most readily, *i.e.* the proton is most readily transferred to the water molecule.

**2** Describe briefly a laboratory preparation of dry chlorine. Outline the most important chemical properties of this element.

Comment on the following statements relating to the chemistry of the halogens.
(*a*) When chromium is made to react with different halogens, the main product of the reaction with fluorine is $CrF_4$, with chlorine it is $CrCl_3$ and with iodine it is $CrI_2$.
(*b*) The boiling points of some simple halides are HF 19°C, HCl $-85$°C, $CF_4$ $-128$°C, $CCl_4$ 76°C, $AlF_3$ 1290°C, $AlCl_3$ 180°C.

*(Oxford and Cambridge)*

Chlorine is prepared by heating concentrated hydrochloric acid with manganese(IV) oxide.

$$MnO_2(s) + 4HCl(aq) \rightarrow MnCl_2(aq) + 2H_2O(l) + Cl_2(g)$$
$$\text{manganese(II) chloride}$$

NB Manganese(IV) oxide often acts as a catalyst but in this reaction it is *not* a catalyst. It is an oxidizing agent and oxidises the hydrochloric acid to chlorine.

The gas is passed through cold water to remove any hydrogen chloride fumes, dried by passing through concentrated sulphuric acid, and collected by downward delivery.

The properties of chlorine are given in Unit 27.3. In addition, the reaction of chlorine with metals such as iron to form chlorides (Unit 27.5) should be included.
(*a*) Fluorine is the most electronegative and iodine is the least electronegative of the three. Fluorine is a stronger oxidizing agent than chlorine and chlorine is a stronger oxidizing agent than iodine. Fluorine oxidizes chromium to oxidation state $+4$ but chlorine is only capable of oxidizing chromium to $+3$ and iodine to $+2$.
(*b*) $CF_4$ is composed of smaller, lighter molecules than $CCl_4$. It therefore has a lower boiling point as less energy is required to separate these molecules. HF should have a lower boiling point than HCl for the same reason. However hydrogen bonding causes association of hydrogen fluoride molecules (Unit 27.4).

$AlF_3$ – the high boiling point is because the compound is ionic.
$AlCl_3$ is predominately covalent.

**3** The halogen family includes chlorine, bromine and iodine. This question concerns these three elements and their compounds.
(*a*) (*i*) Write an ionic equation for the reaction which takes place when chlorine is bubbled through a solution of potassium iodide.
(*ii*) Using this as an example, explain the terms oxidation and reduction.
(*b*) Explain why, although hydrogen chloride can be prepared by warming a mixture of sodium chloride and concentrated sulphuric acid, a similar method cannot be used for preparing hydrogen iodide.
(*c*) Disproportionation occurs when chlorine reacts with hot aqueous sodium hydroxide solution.
(*i*) What is meant by 'disproportionation'?
(*ii*) Write an equation for the reaction.

(*a*) (*i*)
$$Cl_2(g) + 2I^-(aq) \rightarrow 2Cl^-(aq) + I_2(aq)$$

(*ii*) oxidation and reduction Unit 6. Chlorine molecules gain electrons to form chloride ions and are reduced. Iodide ions lose electrons to form iodine molecules and are oxidized.
(*b*) Unit 26.10

(c) (i) Disproportionation. A reaction in which a single substance reacts to form two products. One product is obtained by the oxidation of the original substance and the other by reduction. Other examples will be found in Units 6 and 18.

(ii) When chlorine reacts with cold alkali

$$Cl_2(g) + 2OH^-(aq) \rightarrow Cl^-(aq) + ClO^-(aq) + H_2O(l)$$

On heating the chlorate(I) disproportionates

$$3ClO^-(aq) \rightarrow ClO_3^-(aq) + 2Cl^-(aq)$$

# 28 The $d$-block elements

## 28.1 INTRODUCTION

The first row '$d$-block' elements are a series of metals, from scandium to zinc, which are positioned in the Periodic Table between Groups II and III. The Periodic Table is very closely linked with electron arrangements (Units 1 and 2). Orbitals fill up in order of increasing energy. Calcium has an electron arrangement of $1s^2 2s^2 2p^6 3s^2 3p^6 4s^2$. The next orbitals to be filled are the five $3d$ orbitals, each of which can hold two electrons. The $3d$ orbitals are smaller than the $4s$ and, therefore, electrons are being added but the outer electron arrangement, which determines chemical properties, is little changed. The $d$-block elements (scandium–zinc) represents a family of metals with similar properties.

Questions on $d$-block elements figure frequently on A-level papers. The scope of this topic is very wide and, in this Unit, the basic facts and patterns are discussed. Also, some attempt is made to identify the areas in which questions may be set. After studying this unit, it might be useful to consult your A-level notes or an A-level textbook. These may make more sense with the back-ground of this Unit.

## 28.2 THE $d$-BLOCK ELEMENTS

The elements in the first row of $d$-block elements and their electron arrangements appear in Table 28.1.

**Table 28.1 The first row $d$-block elements**

| Element | Symbol | Atomic number | Electron arrangement |
|---------|--------|---------------|----------------------|
| Scandium | Sc | 21 | $1s^2 2s^2 2p^6 3s^2 3p^6 3d^1 4s^2$ |
| Titanium | Ti | 22 | $1s^2 2s^2 2p^6 3s^2 3p^6 3d^2 4s^2$ |
| Vanadium | V | 23 | $1s^2 2s^2 2p^6 3s^2 3p^6 3d^3 4s^2$ |
| Chromium | Cr | 24 | $1s^2 2s^2 2p^6 3s^2 3p^6 3d^5 4s^1$ |
| Manganese | Mn | 25 | $1s^2 2s^2 2p^6 3s^2 3p^6 3d^5 4s^2$ |
| Iron | Fe | 26 | $1s^2 2s^2 2p^6 3s^2 3p^6 3d^6 4s^2$ |
| Cobalt | Co | 27 | $1s^2 2s^2 2p^6 3s^2 3p^6 3d^7 4s^2$ |
| Nickel | Ni | 28 | $1s^2 2s^2 2p^6 3s^2 3p^6 3d^8 4s^2$ |
| Copper | Cu | 29 | $1s^2 2s^2 2p^6 3s^2 3p^6 3d^{10} 4s^1$ |
| Zinc | Zn | 30 | $1s^2 2s^2 2p^6 3s^2 3p^6 3d^{10} 4s^2$ |

NB The electrons are being added to $3d$ orbitals but the irregularities at chromium and copper exist because half filled orbitals and fully filled orbitals are slightly more stable.

The $d$-block elements from scandium to zinc are silvery metals (apart from copper). They all (except zinc) have high melting points and densities because the atoms are strongly bonded together and closely packed. Zinc has a low melting point because it has full $3d$ and $4s$ orbitals, and these electrons are not used in the same way in bonding the atoms together. The $d$-block elements are good conductors of electricity because the $3d$ and $4s$ electrons are free to move through the structure.

These elements readily form alloys. For example, brass is an alloy of copper and zinc. Since the atoms are similar in size and chemical nature, they can be incorporated in each other's lattices.

Across the first row of the *d*-block elements there is a slight decrease in atomic and ionic radii. Moving from scandium to titanium, there is an additional proton in the nucleus. The extra electron goes into an existing $3d$ orbital causing no increase in radius. Due to the extra nuclear attraction, the radius is slightly smaller.

The first ionization energy is approximately the same but increasing slightly across the series due to extra nuclear attraction.

The $E^\circ$ values (*i.e.* $M^{2+}/M$) for the *d*-block elements except copper are negative. From this it can be concluded that these elements will react with dilute hydrochloric acid or sulphuric acid to produce hydrogen.

*E.g.* $$Mn(s) + H_2SO_4(aq) \rightarrow MnSO_4(aq) + H_2(g)$$
$$\text{manganese(II) sulphate}$$

The *d*-block elements form **interstitial compounds** with hydrogen, carbon etc. The small atoms fit into spaces in the lattice of the *d*-block elements. Steel is an interstitial compound with carbon atoms in spaces in the iron lattice.

The *d*-block elements are much less reactive than alkali and alkaline earth metals, *i.e.* the *s*-block elements.

The *d*-block elements have certain characteristic properties. These include:
(*i*) formation of compounds in a wide variety of oxidation states,
(*ii*) formation of coloured compounds,
(*iii*) existence of paramagnetism
(*iv*) formation of coordination compounds.
Each of these will be considered in Units 28.3–28.6.

## 28.3 VARIABLE OXIDATION STATES

When a first row *d*-block element loses electrons to form positive ions, the $4s$ electrons are lost first, followed by the $3d$ electrons. In no case are electrons lost from the $3p$ orbitals.

Table 28.2 lists the common oxidation states of the first row of *d*-block elements. The most common oxidation state for each element is given in bold type.

**Table 28.2  Common oxidation states of the first row *d*-block elements**

| Element | Oxidation states | Examples |
|---|---|---|
| Scandium | **+3** | $Sc_2O_3$ |
| Titanium | +2, +3, **+4** | $TiO$, $Ti_2O_3$, $TiO_2$ |
| Vanadium | +2, +3, **+4**, +5 | $VO$, $V_2O_3$, $VO_2$, $V_2O_5$ |
| Chromium | +2, **+3**, +6 | $CrO$, $Cr_2O_3$, $CrO_3$ |
| Manganese | **+2**, +3, +4, +6, +7 | $MnO$, $Mn_2O_3$, $MnO_2$, $K_2MnO_4$, $KMnO_4$ |
| Iron | **+2**, +3 | $FeO$, $Fe_2O_3$ |
| Cobalt | **+2**, +3 | $CoO$, $Co_2O_3$ |
| Nickel | **+2**, +3, +4 | $NiO$, $Ni_2O_3$, $NiO_2$ |
| Copper | +1, **+2** | $Cu_2O$, $CuO$ |
| Zinc | **+2** | $ZnO$ |

The maximum oxidation states for scandium, titanium, vanadium, chromium and manganese increase to a peak. For these elements, the maximum oxidation state corresponds to the use of all $3d$ and $4s$ electrons. An oxidation state greater than the number of $3d$ and $4s$ electrons would involve electrons from the $3p$ orbitals which are of much lower energy.

For any element, the higher oxidation states give rise to covalent compounds.

The existence of a variety of oxidation states for each element explains the catalytic properties of the *d*-block elements. For example, the decomposition of hydrogen peroxide with manganese-(IV) oxide can occur in two stages. The manganese(IV) oxide is oxidized by hydrogen peroxide to manganese(VII) oxide. The manganese(VII) oxide then decomposes to reform manganese(IV) oxide and produce oxygen.

## 28.4 FORMATION OF COLOURED COMPOUNDS

Hydrated *d*-block metal ions are coloured where the $3d$ orbitals contain unpaired electrons. Light of a certain colour is removed when electrons move from one *d*-orbital to another. For example, hydrated copper(II) ions absorb red light when certain electrons move from one orbital to another and so the residual blue light is transmitted. This gives hydrated copper(II) compounds their characteristic blue colour.

Table 28.3 lists the colours of some hydrated ions of *d*-block metals.

**Table 28.3  Colours of some hydrated ions of *d*-block elements**

| Hydrated ion | Electron arrangement of ion | Colour |
|---|---|---|
| $Sc^{3+}$ | [Ar] $3d^0$ | colourless |
| $Ti^{3+}$ | [Ar] $3d^1$ | violet |
| $V^{3+}$ | [Ar] $3d^2$ | green |
| $Cr^{3+}$ | [Ar] $3d^3$ | green |
| $Mn^{3+}$ | [Ar] $3d^4$ | violet |
| $Mn^{2+}$ | [Ar] $3d^5$ | pale pink |
| $Fe^{3+}$ | [Ar] $3d^5$ | yellow |
| $Fe^{2+}$ | [Ar] $3d^6$ | green |
| $Co^{2+}$ | [Ar] $3d^7$ | pink |
| $Ni^{2+}$ | [Ar] $3d^8$ | green |
| $Cu^{2+}$ | [Ar] $3d^9$ | blue |
| $Zn^{2+}$ | [Ar] $3d^{10}$ | colourless |

*Note:* [Ar] represents $1s^2 2s^2 2p^6 3s^2 3p^6$.

**Table 28.4  Unpaired electrons in ions of first row *d*-block elements**

| Ion | Electron arrangement of ion | | No. of unpaired electrons |
|---|---|---|---|
| $Sc^{3+}$ | [Ar] $3d^0$ | | 0 |
| $Ti^{3+}$ | [Ar] $3d^1$ | ↑ | 1 |
| $V^{3+}$ | [Ar] $3d^2$ | ↑ ↑ | 2 |
| $Cr^{3+}$ | [Ar] $3d^3$ | ↑ ↑ ↑ | 3 |
| $Mn^{3+}$ | [A4] $3d^4$ | ↑ ↑ ↑ ↑ | 4 |
| $Fe^{3+}$ | [Ar] $3d^5$ | ↑ ↑ ↑ ↑ ↑ | 5 |
| $Co^{3+}$ | [Ar] $3d^6$ | ↑↓ ↑ ↑ ↑ ↑ | 4 |
| $Co^{2+}$ | [Ar] $3d^7$ | ↑↓ ↑↓ ↑ ↑ ↑ | 3 |
| $Ni^{2+}$ | [Ar] $3d^8$ | ↑↓ ↑↓ ↑↓ ↑ ↑ | 2 |
| $Cu^{2+}$ | [Ar] $3d^9$ | ↑↓ ↑↓ ↑↓ ↑↓ ↑ | 1 |
| $Zn^{2+}$ | [Ar] $3d^{10}$ | ↑↓ ↑↓ ↑↓ ↑↓ ↑↓ | 0 |

Where $3d^0$ or $3d^{10}$ arrangements exist ($3d$ orbitals either empty or completely full) the ion is colourless.

## 28.5  PARAMAGNETISM

Those *d*-block metal ions that contain unpaired electrons show **paramagnetism**. The movement of an unpaired negatively charged electron produces a small magnetic field. If a particle containing unpaired electrons is subjected to an external magnetic field, there is a positive interaction which is called paramagnetism. This causes the lines of force to become more concentrated and a measurable paramagnetic moment can be obtained. An extreme case of this is **ferromagnetism**.

A substance containing no unpaired electrons is repelled by the magnetic field and is said to be **diamagnetic**.

Table 28.4 shows the unpaired electrons in ions of first row *d*-block elements.

Fig. 28.1 shows a graph of the relative paramagnetic moments of ions in the first row against the number of unpaired electrons.

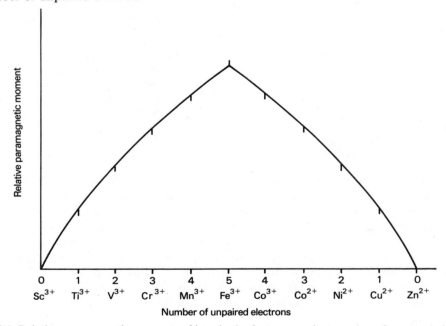

**Fig. 28.1** Relative paramagnetic moments of ions in the first row against number of unpaired electrons

## 28.6  FORMATION OF COORDINATION COMPOUNDS

A **coordination compound** is formed when a number of molecules or negatively charged ions combine with a central *d*-block atom or ion to form a complex ion or molecule. The molecules or ions which bond onto the central metal ion or atom are called **ligands** and the type of bonding involved is **coordinate** (or dative) bonding (Unit 4.7). For example, when a copper(II) compound such as anhydrous copper(II) sulphate is dissolved in water a hexaaquacopper(II) ion is formed.

$$Cu^{2+} + 6H_2O(l) \rightarrow \left[ \begin{array}{c} \text{(hexaaquacopper structure)} \end{array} \right]^{2+}$$

The water molecules are the ligands. Each water molecule acts as a Lewis base (Unit 16.2) donating a pair of electrons to form a coordinate bond. Energy is evolved during this process due to the formation of six new bonds.

Another example of a complex ion is hexacyanoferrate(II) ion. When potassium cyanide solution is added to a solution of iron(II) sulphate, a precipitate of iron(II) cyanide is produced which redissolves to form a solution containing hexacyanoferrate(II) ion.

$$FeSO_4(aq) + 2KCN(aq) \rightarrow Fe(CN)_2(s) + K_2SO_4(aq)$$

$$Fe(CN)_2(s) + 4CN^-(aq) \rightarrow \left[ \begin{array}{c} \text{(hexacyanoferrate structure)} \end{array} \right]^{4-}$$

hexacyanoferrate(II) ion

The hexacyanoferrate(II) ion was previously called the ferrocyanide ion. This should be distinguished from the hexacyanoferrate(III) or ferricyanide which is similar in structure but contains iron in oxidation state $+3$.

In these complex ions the cyanide ($CN^-$) ions are the ligands. In the example hexaaquacopper(II) ion, the resulting complex ion is positively charged but in hexacyanoferrate(II) and hexacyanoferrate(III) the resulting ions are negatively charged. The resulting complexes can be neutral, *e.g.* $Ni(CO)_4$ tetracarbonylnickel(0) or positively charged (called **cationic complexes**) or negatively charged (called **anionic complexes**).

$Fe^{2+}$ ions contain six electrons in the $3d$ sub-shell, and require 12 electrons to achieve the noble gas electron arrangement of krypton. Twelve electrons are supplied by accepting six lone pairs. In all complexes a similar tendency to achieve a noble gas electron arrangement exists, but this is not always achieved.

The properties of a complex ion are different from the simple ions from which it is composed. For example, it is impossible to obtain positive tests for $Fe^{2+}$ and $CN^-$ ions in the hexacyanoferrate(II) complex. Instead, there are distinct tests for the hexacyanoferrate(II) ion. When copper(II) sulphate solution is added to a solution containing hexacyanoferrate(II) ions, a brown precipitate of copper hexacyanoferrate(II) is produced.

$$2Cu^{2+}(aq) + Fe(CN)_6^{4-} (aq) \rightarrow CuFe(CN)_6(s)$$

Ligands such as $H_2O$, $NH_3$ and $Cl^-$ form only a single coordinate bond with a central metal ion. They are said to be **unidentate**. Some ligands are able to form two or more coordinate bonds to the central metal ion. They are said to be **polydentate**. Some ligands such as EDTA can form as many as six coordinate bonds with the central metal ion. The resulting complexes are called **chelates** or **chelated complexes**.

Some polydentate ligands are

ethanedioate
ion

butanedione dioxime
(dimethylglyoxime)

EDTA

There is a system for naming complexes. Negatively charged ligands end in '-o'.
*E.g.* chloro- $Cl^-$, cyano- $CN^-$
Neutral ligands include water and ammonia. The name **aqua-** is used for water and **ammine** for ammonia. (NB It is important to distinguish **ammine** from **amine**. Amines (Unit 35) are organic derivatives obtained by replacing hydrogens of ammonia with alkyl and aryl groups.)

If the complex ion is negatively charged (anionic complex) the name ends in '-ate', *e.g.* hexacyanoferrate(II). (NB Some metals are given the Latin name, *e.g.* iron – ferrate, lead – plumbate, silver – argentate.)

The number in Roman numerals is the oxidation state of the central metal ion. In hexacyanoferrate(II), the central metal ion is $Fe^{2+}$ in oxidation state $+2$.

Where the complex is positively charged or neutral, the name does not end in -ate, *e.g.* tetracarbonylnickel(0) and hexaaquacopper(II).

The shape of a complex is determined by the coordination number of the central metal atom or ion. The coordination number is the number of ligands which will bond to a central metal atom or ion. In hexacyanoferrate(II) and hexaaquacopper(II) ions the coordination number is 6 but in tetracarbonylnickel(0) the coordination number is 4. The common coordination numbers are 2, 4 and 6.

### Coordination number 2

The complex is linear, *e.g.* dichlorocuprate(I) ion $[CuCl_2]^-$

The central ion is *sp* hybridized. $[Cl \rightarrow Cu \leftarrow Cl]^-$

### Coordination number 4

The usual shape for the complex ion is tetrahedral, *e.g.* tetracarbonylnickel(0).

$$
\begin{array}{c}
O \\
C \\
\downarrow \\
Ni \leftarrow CO \\
\nearrow \quad \nwarrow \\
C \qquad C \\
O \qquad O
\end{array}
$$

The central atom is $sp^3$ hybridised. In cases where the central metal ion has a $d^8$ arrangement, the shape is square planar, *e.g.* tetracyanonickelate(II)

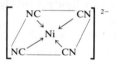

The central ion is $dsp^2$ hybridized.

### Coordination number 6

The coordination number 6 is extremely common and the octahedral shape is preferred, *e.g.* hexacyanoferrate(II), with $d^2sp^3$ hybridization of the central ion.

Ligands may change during a reaction and, rarely, the coordination number may change.

*E.g.*
$$[Co(H_2O)_6]^{2+}(aq) + 4Cl^-(aq) \rightleftharpoons [CoCl_4]^{2-}(aq) + 6H_2O(l)$$

| octahedral | tetrahedral |
| :---: | :---: |
| pink | blue |
| hexaaquacobalt(II) ion | tetrachlorocobaltate(II) |

In this case the water ligands are replaced by chloride ligands and the coordination number changes from 6 to 4.

In the sections 28.7 – 28.13 some of the *d*-block elements will be considered separately. As you work through these sections you will see some of the above principles in operation.

### 28.7 CHROMIUM

Chromium is a hard bluish-white metal used to electroplate steel (*e.g.* bicycle handlebars) because it is shiny and prevents corrosion.

It reacts with steam at red heat to produce chromium(III) oxide:

$$2Cr(s) + 3H_2O(g) \rightarrow Cr_2O_3(s) + 3H_2(g)$$

Chromium reacts slowly with dilute hydrochloric acid to produce chromium(II) chloride.

$$Cr(s) + 2HCl(aq) \rightarrow CrCl_2(aq) + H_2(g)$$

Chromium can be extracted from chromium(III) oxide by heating a mixture of chromium(III) oxide and aluminium powder.

$$Cr_2O_3(s) + 2Al(s) \rightarrow Al_2O_3(s) + 2Cr(s)$$

Chromium has an electronic arrangement $[Ar]3d^54s^1$. There are three possible oxidation states ($+2$, $+3$ and $+6$).

### Chromium(II)

This is an unstable oxidation state and is readily oxidized to chromium(III). Chromium(II) oxide, CrO, is a basic oxide.

## Chromium(III)

Chromium(III) is the most common oxidation state of chromium. If a chromium(III) salt contains water of crystallization the crystals are usually violet.

In a solution of a chromium(III) salt, the hexaaquachromium(III) ion $[Cr(H_2O)_6]^{3+}$ exists. This complex ion readily loses protons due to the high charge density of the $Cr^{3+}$ ion (see Unit 23.5).

$$[Cr(H_2O)_2(OH)_4]^- \xrightarrow{\text{addn of } OH^-} [Cr(H_2O)_3(OH)_3] \xrightarrow{\text{addn of } H_3O^+} [Cr(H_2O)_4(OH)_2]^+$$

The action of ammonia solution on a chromium(III) salt solution is different. The chromium(III) hydroxide dissolves in excess ammonia solution to form a yellow solution of hexamminechromium(III) ion, $[Cr(NH_3)_6]^{3+}$.

Chromium(III) salts are oxidized to chromium(VI) by warming a solution with hydrogen peroxide in an alkaline solution:

$$2Cr^{3+}(aq) + 5H_2O_2(aq) \rightarrow 2CrO_4^{2-}(aq) + 10H^+(aq) + O_2(g)$$
$$\text{chromate(VI) yellow}$$

## Chromium(VI)

Chromium(VI) compounds are strong oxidizing agents. The two common chromium(VI) compounds are potassium chromate(VI) ($K_2CrO_4$ containing the $CrO_4^{2-}$ ions) and potassium dichromate(VI) ($K_2Cr_2O_7$ containing $Cr_2O_7^{2-}$ ions).

The chromate(VI) and dichromate(VI) ions are both in the same oxidation state and are readily interconverted. If an acid is added to yellow chromate(VI) solution, the orange dichromate(VI) is produced.

$$2CrO_4^{2-}(aq) + 2H_3O^+(aq) \rightleftharpoons Cr_2O_7^{2-}(aq) + 3H_2O(l)$$

Addition of an alkali to a dichromate(VI) solution produces the chromate(VI). The structures of the two ions are:

chromate(VI)        dichromate(VI)

NB The arrangement of oxygens around each chromium atom is approximately tetrahedral.

The dichromate(VI) ion is a strong oxidizing agent in acid solution (Unit 6.3). The dichromate(VI) ion does not, however, oxidize hydrogen chloride to chlorine (c.f. with manganate(VII)).

Chromium(VI) oxide is an acidic oxide. It is produced as bright red, needle-shaped crystals when concentrated sulphuric acid is added to concentrated solutions of chromate(VI) or dichromate(VI):

$$CrO_4^{2-}(aq) + 2H_2SO_4(l) \rightarrow CrO_3(s) + 2HSO_4^-(aq) + H_2O(l)$$
$$Cr_2O_7^{2-}(aq) + 2H_2SO_4(l) \rightarrow 2CrO_3(s) + 2HSO_4^-(aq) + H_2O(l)$$

This chromium(VI) oxide dissolves in water to form chromic(VI) acid ('chromic acid')

$$CrO_3(s) + H_2O(l) \rightarrow H_2CrO_4(aq)$$

## 28.8   MANGANESE

Manganese is a hard, grey metal. It reacts readily with hot water and dilute hydrochloric acid.

$$Mn(s) + 2HCl(aq) \rightarrow MnCl_2(aq) + H_2(g)$$

(Some confusion exists especially with weaker candidates between manganese (Mn – a dense metal in the d-block) and magnesium (Mg – a less dense metal in group II).

Manganese has an electron arrangement $[Ar]3d^5 4s^2$ and can exist in oxidation states $+2$, $+3$, $+4$, $+6$ and $+7$.

## Manganese(II)

Manganese(II) is the most stable oxidation state corresponding to a $3d^5$ electron arrangement, *i.e.* 1 electron in each d orbital. Manganese(II) oxide is a basic oxide. Solutions of manganese(II) ions in water are pale pink due to the presence of $[Mn(H_2O)_6]^{2+}$ ions.

## Manganese(III)

This is an unstable oxidation state existing only in complexes. In acid solution manganese(III) disproportionates.

$$2Mn(III) \rightarrow Mn(IV) + Mn(II)$$

## Manganese(IV)

Manganese(IV) exists only in insoluble compounds (*e.g.* manganese(IV) oxide) and complexes.

Manganese(IV) oxide is a brown-black covalent oxide with feebly amphoteric properties. It is an oxidizing agent, oxidizing concentrated hydrochloric acid to chlorine:

$$MnO_2(s) + 4HCl(aq) \rightarrow MnCl_2(aq) + 2H_2O(l) + Cl_2(g)$$

## Manganese(VI)

The manganese(VI) oxidation state is an unstable oxidation state. The manganate(VI) ion (**manganate**) is produced when manganese(IV) oxide is fused with a strong oxidizing agent such as potassium chlorate(V):

$$3MnO_2(s) + KClO_3(l) + 6KOH(l) \rightarrow 3K_2MnO_4(l) + KCl(l) + 3H_2O(g)$$
$$\text{potassium manganate(VI)}$$

Potassium manganate(VI) is a green salt. In neutral or acid solution, it disproportionates to form manganese(IV) oxide and manganate(VII) (**permanganate**):

$$3MnO_4^{2-}(aq) + 2H_2O(l) \rightarrow 2MnO_4^-(aq) + MnO_2(s) + 4OH^-(aq)$$

Potassium manganate(VI) is produced when potassium manganate(VII) is heated.

$$2KMnO_4(s) \rightarrow K_2MnO_4(s) + MnO_2(s) + O_2(g)$$

The structure of the manganate(VI) ion is:

$$\left[ \begin{array}{c} \overset{\textstyle \ddot{O}}{\underset{\textstyle \ddot{O}}{\overset{\|}{\underset{\|}{\ddot{O}-Mn-\ddot{O}}}}} \end{array} \right]^{2-}$$

## Manganese(VII)

Manganese(VII) exists in the oxide $Mn_2O_7$ which is strongly acidic. Manganate(VII) can be produced directly from manganese(II) by warming with a strong oxidizing agent, *e.g.* sodium bismuthate $NaBiO_3$. The manganate(VII) (permanganate) ion is produced by acidification of manganate(VI) ions, or passing chlorine through a solution of potassium manganate(VI).

$$2MnO_4^{2-}(aq) + Cl_2(g) \rightarrow 2MnO_4^-(aq) + 2Cl^-(aq)$$

Potassium manganate(VII) is a powerful oxidizing agent in inorganic and organic chemistry (Unit 6.3).

## 28.9 IRON

### Extraction of iron

Iron exists in the earth's crust in ores including **haematite** $Fe_2O_3$, **magnetite** $Fe_3O_4$ and **iron pyrites** $FeS_2$. Iron is obtained by reducing the ores in a blast furnace.

The blast furnace is loaded with a charge of iron ore, coke and limestone through the top of the furnace. The furnace is heated by blasts of hot air into the base. Various reactions take place in the furnace.

(*i*) The burning of the coke in the air produces temperatures in excess of 1500°C.

$$C(s) + O_2(g) \rightarrow CO_2(g)$$

(*ii*) The reduction of carbon dioxide to carbon monoxide.

$$CO_2(g) + C(s) \rightarrow 2CO(g)$$

(*iii*) The reduction of iron ore takes place in the furnace. At the top of the furnace:

$$Fe_2O_3(s) + 3CO(g) \rightarrow 2Fe(l) + 3CO_2(g)$$
$$Fe_2O_3(s) + CO(g) \rightarrow 2FeO(s) + CO_2(g)$$

and lower in the furnace where the temperature is higher.

$$FeO(s) + C(s) \rightarrow Fe(l) + CO(g)$$

(*iv*) The limestone is added to the furnace to remove impurities of silicon(IV) oxide in the ore. The calcium carbonate decomposes to form calcium oxide.

$$CaCO_3(s) \rightarrow CaO(s) + CO_2(g)$$

(*v*) The calcium oxide reacts with silicon(IV) oxide to form calcium silicate (**slag**).

$$CaO(s) + SiO_2(s) \rightarrow CaSiO_3(l)$$

The slag floats on top of the molten iron and both can be tapped off separately. The iron produced is called pig iron and contains about 4% carbon plus other impurities including phosphorus, silicon and manganese.

Most of the pig iron is converted into steel. The molten iron is loaded into a furnace and a blast of hot air or oxygen is blown through the iron to burn off the impurities. Calculated quantities of carbon and other elements are then added to give steel of the desired composition.

Steel is basically iron with 0.5–1.5% carbon added. The properties of steel depend upon the percentage of carbon and other elements added. Steel containing 1.5% carbon is very hard. Steel containing chromium and nickel is called stainless steel and shows resistance to corrosion.

## Rusting of iron

The rusting of iron requires the presence of water and oxygen. The product, rust, is essentially a hydrated iron(III) oxide. The rusting process is accelerated by carbon dioxide and electrolytes such as salt.

Rust can be prevented by mechanical means such as painting and oiling which prevent the metal coming in contact with oxygen and water. Rusting can also be prevented by sacrificial protection and coating with metals.

## Properties of iron and its compounds

Iron is a grey metal which rusts in contact with moist air. It reacts with steam to produce iron(II) diiron(III) oxide (ferrosoferric oxide).

$$3Fe(s) + 4H_2O(g) \rightleftharpoons 3Fe_3O_4(s) + 4H_2(g)$$

It reacts with dilute hydrochloric and sulphuric acids to produce hydrogen.

$$Fe(s) + 2H_3O^+(aq) \rightarrow Fe^{2+}(aq) + H_2(g) + 2H_2O(l)$$

Compounds and complexes of iron exist in oxidation states $+2$ and $+3$. Since an iron atom has an electron arrangement [Ar] $3d^6\ 4s^2$, the electron arrangements in the two oxidation states are [Ar] $3d^6$ (Fe(II)) and [Ar]$3d^5$ (Fe(III)). Iron(III) is more stable than iron(II) because of the stability of the $d^5$ arrangement.

Both iron(II) and iron(III) ions form hexaaquacomplexes in aqueous solution.

$$\left[ \begin{array}{c} OH_2 \\ | \quad OH_2 \\ H_2O \rightarrow Fe \leftarrow OH_2 \\ H_2O \quad | \\ OH_2 \end{array} \right]^{2+} \rightarrow \left[ \begin{array}{c} OH_2 \\ | \quad OH_2 \\ H_2O \rightarrow Fe \leftarrow OH_2 \\ H_2O \quad | \\ OH_2 \end{array} \right]^{3+}$$

**very pale green**
**hexaaquairon(II)**

**yellow**
**hexaaquairon(III)**

The hexaaquairon(II) complex is relatively stable in acid solution, but in neutral or alkaline solution it is oxidized by air to hexaaquairon(III).

The hexaaquairon(III) complex is acidic in solution because of hydrolysis:

$$[Fe(H_2O)_6]^{3+}(aq) + H_2O(l) \rightleftharpoons [Fe(H_2O)_5(OH)]^{2+}(aq) + H_3O^+(aq)$$

It is important to know how to distinguish iron(II) and iron(III) by chemical tests. One method of distinguishing them is the action of aqueous solution of sodium hydroxide on solutions of iron(II) and iron(III) salts. With iron(II) salts, a green precipitate of iron(II) hydroxide is formed which slowly turns brown due to oxidation:

$$Fe^{2+}(aq) + 2OH^-(aq) \rightarrow Fe(OH)_2(s)$$

Iron(III) hydroxide is formed as a reddish brown precipitate when sodium hydroxide solution is added to a solution of an iron(III) salt.

$$Fe^{3+}(aq) + 3OH^-(aq) \rightarrow Fe(OH)_3(s)$$

Another method of distinguishing the two is to add a solution of potassium thiocyanate (KCNS) to solutions of iron(II) and iron(III) ions. A blood-red complex is formed with iron(III), and no colouration with the iron(II).

When potassium hexacyanoferrate(II) (potassium ferrocyanide) is added to a solution containing iron(III) ions, a dark blue or Prussian blue precipitate is formed. This is a good test for iron(III) ions as iron(II) does not form this precipitate.

$$K^+(aq) + [Fe(CN)_6]^{4-}(aq) + Fe^{3+}(aq) \rightarrow KFe[Fe(CN)_6]$$

potassium iron(III) hexacyano-
ferrate(II)

A similar precipitate is formed, however, if potassium hexacyanoferrate(III) (potassium ferricyanide) is added to a solution containing iron(II) ions. This occurs in two steps. The iron(II) ions are oxidized to iron(III) ions while the potassium hexacyanoferrate(III) is reduced to potassium hexacyanoferrate(II). Then the precipitation occurs as before.

$$Fe^{2+}(aq) + [Fe(CN)_6]^{3-}(aq) \rightarrow Fe^{3+}(aq) + [Fe(CN)_6]^{4-}(aq)$$
$$K^+(aq) + [Fe(CN)_6]^{4-}(aq) + Fe^{3+}(aq) \rightarrow KFe[Fe(CN)_6]$$

The distinction between a **complex** or **coordinate compound** and a **double salt** is also important. Iron(II) ammonium sulphate, $FeSO_4.(NH_4)_2SO_4.7H_2O$ is a double salt. It is prepared by mixing solutions of iron(II) sulphate and ammonium sulphate in the correct proportions. The solution is then crystallized. Only one type of crystal is obtained; it contains $Fe^{2+}$, $SO_4^{2-}$, and $NH_4^+$ ions and water molecules. The double salt behaves as you would expect, each ion being detectable. Potassium aluminium sulphate (called alum) $KAl(SO_4).12H_2O$ is another double salt.

A complex salt such as $K_4[Fe(CN)_6]$, potassium hexacyanoferrate(II), does not give all of the tests for the metal present. In this case, addition of sodium hydroxide solution would not precipitate iron(II) hydroxide.

### 28.10  COBALT

Cobalt does not react with air or water at room temperature. It reacts very slowly with dilute hydrochloric and sulphuric acids.

$$Co(s) + 2HCl(aq) \rightarrow CoCl_2(aq) + H_2(g)$$
$$\text{cobalt(II) chloride}$$

The electron arrangement of cobalt is $[Ar]3d^74s^2$. It can form compounds in oxidation states $+2$ and $+3$. The $+3$ oxidation state is stable in complexes, and cobalt(III), which reduces readily to cobalt(II), is a strong oxidizing agent.

In aqueous solution cobalt(II) compounds exist as hexaaquacobalt(II) ions $[Co(H_2O)_6]^{2+}$. This solution is pink in colour. On addition of sodium hydroxide solution, a blue precipitate of cobalt(II) hydroxide is formed which does not dissolve in excess sodium hydroxide solution.

$$[Co(H_2O)_6]^{2+}(aq) + 2OH^-(aq) \rightarrow [Co(H_2O)_4(OH)](s) + 2H_2O(l)$$

Addition of ammonia solution to hexaaquacobalt(II) ions precipitates cobalt(II) hydroxide which dissolves in excess forming a hexaammine complex.

$$[Co(H_2O)_6]^{2+}(aq) + 6NH_3(aq) \rightarrow [Co(NH_3)_6]^{2+}(aq) + 6H_2O(l)$$
$$\text{hexaamminecobalt(II) ion}$$
$$\text{pale yellow.}$$

On passing air through this solution or by adding hydrogen peroxide, the cobalt in the complex is oxidized from $+2$ to $+3$.

$$[Co(NH_3)_6]^{2+}(aq) \rightarrow [Co(NH_3)_6]^{3+}(aq) + e^-$$

In the hexaamminecobalt(III) ion, the central cobalt ion has achieved the noble gas electron arrangement of krypton. For this reason it is more stable than hexaamminecobalt(II).

When concentrated hydrochloric acid is added to a solution of cobalt(II) chloride, the solution turns blue due to the formation of the tetrachlorocobaltate(II) ion.

$$\begin{bmatrix} & OH_2 & OH_2 \\ H_2O \rightarrow & Co & \leftarrow OH_2 \\ H_2O & OH_2 \end{bmatrix}^{2+} (aq) + 4Cl^-(aq) \rightleftharpoons \begin{bmatrix} & Cl \\ & Co \leftarrow Cl \\ Cl & Cl \end{bmatrix}^{2-} (aq) + 6H_2O(l)$$

*octahedral*                                    *tetrahedral*

One reason given for this change in coordination number is that the chloride ion is larger than a water molecule, and only four will fit around the central cobalt ion.

### 28.11  NICKEL

Nickel as a metal resembles cobalt in its chemical properties. Nickel has an electron arrangement of $[Ar]3d^84s^2$ and can form compounds in oxidation states $+2$, $+3$ and $+4$. Of these $+2$ is far more stable.

Nickel can form complexes in oxidation state 0, *e.g.* tetracarbonylnickel(0)

$$\begin{matrix} & O \\ & C \\ & \downarrow \\ & Ni \leftarrow CO \\ C & & C \\ O & & O \end{matrix}$$

In this complex nickel has received eight electrons, sufficient to give nickel the electron arrangement of krypton. Tetracarbonylnickel(0) is a key compound in the production of metallic nickel.

Aqueous solutions of nickel(II) salts contain the hexaaquanickel(II) ion $[Ni(H_2O)_6]^{2+}(aq)$, and this solution is green. On addition of sodium hydroxide solution a green precipitate of nickel(II) hydroxide is formed.

$$[Ni(H_2O)_6]^{2+}(aq) + 2OH^-(aq) \rightarrow Ni(H_2O)_4(OH)_2(aq) + 2H_2O(l)$$

Addition of ammonia solution to an aqueous solution of nickel(II) salt precipitates nickel(II) hydroxide, which redissolves to form the bluish violet $[Ni(NH_3)_6]^{2+}(aq)$ complex.

Square planar complexes are known with nickel(II) as the central ion. An example is tetracyanonickelate(II), $[Ni(CN)_4]^{2-}(aq)$.

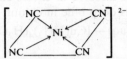

A common test for nickel(II) ions in solution is the addition of a solution of butanedione dioxime (dimethylglyoxime) in slightly alkaline conditions. This produces a pink precipitate.

## 28.12 COPPER

A copper atom has an electron arrangement of $[Ar]3d^{10}4s^1$. Sometimes questions make a comparison between copper and, for example, the alkali metal potassium ($[Ar]4s^1$). Both contain a single electron in a $4s$ orbital. However, in copper there are extra protons in the nucleus and the poor screening of the $3d$ orbitals decreases the radius of the atom. As a result, the first ionization energy of copper is much greater than that of potassium. Copper is, therefore, much less reactive than potassium.

There are two common oxidation states of copper $+1$ and $+2$. The $+2$ oxidation state is more stable than the $+1$ because the higher charge produced gives stronger bonding and this compensates for the extra energy required to remove an electron from the $3d$ orbital in addition to the electron from the $4s$ orbital.

### Copper(I)

Copper(I) is unstable in aqueous solution and exists only

(*i*) at high temperatures, (*ii*) when insoluble and precipitated, (*iii*) in complexes.

In aqueous solution, copper(I) compounds disproportionate, *e.g.* a solution of copper(I) sulphate:

$$Cu_2SO_4(aq) \rightarrow Cu(s) + CuSO_4(aq)$$
$$Cu(I) \rightarrow Cu(0) + Cu(II)$$

The copper(I) ion can be stabilized in solution by adding concentrated hydrochloric acid to form a complex.

$$CuCl(s) + Cl^-(aq) \rightarrow [CuCl_2]^-(aq)$$
$$\text{dichlorocuprate(I)}$$

Copper(I) chloride is prepared by boiling copper(II) chloride with copper and concentrated hydrochloric acid. The tetrachlorocuprate(II) ion is reduced to the dichlorocuprate(I) ion.

$$[Cu(H_2O)_6]^{2+}(aq) + 4Cl^-(aq) \rightarrow [CuCl_4]^{2-}(aq) + 6H_2O(l)$$
$$[CuCl_4]^{2-}(aq) + Cu(s) \rightarrow 2[CuCl_2]^-(aq)$$

When this solution is poured into water a white precipitate of copper(I) chloride is formed.

$$[CuCl_2]^-(aq) \rightarrow CuCl(s) + Cl^-(aq)$$

Copper(I) iodide is precipitated as a white precipitate (coloured by the iodine also formed) when potassium iodide solution is added to copper(II) sulphate solution:

$$2CuSO_4(aq) + 4KI(aq) \rightarrow 2CuI(s) + 2K_2SO_4(aq) + I_2(aq)$$

ionic equation:  $$2Cu^{2+}(aq) + 4I^-(aq) \rightarrow 2CuI(s) + I_2(aq)$$

Some of the iodide ions reduce the copper(II) to copper(I). The iodide ions are oxidized to iodine. This is an extremely important reaction for A level students. It frequently appears and many candidates in ignorance write:

$$CuSO_4 + 2KI \rightarrow CuI_2 + K_2SO_4$$

This reaction between copper(II) and iodide ions can be used in volumetric analysis to estimate the concentration of copper(II) by titration of the resulting solution with standard sodium thiosulphate solution.

Copper(I) oxide, $Cu_2O$, is precipitated as a red precipitate when an alkaline solution of copper(II) sulphate is reduced by glucose (or an aldehyde) on warming, this is Fehling's test.

The copper(II) ions are complexed with 2,3-dihydroxybutane dioates (tartrates) to prevent precipitation of copper(II) hydroxide.

$$2Cu^{2+}(aq) + 2OH^-(aq) \rightarrow Cu_2O(s) + H_2O(l)$$

**Copper(II)**

This is the more stable oxidation state of copper. Solutions of copper(II) ions in water are blue due to the presence of hexaaquacopper(II) ion $[Cu(H_2O)_6]^{2+}$.

Copper(II) hydroxide is precipitated as a pale blue precipitate when sodium hydroxide solution or ammonia solution is added to an aqueous solution of hexaaquacopper(II) ions

$$[Cu(H_2O)_6]^{2+}(aq) + 2OH^-(aq) \rightarrow [Cu(H_2O)_4(OH)_2](s) + 2H_2O(l)$$

The copper(II) hydroxide does not dissolve in excess sodium hydroxide solution to form the tetraamminecopper(II) ions. This is deep blue in colour.

$$[Cu(H_2O)_4(OH)_2](s) + 4NH_3(aq) \rightarrow [Cu(NH_3)_4(H_2O)_2]^{2+}(aq) + 2OH^-(aq) + 2H_2O(l)$$

### 28.13    Isomerism of complex ions

Examples of geometric (or *cis-trans* isomers) and of structural isomers will be found in Question 4 at the end of this Unit. Optical isomerism is possible where three bidentate ligands are coordinated with a central metal ion.

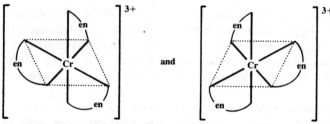

en represents 1,2-diaminoethane $NH_2CH_2CH_2NH_2$

These two forms are mirror images and have differing effects on plane polarized light (see Unit 7.12)

1 (*a*) For the compounds whose formulae are

write down         $Mn_2O_7 \; Cu_2O \; CrO_3 \; FeO \; V_2O_5 \; P_2O_3(P_4O_6)$

(*i*) two formulae which represent basic oxides

(*ii*) four formulae which represent acidic oxides giving reasons for both answers (*i*) and (*ii*).

(*b*) Name and give an ionic formula for

(*i*) a four coordinated complex of Cu(I)

(*ii*) a six coordinated complex of Fe(III)

Sketch diagrams to show the bond type and shape of each complex.

(*c*) The ion $VO_2^+$(aq) is reduced by sulphur dioxide to the ion $VO^{2+}$(aq) in acidic solution. Balance the half equations or partial ionic equations.

(*i*) $VO_2^+(aq) \rightarrow VO^{2+}(aq)$

and (*ii*) $SO_2(g) \rightarrow SO_4^{2-}(aq)$

Finally, construct the redox equation from the two half equations.

(*d*) Cite briefly FOUR general characteristics of transition metals.    (*Southern Universities Joint Board*)

(*a*) (*i*) Two basic oxides: $Cu_2O$, FeO

(*ii*) Four acidic oxides $Mn_2O_7 \; CrO_3 \; V_2O_5 \; P_2O_3(P_4O_6)$

Acidic oxides are oxides of non metals (*e.g.* P) or less electropositive metals in high oxidation states.

(Oxidation states in these oxides: Mn +7, Cr +6, V +5, Cu +1, Fe +2.)

(*b*) (*i*) A suitable four coordinated complex of copper(I) is tetracyanocuprate(I) ion, $[Cu(CN)_4]^{3-}$

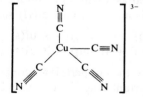

*tetrahedral*

(*ii*) A suitable six coordinated complex of iron(III) is hexacyanoferrate(III) ion, $[Fe(CN)_6]^{3-}$

*octahedral*

(c) (i) $$VO_2^+(aq) + 2H^+(aq) + e^- \rightarrow VO^{2+}(aq) + H_2O(l)$$

(ii) $$SO_2(g) + 2H_2O(l) \rightarrow SO_4^{2-}(aq) + 4H^+(aq) + 2e^-$$

Multiply equation (i) by 2 and add two equations together.

$$2VO_2^+(aq) + SO_2(g) \rightarrow 2VO^{2+}(aq) + SO_4^{2-}(aq)$$

(d) Four characteristics of transition metals

*1* Formation of coloured compounds, *e.g.* copper(II) compounds (hydrated) are blue.

*2* Variable oxidation states, *e.g.* copper has oxidation states +1 and +2.

*3* Shows paramagnetism due to unpaired electrons, *e.g.* manganese(II) contains five unpaired electrons.

*4* Forms complex ions, *e.g.* hexacyanoferrate(III).

**2** (*a*) Four compounds A, B, C, and D have the following percentage composition by mass

| Compound | Co | Cl | NH₃ |
|---|---|---|---|
| | | | *NH₃* |
| A | 25.22 | 45.59 | 29.19 |
| B | 25.22 | 45.59 | 29.19 |
| C | 23.53 | 42.46 | 34.01 |
| D | 22.03 | 39.76 | 38.21 |

Calculate the empirical formula of A, B, C, and D

When $10^{-2}$ mol of compounds A and B were separately dissolved in water and treated with an excess of aqueous silver nitrate, $10^{-2}$ mol of silver chloride was precipitated.

When $10^{-2}$ mol of compound C was dissolved in water and treated with an excess of aqueous silver nitrate, $2 \times 10^{-2}$ mol of silver chloride was precipitated.

When $10^{-2}$ mol of compound D was similarly treated $3 \times 10^{-2}$ mol of silver chloride was precipitated. What conclusions can be drawn from these results concerning A, B, C and D?

(*b*) Describe what is seen of the following and write appropriate equations for the reactions which occur.

(*i*) A solution of iron(II) sulphate is added to an acidified solution of potassium manganate(VII) (permanganate).

(*ii*) A solution of potassium iodide is added to a solution of copper(II) sulphate.

(*iii*) A solution of barium chloride is added to a solution of potassium chromate(VI).

(relative atomic masses H = 1, N = 14, Cl = 35.5, Co = 59)

*(Associated Examining Board)*

(*a*) Empirical formula of A

| | Co | Cl | NH₃ |
|---|---|---|---|
| | *Co* | *Cl* | *NH₃* |
| **Percentage** | 25.22 | 45.59 | 29.19 |
| **divide by relative atomic mass (or relative molecular mass)** | 59 | 35.5 | 17 |
| | 0.42 | 1.28 | 1.71 |
| **Divide by smallest** | 1 | 3 | 4 |

Empirical formula of A is $CoCl_3(NH_3)_4$

Since B has identical composition, its empirical formula is $CoCl_3(NH_3)_4$.

Using the same type of calculation, the empirical formulae of C and D can be found

| C | $CoCl_3(NH_3)_5$ |
|---|---|
| D | $CoCl_3(NH_3)_6$ |

In A and B, $10^{-2}$ mol of AgCl is produced from $10^{-2}$ mol of A and B. Only one chloride ion is free to react with silver nitrate:

A and B $[CoCl_2(NH_3)_4]^+Cl^-$

In C, $2 \times 10^{-2}$ mol of silver chloride is produced from $10^{-2}$ mol of C, therefore C contains 2 free chloride ions

$$[CoCl(NH_3)_5]^{2+}2Cl^-$$

Similarly D contains three free chloride ions

$$[Co(NH_3)_6]^{3+}3Cl^-$$

A and B are isomers (Unit 7.10). They have the same molecular formula but the atoms are differently arranged. Isomerism is common in transition metal complexes

A and B

*cis*                   *trans*

C                         D

(*b*) In this part of the question it is important to describe what is seen and this is often missed by candidates.

(*i*) (see Unit 6)    $MnO_4^-(aq) + 8H^+(aq) + 5e^- \rightarrow Mn^{2+}(aq) + 4H_2O(l)$

$$Fe^{2+}(aq) \rightarrow Fe^{3+}(aq) + e^-$$

$$MnO_4^-(aq) + 8H^+(aq) + 5Fe^{2+}(aq) \rightarrow Mn^{2+}(aq) + 4H_2O(l) + 5Fe^{3+}(aq)$$

On addition of iron(II) sulphate, the purple colour is removed when all the potassium manganate(VII) is used up. The products are almost colourless.

(*ii*)    $$2Cu^{2+}(aq) + 4I^-(aq) \rightarrow 2CuI(s) + I_2(aq)$$

When potassium iodide solution is added to blue copper(II) sulphate solution, a precipitate of copper(I) iodide is formed. The precipitate is white but is coloured brown by the iodine produced.

(*iii*) This is not included in Unit 28. When a solution of barium chloride is added to a yellow solution of potassium chromate(VI), a bright yellow precipitate of barium chromate(VI) is produced:

$$BaCl_2(aq) + K_2CrO_4(aq) \rightarrow BaCrO_4(s) + 2KCl(aq)$$

**3** A transition metal M forms a complex ion $[M(NH_3)_n(H_2O)_{6-n}]^{2+}$ with ammonia. An experiment was carried out to determine the composition of the complex.

Aqueous solutions of ammonia and sulphate of M (both 1 mol dm$^{-3}$) were mixed in varying proportions. Any precipitate formed on mixing the solutions was removed by filtration, dried and weighed. The results are shown in Fig. 28.2.

(*i*) Write an equation for the reaction between ammonia solution and the metal ion solution to produce a precipitate.

(*ii*) From the graph, what is the ratio (mols of $NH_3$ : mols of $M^{2+}$) when all the precipitate dissolves in the ammonia solution?

(*iii*) Suggest a formula for the complex ion which is formed and indicate a likely structure.

(*iv*) Name a metal ion which would behave in this way.

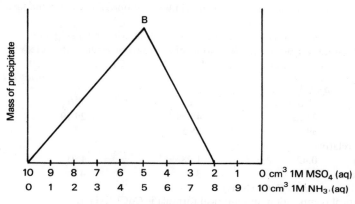

**Fig. 28.2** Question

(*i*)    $$[M(H_2O)_6]^{2+}(aq) + nNH_3(aq) \rightarrow [M(NH_3)_n(H_2O)_{6-n}]^{2+}(aq) + nH_2O(l)$$

(*ii*) Precipitate completely dissolves when

$2cm^3$ of $MSO_4$ reacts with $8cm^3$ of ammonia solution.
ratio of no. of mols of ammonia to no. of mols of $M^{2+}$
$= 4:1$

(*iii*) $[M(NH_3)_4(H_2O)_2]^{2+}$

(*iv*) $Cu^{2+}$ ion would behave in this way.

**4** Properties of chromyl chloride: formula $Cr?_2Cl_2$; boiling point 116°C; red liquid, hydrolysed rapidly in contact with water.
Chromyl chloride is prepared by distilling a mixture of potassium dichromate (VI), sodium chloride and concentrated sulphuric acid. The equation is:

$K_2Cr_2O_7(s) + 4NaCl(s) + 3H_2SO_4(l) \rightarrow 2CrO_2Cl_2(l) + 3H_2O(l) + 2Na_2SO_4(s) + K_2SO_4(s)$

(*a*) Write the simplest ionic equation for the reaction taking place.

(*b*) What change in oxidation state of chromium, if any, occurs during the reaction?

(*c*) (*i*) Draw a labelled diagram of apparatus suitable for preparing a sample of chromyl chloride in the laboratory. Include in your diagram: **1** a suitable means of mixing the reagents safely; **2** a method of removing chromyl chloride from the final mixture.

(*ii*) What precautions could be taken to prevent chromyl chloride hydrolysing during during the preparation?

$$CrO_2Cl_2$$

↓ + H₂O/steamy fumes produced

solution

↓ neutralize with NaOH (aq)

yellow solution → add silver nitrate soln. → red precipitate

Name: (*i*) the steamy fumes when water is added to chromyl chloride;
(*ii*) the yellow solution;
(*iii*) the red precipitate.

(*a*) $Cr_2O_7^{2-} + 4Cl^- + 6H^+ \rightarrow 2CrO_2Cl_2 + 3H_2O$
(*b*) No change −oxidation state +6 throughout.
(*c*) (*i*)

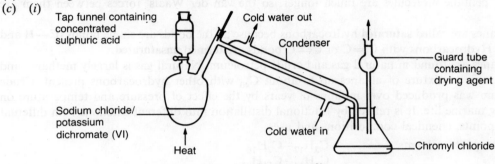

**Fig. 28.3** Question

(*ii*) Dry reagents, dry apparatus in oven before use, guard tubes containing drying agent to prevent water entering.
(*d*) (*i*) Hydrogen chloride HCl
(*ii*) chromate ions, $CrO_4^{2-}$, or sodium chromate, $Na_2CrO_4$.
(*iii*) Silver chromate, $Ag_2CrO_4$.

# 29 Hydrocarbons

## 29.1 INTRODUCTION

Hydrocarbons are organic compounds composed of carbon and hydrogen only. In this Unit aliphatic and aromatic hydrocarbons will be discussed.

**Aliphatic hydrocarbons** consist of alkanes, alkenes, alkynes and cycloalkanes. **Aromatic hydrocarbons** include benzene, methylbenzene, naphthalene etc.

During a GCSE course you have probably studied some of these at an elementary level.

## 29.2 ALKANES

Alkanes are a series of hydrocarbons which all have molecular formulae which fit the general formula $C_nH_{2n+2}$. The simplest alkanes are shown in Table 29.1.

**Table 29.1 The alkanes**

| | | | |
|---|---|---|---|
| CH₄ | methane | C₆H₁₄ | hexane |
| C₂H₆ | ethane | C₇H₁₆ | heptane |
| C₃H₈ | propane | C₈H₁₈ | octane |
| C₄H₁₀ | butane | C₉H₂₀ | nonane |
| C₅H₁₂ | pentane | C₁₀H₂₂ | decane |

Any group of compounds which fit the general formula and differ only by different numbers of CH₂ units is called a **homologous series**. Within a homologous series the compounds have the same chemical properties and there are trends in physical properties such as increasing melting and boiling points. These increases are due to increasing relative molecular masses and van der Waals' forces (Unit 4.9).

In addition to the alkanes in Table 29.1, other isomers are possible with alkanes containing more than three carbon atoms. This isomerism is due to branching of chains. The possible isomers of $C_5H_{12}$ are

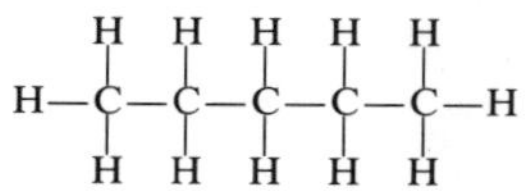

pentane
boiling point
36°C

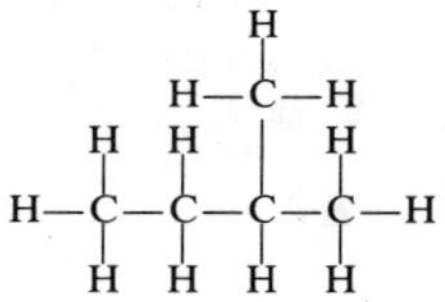

2-methylbutane
boiling point
28°C

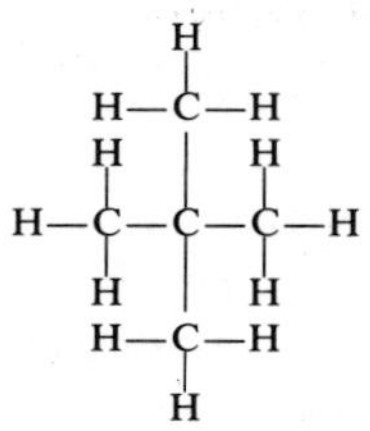

2,2 dimethylpropane
boiling point
9°C

These three isomers have identical relative molecular mass but differ in boiling point. This is because pentane molecules are much longer, so the van der Waals' forces between them are greater.

Alkanes are called **saturated** hydrocarbons because all the bonds are single bonds (C—H and C—C). Hydrocarbons with C=C or C≡C bonds are termed **unsaturated**.

Alkanes are found in natural gas and crude petroleum. Natural gas is largely methane, and petroleum is a mixture of alkanes up to about $C_{40}$ with other hydrocarbons present. Crude petroleum was produced over millions of years by the effect of pressure and temperature on decaying marine life. It is refined by fractional distillation into different fractions with different boiling points, chemical compositions and uses.

| Aviation spirit – solvents | $C_5H_{12} - C_7H_{16}$ |
|---|---|
| Petrol | $C_6H_{14} - C_{11}H_{24}$ |
| Paraffin | $C_{12}H_{26} - C_{16}H_{34}$ |
| Fuel oil | $C_{13}H_{28} - C_{18}H_{38}$ |
| Lubricating oils | $C_{19}H_{40} - C_{40}H_{82}$ |

The remaining fractions are used for heavy oils, greases and waxes.

## 29.3   REACTIONS OF ALKANES

Alkanes are comparatively unreactive and there are few chemical reactions of importance.

### (i)  Combustion of alkanes

Many of the uses of alkanes rely upon their ready combustion or oxidation. The products depend upon the amount of oxygen available.

*Unlimited supply of oxygen*

Complete combustion of alkanes produces carbon dioxide and water.

$E.g.$ $\qquad 2C_2H_6(g) + 7O_2(g) \rightarrow 4CO_2(g) + 6H_2O(l)$

*Limited supply of oxygen*

Carbon monoxide and water are produced. Carbon monoxide is very poisonous, which explains the danger of running a car engine in an enclosed space.

$E.g.$ $\qquad 2C_2H_6(g) + 5O_2(g) \rightarrow 4CO(g) + 6H_2O(l)$

### (ii)  Chlorination of alkanes

A mixture of methane and chlorine vapours subjected to ultraviolet light undergoes a series of substitution reactions. The mechanism of this reaction is important and is discussed in Unit 38.4.

$$CH_4 + Cl_2 \rightarrow CH_3Cl \qquad + HCl$$
$$\text{chloromethane}$$
$$CH_3Cl + Cl_2 \rightarrow CH_2Cl_2 \qquad + HCl$$
$$\text{dichloromethane}$$
$$CH_2Cl_2 + Cl_2 \rightarrow CHCl_3 \qquad + HCl$$
$$\text{trichloromethane}$$
$$CHCl_3 + Cl_2 \rightarrow CCl_4 + \qquad + HCl$$
$$\text{tetrachloromethane}$$

In practice a complex mixture of all products will be produced.

### (iii)  Cracking of hydrocarbons

Higher fractions can be cracked into shorter chain molecules by passing over a heated catalyst. The products of cracking include alkenes which are suitable for making polymers (Unit 36).

## 29.4 ALKENES

Alkenes are a homologous series of hydrocarbons containing a double bond. They all have molecular formulae which fit a general formula $C_nH_{2n}$.

The simplest members of the homologous series are

ethene $C_2H_4$

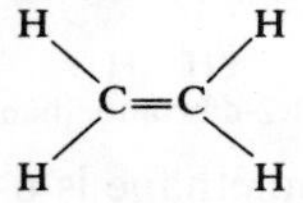

propene $C_3H_6$

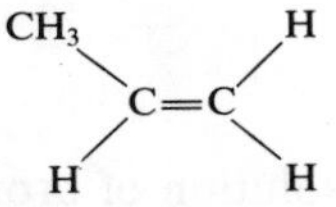

but-l-ene

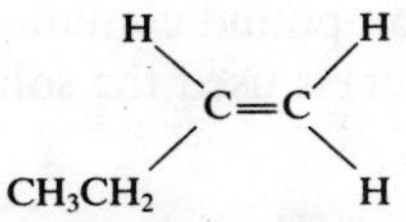

but-2-ene

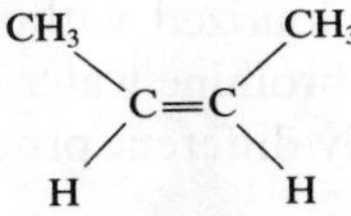

Alkenes are obtained by cracking alkanes derived from petroleum. In the laboratory, alkenes can be prepared by dehydration of alcohols or by dehydrohalogenation of haloalkanes (alkyl halides).

## 29.5 LABORATORY PREPARATION OF ALKENES

Ethene is prepared by the dehydration of ethanol

$$CH_3CH_2OH \rightarrow CH_2 = CH_2 + H_2O$$

The dehydration can be carried out using concentrated sulphuric acid added slowly in excess to ethanol. The flask containing the mixture is cooled during this step to reduce charring. Ethene is produced when this mixture is heated to 170°C.

$$CH_3CH_2OH + H_2SO_4 \rightarrow CH_3CH_2OSO_2OH + H_2O$$
$$CH_3CH_2OSO_2OH \rightarrow CH_2 = CH_2 + H_2SO_4$$

The mechanism for this process is discussed in Unit 38.

With excess ethanol present and at a lower temperature, ethoxyethane (an ether) is produced.

$$CH_3CH_2OSO_2OH + CH_3CH_2OH \rightarrow CH_3CH_2OCH_2CH_3$$

Less charring is obtained if concentrated phosphoric(V) acid is used. Alternatively the dehydration can be carried out by passing the ethanol vapour over heated aluminium oxide at about 350°C.

Alkenes can also be produced by dehydrohalogenation of haloalkanes by refluxing with a solution of potassium hydroxide dissolved in ethanol. The mechanism for this reaction will be found in Unit 38. It is an elimination reaction.

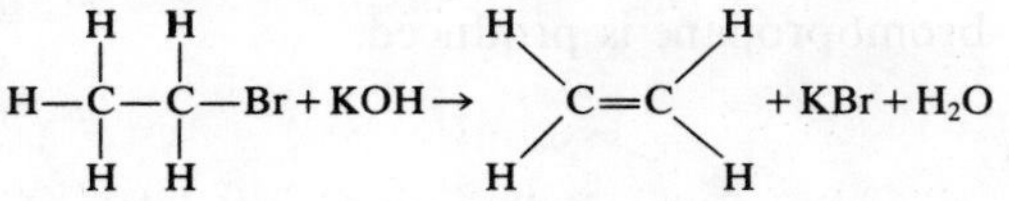

## 29.6 REACTIONS OF ALKENES

Most of the reactions of alkenes involve addition reactions to the carbon–carbon double bond. The mechanism called electrophilic addition is discussed in Unit 38.5.

### (i) Hydrogenation

When an alkene is mixed with hydrogen and passed over a platinum catalyst at room temperature or a nickel catalyst at between 140 and 200°C, an addition reaction takes place forming the corresponding alkane

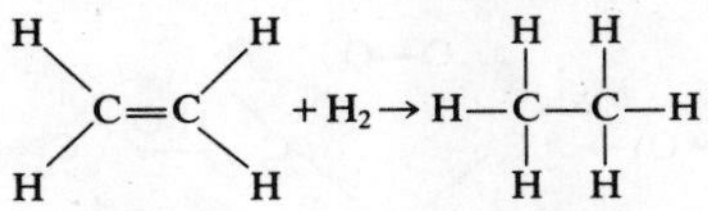

ethene          ethane

This process is important in the 'hardening' of vegetable and animal oils to produce solid fats used to make margarine. The oils are unsaturated and addition of hydrogen removes the double bonds.

### (ii) Hydration of alkenes

This is the reverse of the method used to produce ethene in the laboratory from ethanol and concentrated sulphuric acid. It is widely used to produce ethanol from ethene. The ethene is passed into almost concentrated sulphuric acid and then water is added to decompose the product.

$$CH_2 = CH_2 + H_2SO_4 \rightarrow CH_3CH_2O.SO_2OH \xrightarrow{+H_2O} CH_3CH_2OH + H_2SO_4$$

### (iii) Addition of a halogen

Chlorine and bromine are readily added to an alkene to form a dichloro- or dibromo- compound.

$$\begin{matrix} H \\ \diagdown \\ C=C \\ \diagup \\ H \end{matrix} \begin{matrix} H \\ \diagup \\ \diagdown \\ H \end{matrix} + Br_2 \rightarrow \begin{matrix} H & H \\ | & | \\ Br-C-C-Br \\ | & | \\ H & H \end{matrix}$$

1,2-dibromoethane

A solution of bromine in hexane or tetrachloromethane is used as a test for unsaturation. It is decolourized without heating when added to a compound containing a double or triple bond.

If bromine water (a solution of bromine in water) is used the solution is decolourized, but a slightly different product is obtained.

$$\begin{matrix} H \\ \diagdown \\ C=C \\ \diagup \\ H \end{matrix} \begin{matrix} H \\ \diagup \\ \diagdown \\ H \end{matrix} + Br_2/H_2O \rightarrow \begin{matrix} H & H \\ | & | \\ H-C-C-H \\ | & | \\ Br & OH \end{matrix} + HBr$$

2-bromoethanol

The mechanisms for these reactions are given in Unit 38.5.

### (iv) Addition of hydrogen halides

Hydrogen halides can be added to alkenes to produce haloalkanes.

$$\begin{matrix} H \\ \diagdown \\ C=C \\ \diagup \\ H \end{matrix} \begin{matrix} H \\ \diagup \\ \diagdown \\ H \end{matrix} + HBr \rightarrow \begin{matrix} H & H \\ | & | \\ H-C-C-Br \\ | & | \\ H & H \end{matrix}$$

ethene                      bromoethane

Ethene is a symmetrical alkene because the same types of atoms are bonded to both carbon atoms. If an asymmetrical alkene – one where one carbon atom is bonded to a different atom – is used, two products are possible:

$$\begin{matrix} CH_3 \\ \diagdown \\ C=C \\ \diagup \\ H \end{matrix} \begin{matrix} H \\ \diagup \\ \diagdown \\ H \end{matrix} + HBr \rightarrow \begin{matrix} CH_3 & H \\ | & | \\ Br-C——C-H \\ | & | \\ H & H \end{matrix}$$

2-bromopropane

$$\begin{matrix} CH_3 \\ \diagdown \\ C=C \\ \diagup \\ H \end{matrix} \begin{matrix} H \\ \diagup \\ \diagdown \\ H \end{matrix} + HBr \rightarrow \begin{matrix} CH_3 & H \\ | & | \\ H-C——C-Br \\ | & | \\ H & H \end{matrix}$$

1-bromopropane

When the reaction is carried out, the product is 2-bromopropane rather than 1-bromopropane. This is predicted by Markownikoff's rule (Unit 38.5). If the reaction is carried out in the presence of peroxides, 1-bromopropane is produced.

### (v) Oxidation of alkenes

*1* Alkenes burn in air or oxygen to form similar products to those from alkanes (Unit 29.3)

*e.g.*  
$$C_2H_4(g) + 3O_2(g) \rightarrow 2CO_2(g) + 2H_2O(l) \quad \text{excess oxygen}$$
$$C_2H_4(g) + 2O_2(g) \rightarrow 2CO(g) + 2H_2O(l) \quad \text{limited oxygen}$$

Combustion usually occurs with a slightly smoky flame caused by the greater proportion of carbon compared with alkanes.

*2* The reaction of alkenes with trioxygen (ozone) is important as it is the basis of the technique called **ozonolysis**. When trioxygen is passed into a solution of an alkene in an inert solvent, an ozonide is formed. When the ozonide is hydrolysed by boiling with water, aldehydes or ketones are produced (Unit 32).

$$\begin{matrix} \diagdown \\ C=C \\ \diagup \end{matrix} + O_3 \rightarrow \begin{matrix} O-O \\ \diagdown \quad \diagup \\ C \quad C \\ \diagdown \quad \diagup \\ O \end{matrix} \xrightarrow{H_2O} \begin{matrix} \diagdown \\ C=O \quad O=C \\ \diagup \end{matrix}$$

ozonide

An example of a question using ozonolysis will be found in the questions at the end of this unit.

*\*3* Alkenes are oxidised by an alkaline solution of potassium manganate(VII) (permanganate). The oxidation does not require heating and results in the formation of a diol or glycol.

$$\begin{matrix} H \\ \diagdown \\ C=C \\ \diagup \\ H \end{matrix} \begin{matrix} H \\ \diagup \\ \diagdown \\ H \end{matrix} + H_2O + [O] \rightarrow \begin{matrix} H & H \\ | & | \\ H-C-C-H \\ | & | \\ OH & OH \end{matrix}$$

ethane-1,2-diol  
(ethylene glycol)

Ethane-1,2-diol is the major ingredient of antifreeze used in car cooling systems, and is also used in making Terylene. The decolourization of an alkaline potassium manganate(VII) solution is another test for unsaturation.

When an alkene is treated with peroxotrifluoroethanoic acid, the corresponding alkene oxide or epoxide is formed:

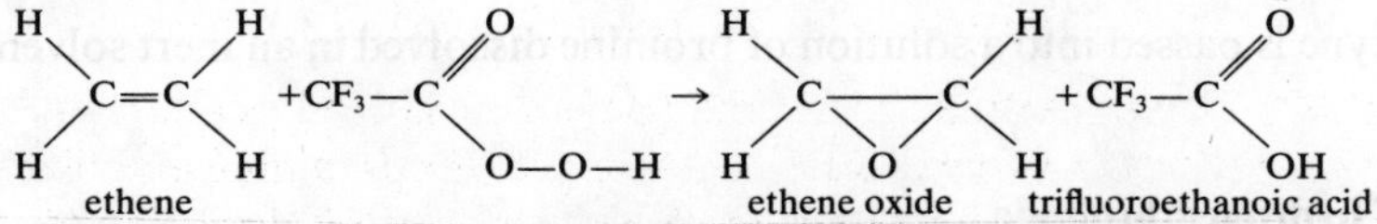

ethene ethene oxide trifluoroethanoic acid

Alkene oxides are useful organic intermediates. For example, warming ethene oxide with very dilute hydrochloric acid produces ethane-1,2-diol:

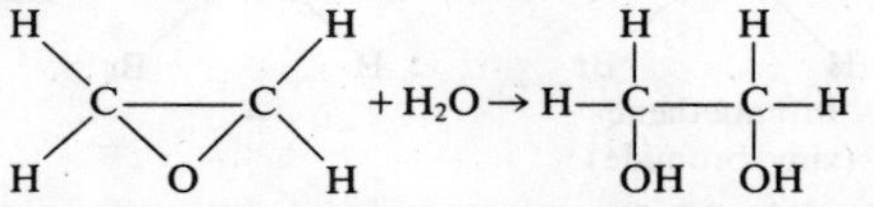

Alkene oxides can also be produced by passing a mixture of alkene and oxygen over a heated silver catalyst at 300°C.

**(vi) Polymerization of alkenes**

See Unit 36.

## 29.7 ALKYNES

The homolgous series of alkynes has a general formula $C_nH_{2n-2}$. All alkynes contain a triple bond between two carbon atoms. The simplest alkyne is called ethyne (sometimes called acetylene). It has a molecular formula of $C_2H_2$ and a structural formula of:

$$H-C\equiv C-H$$

## *29.8 LABORATORY PREPARATION OF ALKYNES

Ethyne can be prepared by the action of cold water on calcium dicarbide (calcium carbide)

$$CaC_2(s) + 2H_2O(l) \rightarrow Ca(OH)_2(s) + C_2H_2(g)$$

(This equation is frequently written by candidates as

$$CaC_2(s) + H_2O(l) \rightarrow CaO(s) + C_2H_2(g)$$

By writing this the candidate fails to appreciate that calcium oxide reacts with water).

Calcium dicarbide is manufactured by heating calcium oxide and carbon strongly in an electric furnace.

$$CaO(s) + 3C(s) \rightarrow CaC_2(s) + CO(g)$$

Other alkynes cannot be prepared by a similar method. The usual method is to reflux a dihaloalkane with a solution of potassium hydroxide in ethanol. A dehydrohalogenation reaction takes place to produce the alkyne. Two molecules of hydrogen halide are eliminated.

$$CH_3-CH(Br)-CH_2(Br) + 2KOH \rightarrow CH_3-C\equiv CH + 2KBr + 2H_2O$$

propyne

## 29.9 REACTIONS OF ALKYNES

Alkynes undergo addition reactions similar to those of alkenes. These reactions are electrophilic addition reactions (Unit 38.5). It is also possible for the terminal hydrogen atoms to be substituted.

**(i) Addition of hydrogen**

A mixture of hydrogen and alkyne is passed over a platinum catalyst at room temperature, or a nickel catalyst at 250°C. The reaction takes place in two stages.

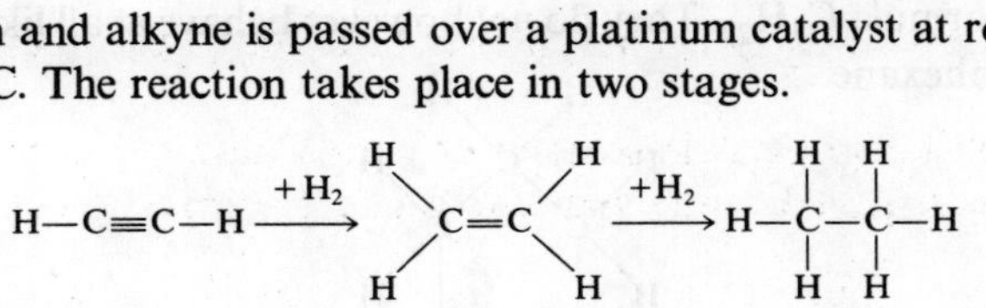

ethyne ethene ethane

The reaction usually goes directly to the alkane, and the alkene is not isolated. The reaction can be stopped at the alkene stage by adding poisons to the catalyst which prevent the second stage.

### (ii) Addition of bromine

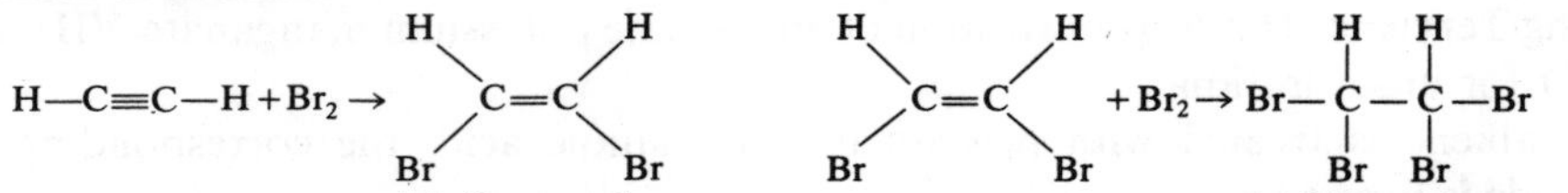

$$H—C≡C—H + Br_2 →$$

1,2-dibromoethene

$$+ Br_2 → Br—\underset{Br}{\overset{H}{C}}—\underset{Br}{\overset{H}{C}}—Br$$

1,1,2,2-tetrabromoethane

When an alkyne is passed into a solution of bromine dissolved in an inert solvent, the bromine is decolourized.

### (iii) Addition of hydrogen bromide

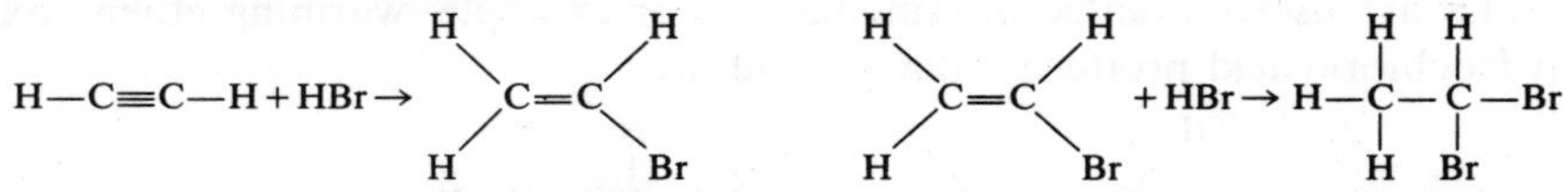

$$H—C≡C—H + HBr →$$

bromoethene
(vinyl bromide)

$$+ HBr → H—\underset{H}{\overset{H}{C}}—\underset{Br}{\overset{H}{C}}—Br$$

1,1-dibromoethane

(see Markownikoff's rule 38.5)

### (iv) Substitution reactions of alkynes

Ethyne differs from ethene because the hydrogen atoms in ethyne are slightly acidic. If ethyne is passed into an ammonical solution of silver nitrate, a white precipitate of silver dicarbide (silver acetylide) is produced.

$$H − C≡C − H + 2[Ag(NH_3)_2]^+ → Ag^+(C≡C)^{2-}Ag^+ + 2H^+ + 4NH_3$$

This reaction can be used to distinguish ethyne from ethene.

### (v) Reactions of ethyne with dilute sulphuric acid

When ethyne is passed into warm, dilute sulphuric acid in the presence of a catalyst, a reaction takes place producing ethanal.

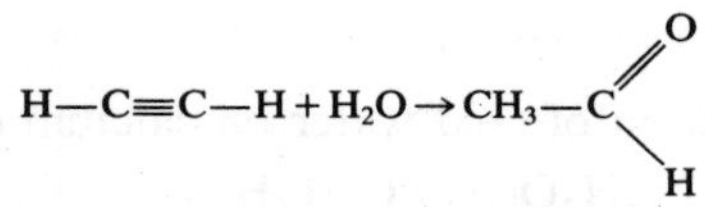

$$H—C≡C—H + H_2O → CH_3—\overset{O}{\underset{H}{C}}$$

ethanal

The catalyst is a mixture of mercury(II) and iron(III) salts.

### (vi) Conversion of ethyne to benzene

When ethyne gas is passed through a copper tube at about 300°C a reaction takes place. On cooling the gas, a colourless liquid condenses. This liquid is **benzene**, produced by joining three molecules of ethyne together.

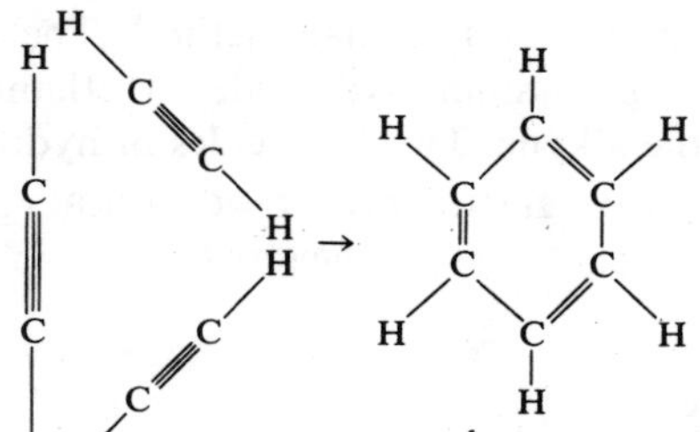

benzene

The copper tube acts as a catalyst.

### 29.10 CYCLOALKANES

It is possible to produce saturated hydrocarbons with a ring structure. These compounds, like alkenes, fit the general formula $C_nH_{2n}$. They do *not* however behave at all like alkenes. An example of a cycloalkane is cyclohexane.

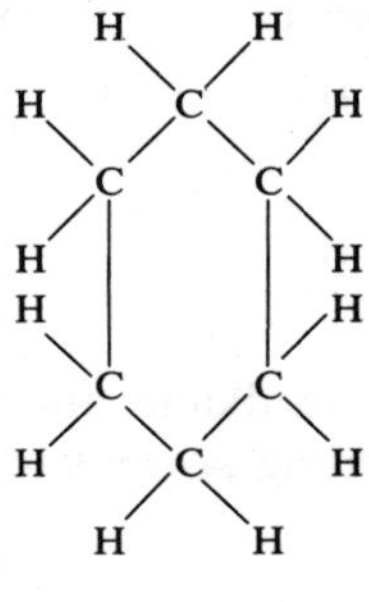

This molecule is not planar but can exist in two readily interconvertible forms. These forms are the chair and boat forms.

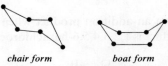

*chair form*      *boat form*

In these two forms the bond angles are close to the tetrahedral angle. Neither of these two forms can be isolated and the chair form, which allows the hydrogen atoms to be further apart, predominates.

The chemical reactions of cycloalkanes closely resemble those of the alkanes.

### 29.11 AROMATIC HYDROCARBONS

The most important aromatic hydrocarbon is benzene, $C_6H_6$. A benzene molecule consists of a planar ring of six carbon atoms. All of the carbon–carbon bond lengths are the same–0.139nm. Since the bond lengths in carbon–carbon single and double bonds are 0.154nm and 0.132nm respectively, it can be concluded that these bonds in benzene are intermediate between single and double bonds.

Benzene may be considered to be a resonance hybrid of a number of extreme structures such as the Kekulé formulae.

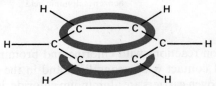

Neither of these extreme structures actually exists.

The modern approach to the structure of benzene explains why the carbon–carbon bonds are all the same length. It also explains the planar nature of the molecule, the bond angles and the fact that benzene does not decolourise a solution of bromine.

Each carbon atom is $sp^2$ hybridised forming three $sp^2$ hybrid orbitals in the same plane but at angles of 120°C. One of these $sp^2$ orbitals on each carbon atom overlaps with the 1s orbital of a hydrogen atom and the other two $sp^2$ orbitals overlap with $sp^2$ orbitals of other carbon atoms. This forms a planar hexagonal framework with one p orbital on each carbon atom unused and at right angles to the ring. These remaining p orbitals can overlap to form a ring of negative charge above and below the ring (Fig. 29.1).

**Fig. 29.1** Structure of benzene

It is usual to represent a benzene ring in a simplified way. The most common method is:

Do not forget that this molecule has six carbon atoms and six hydrogen atoms.

Other aromatic hydrocarbons include:

methylbenzene      naphthalene

### 29.12 REACTIONS OF BENZENE

#### (i) Addition reactions of benzene

Because the benzene ring structure is stable (Unit 14.2) addition reactions, which would result in the destruction of this system, do not readily take place.

If a mixture of hydrogen and benzene is passed over a nickel catalyst at 200°C, a series of reactions takes place forming cyclohexane.

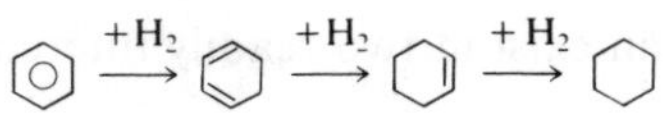

$$\bigcirc \xrightarrow{+H_2} \bigcirc \xrightarrow{+H_2} \bigcirc \xrightarrow{+H_2} \bigcirc$$

Chlorine and benzene react to form an addition product when the mixture of vapours is subjected to ultraviolet light. The product is 1,2,3,4,5,6-hexachlorocyclohexane.

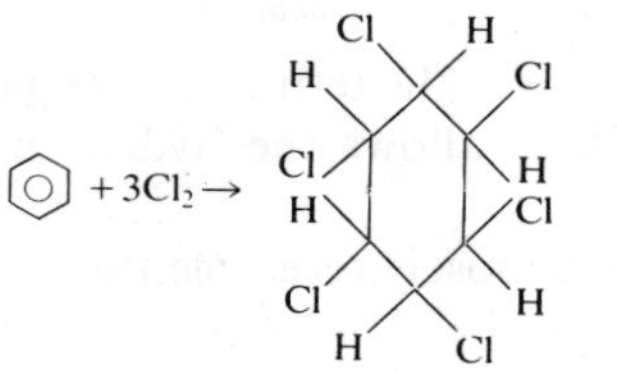

$$\bigcirc + 3Cl_2 \rightarrow$$

1,2,3,4,5,6-hexachlorocyclohexane
benzene hexachloride

## (ii) Substitution reactions of benzene

Benzene undergoes a range of substitution reactions. These reactions are examples of electrophilic substitutions and the mechanisms are discussed in Unit 38.

### Nitration of benzene

When benzene is refluxed with a mixture of concentrated nitric and sulphuric acids, nitrobenzene is produced.

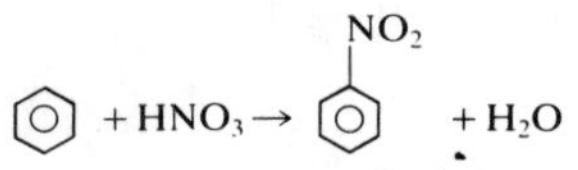

$$\bigcirc + HNO_3 \rightarrow \bigcirc NO_2 + H_2O$$

nitrobenzene

If the temperature rises above 55°C further reaction is possible.

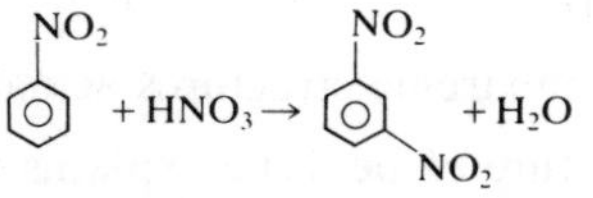

$$\bigcirc NO_2 + HNO_3 \rightarrow \bigcirc \begin{array}{c} NO_2 \\ NO_2 \end{array} + H_2O$$

1,3-dinitrobenzene

### Sulphonation of benzene

Refluxing benzene with concentrated sulphuric acid for 24 hours produces a benzenesulphonic acid. The reaction is, however, reversible.

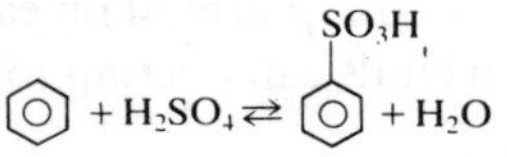

$$\bigcirc + H_2SO_4 \rightleftarrows \bigcirc SO_3H + H_2O$$

benzenesulphonic acid

### *Halogenation of benzene

Benzene undergoes substitution reactions with chlorine and bromine. The reaction takes place on heating the mixture out of contact with strong light and in the presence of catalysts called 'halogen carriers'. Suitable halogen carriers are aluminium chloride, iron(III) chloride and iodine.

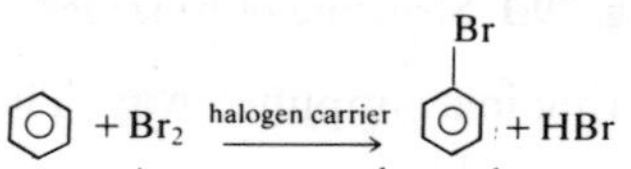

$$\bigcirc + Br_2 \xrightarrow{\text{halogen carrier}} \bigcirc Br + HBr$$

bromobenzene

### Friedel Craft reactions

There are two variations on the Friedel Craft reactions.
(a) Alkylation
  Benzene is refluxed with a haloalkane in the presence of aluminium chloride.

*E.g.*

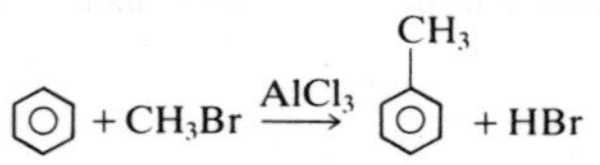

$$\bigcirc + CH_3Br \xrightarrow{AlCl_3} \bigcirc CH_3 + HBr$$

It is difficult to stop the reaction at this stage. Further substitutions take place.

*E.g.*

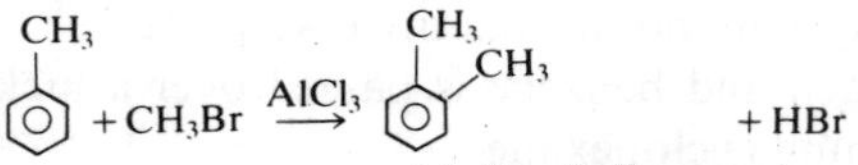

$$\bigcirc CH_3 + CH_3Br \xrightarrow{AlCl_3} \bigcirc \begin{array}{c} CH_3 \\ CH_3 \end{array} + HBr$$

1,2-dimethylbenzene

### (b) Acylation

A similar reaction takes place when benzene is refluxed with an acyl chloride in the presence of aluminium chloride. The product of the reaction is a ketone.

*E.g.*

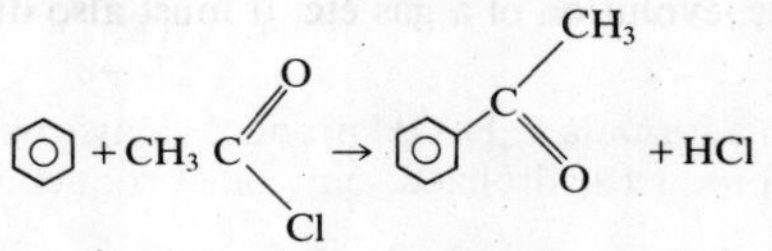

ethanoyl chloride     phenylethanone

The substituent in the benzene ring may speed up or slow down similar reactions. If the substituent is electron supplying, *e.g.* $CH_3$, the reaction is faster. If the substituent is electron-withdrawing the reaction is slower (Unit 38.7).

### 29.13 REACTIONS OF METHYLBENZENE

Since methylbenzene contains a benzene ring, it will undergo the usual electrophilic substitutions of benzene, *i.e.* nitration, sulphonation, halogenation and Friedel Craft reactions. (Note: the reactions are *quicker* with methylbenzene than benzene.)

Methylbenzene also contains a $CH_3$ side-chain which, since it is a saturated hydrocarbon chain, undergoes reactions of alkanes.

The reaction of chlorine with methylbenzene depends upon the conditions. In the dark, on refluxing the mixture with a halogen carrier, a mixture of chloromethylbenzenes is produced.

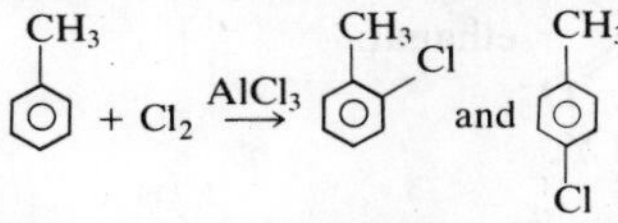

If the mixture is subjected to ultraviolet light, free radical substitution reactions in the side-chain take place.

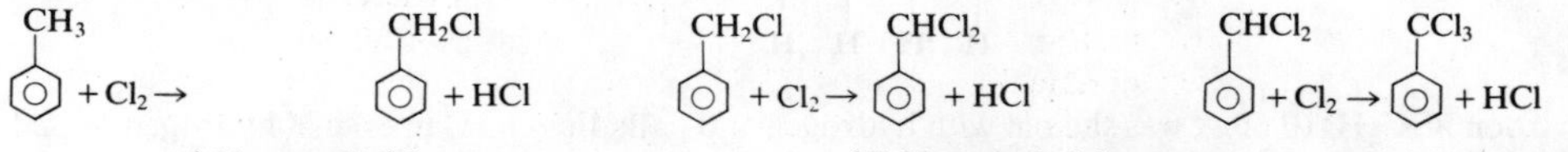

(chloromethyl) benzene          (dichloromethyl) benzene          (trichloromethyl) benzene

## QUESTIONS

Many of the questions in this section include material from Unit 38.

1 A multiple completion question (see page 3).

The products of the reaction of 2-methylpent-2-ene $(CH_3)_2C{=}CHCH_2CH_3$ with trioxygen followed by hydrolysis include

(i) $CH_3CH_2CHO$

(ii) $(CH_3)_2C{-}CH\ CH_3$ (with O bridging)

(iii) $(CH_3)_2C(OH)CH(OH)CH_2CH_3$

(iv) $(CH_3)_2C{=}O$

Ozonolysis splits carbon–carbon double bonds.

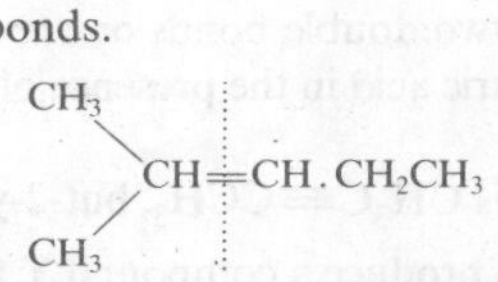

(i) is produced from the right hand end of the molecule and (iv) from the left hand end. E is therefore the correct answer.

2 Give a reagent which gives an observable chemical change with only one compound in each of the following pairs. For each reagent, state what you would observe and give the formula of the appropriate product.

(a) Hexane and hex-1-ene.

(b) Hexane and hex-1-yne.

(a) The reagent is a solution of bromine in an inert solvent. The red-brown bromine colour is removed and a colourless solution results.

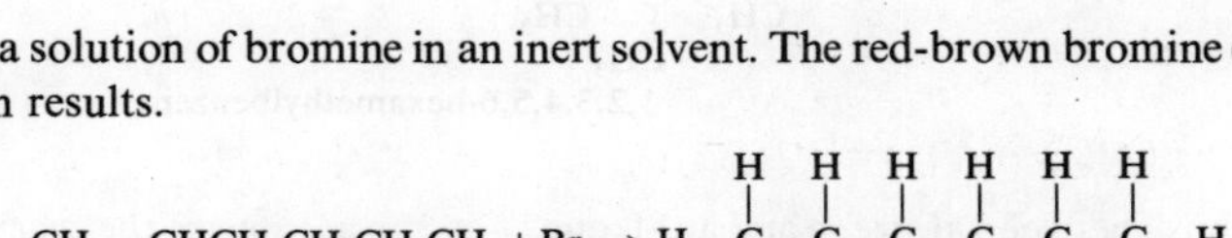

1,2-dibromohexane

(b) The reagent is an ammonical solution of silver nitrate. Dilute ammonia solution is added to silver nitrate solution until the precipitate of silver oxide just redissolves. When this solution is added to hex-1-yne, a white precipitate is observed. This is due to the acidic nature of the terminal hydrogen. The product is $C_4H_9C\equiv C^-Ag^+$. No precipitate is produced with hexane.

In this question it is important to choose a reaction which gives an observable change, *e.g.* change of colour, formation of a precipitate, evolution of a gas etc. It must also distinguish the two compounds.

**3** Dehydration of an alcohol X with a formula $C_4H_9OH$ produces a hydrocarbon Y, $C_4H_8$. When Y is treated with trioxygen (ozone), and the product hydrolized, only one product Z is obtained.

Identify X, Y and Z.

There are three possible structural isomers of Y.

A          B          C

Ozonolysis of A would produce 2 products $CH_3CH_2CHO$ and HCHO.
Ozonolysis of B would produce 2 products $CH_3COCH_3$ and HCHO.
Ozonolysis of C would produce only 1 product $CH_3CHO$ because C is symmetrical about the carbon–carbon double bond.

The product Z is therefore $CH_3-C\!\!\begin{smallmatrix}H\\ \diagdown\\ \diagup\\ O\end{smallmatrix}$  ethanal.

C (but–2–ene) is therefore compound Y.
Working backwards X is butan-2-ol.

**4** A hydrocarbon X, $C_4H_6$ (0.50g), was shaken with hydrogen and palladium until uptake of hydrogen ceased; $415cm^3$ of hydrogen (measured at stp) were absorbed. Reaction with mercury(II) sulphate in dilute sulphuric acid yielded Y, $C_4H_8O$; Y gave a positive iodoform reaction. Polymerization of X gave Z $C_{12}H_{18}$. What are the structures of X, Y and Z?

$A_r(H)=1$, $A_r(C)=12$, $A_r(O)=16$; 1 mole of a gas at stp occupies $22\,400\ cm^3$)

*(Oxford and Cambridge)*

It is important to use the information given in the question.

$$415cm^3 \text{ of hydrogen at stp} = \frac{415}{22\,400} \text{ mole of hydrogen molecules.}$$
$$= 0.018 \text{ mole of hydrogen molecules.}$$
$$\text{Relative molecular mass of X} = 54 \ (i.e.\ 12\times4+6)$$
$$\text{Number of moles of X} = \frac{0.5}{54} = 0.009 \text{ mole}$$

0.009 mole of X reacts with 0.018 mole of hydrogen molecules. 1 mole of X reacts with 2 moles of hydrogen molecules. X therefore contains either two double bonds or one triple bond.

The reaction of X with dilute sulphuric acid in the presence of mercury(II) sulphate and polymerization suggests an alkyne.

X is $CH_3C\equiv CCH_3$, but-2-yne

This reacts with dilute sulphuric acid to produce a compound Y which gives a positive iodoform test.

Y is therefore $CH_3\ \underset{\underset{O}{\|}}{C}\ CH_2\ CH_3$, butanone

Polymerization of X leads to a substituted benzene compound, Z.

$$3CH_3-C\equiv C-CH_3 \rightarrow$$

1,2,3,4,5,6-hexamethylbenzene

**5** How would you distinguish between an arene and an alkene? Give an account of the reactions of arenes. What effect does the introduction of a substituent atom or group have on the reactivity of the benzene ring?

*(Nuffield)*

Distinguish between an arene and ethene using a solution of bromine in hexane. When bromine solution is added to benzene at room temperature there is no decolourization. With ethene the bromine is decolourized. For effect of substitutent, see Unit 29.12.

# 30 Organohalogen compounds

## 30.1 INTRODUCTION

In this Unit a comparison is made between organic compounds containing a single halogen atom. This atom may be a fluorine, chlorine, bromine or iodine atom.

Common organohalogen compounds include:

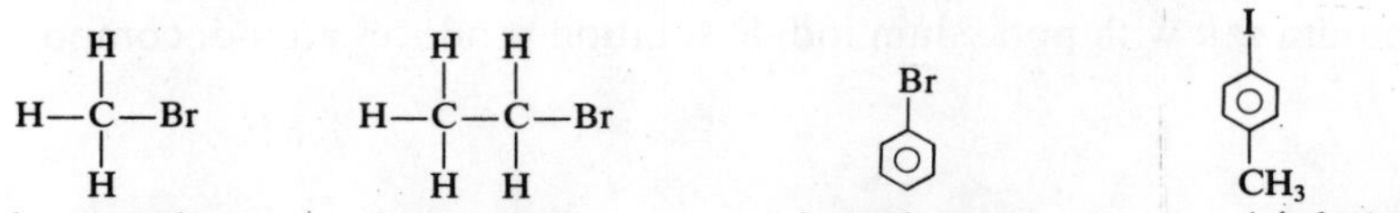

bromomethane    bromoethane    bromobenzene    1-iodo-4-methylbenzene

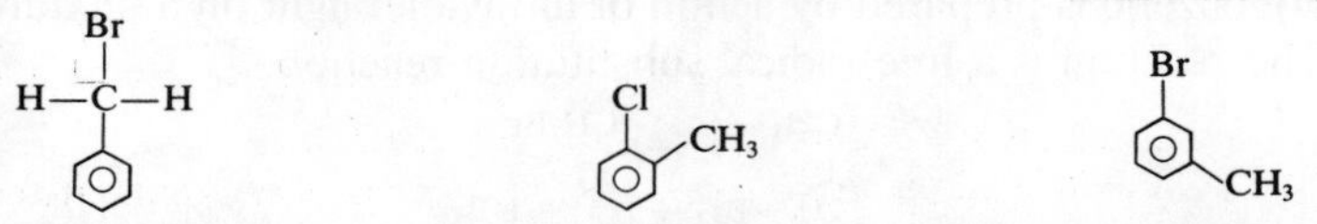

(bromomethyl) benzene    1-chloro-2-methylbenzene    1-bromo-3-methylbenzene

Bromomethane and bromoethane are examples of haloalkanes (sometimes called halogeno-alkanes or alkyl halides). They are useful organic chemicals because they can readily be converted into other products. The halogen atom of the haloalkane is readily replaced.

When a halogen atom is attached directly to an aromatic ring the halogen is difficult to replace. (Bromomethyl) benzene reacts in a similar way to a haloalkane since the bromine atom is not directly attached to the benzene ring.

## 30.2 PREPARATION OF ORGANOHALOGEN COMPOUNDS

(i) Addition of a halogen halide molecule to an alkene

*E.g.*

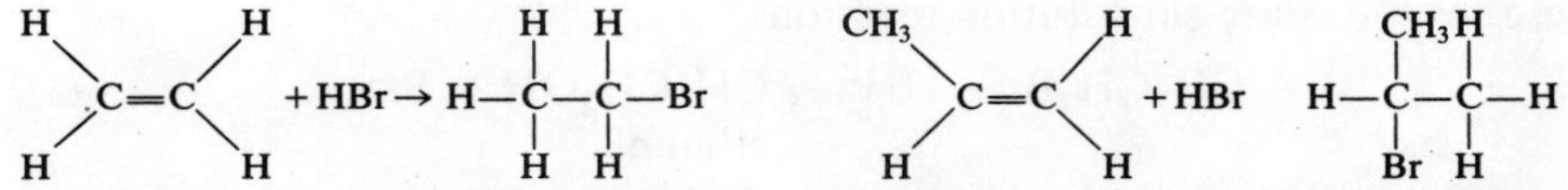

ethene    chloroethane    propene    2-bromopropane

(ii) From an alcohol using a hydrogen halide or phosphorus halide

*E.g.*

$$CH_3CH_2OH + HCl \rightleftarrows CH_3CH_2Cl + H_2O$$

ethanol    chloroethane

$$CH_3CH_2OH + PCl_5 \rightarrow CH_3CH_2Cl + POCl_3 + HCl$$

chloroethane  phosphorus trichloride oxide

Preparing a haloalkane from the corresponding alcohol is often the most convenient method.

The reaction of alcohol with hydrogen halide is most suitable for chlorides and bromides. This reaction is reversible and is usually carried out by bubbling dry hydrogen chloride gas into the alcohol in the presence of a catalyst of anhydrous zinc chloride.

The halide of phosphorus can be prepared *in situ, e.g.* by mixing red phosphorus and iodine together to produce phosphorus triiodide.

$$3CH_3OH + PI_3 \rightarrow 3CH_3I + H_3PO_3$$

methanol    iodo-    phosphonic acid
methane   (phosphorous acid)

Sulphur dichloride oxide (thionyl chloride) is useful for producing a chloroalkane. If halides of phosphorus are used, the task of separating the haloalkane from the phosphorus compound remains. This may require fractional distillation. If sulphur dichloride oxide is used the other products are gases and escape from the solution:

$$CH_3CH_2OH + SOCl_2 \rightarrow CH_3CH_2Cl + SO_2 + HCl$$

ethanol + sulphur dichloride oxide → chloroethane + sulphur dioxide + hydrogen chloride

Aryl halides are prepared by different methods.

(i) From benzene

Halogenation of benzene can be used to prepare aryl halides (29.11 and 38.6). These reactions take place when the benzene and halogen are mixed and the mixture heated out of contact with light in the presence of a halogen carrier.

$$\text{C}_6\text{H}_6 + \text{Br}_2 \rightarrow \text{C}_6\text{H}_5\text{Br} + \text{HBr}$$

## (*ii*) Via a diazonium salt (Unit 35.4)

A diazonium salt is warmed with a copper(I) halide and concentrated hydrogen halide or copper to produce an aryl halide.

*E.g.*

$$\text{C}_6\text{H}_5\text{N}_2^+\text{Cl}^- \rightarrow \text{C}_6\text{H}_5\text{Cl} + \text{N}_2$$

Warming a diazonium salt with potassium iodide solution produces an iodocompound.

(Bromomethyl)benzene is prepared by action of ultraviolet light on a mixture of bromine and methylbenzene. The reaction is a free radical substitution reaction.

$$\text{C}_6\text{H}_5\text{CH}_3 + \text{Br}_2 \rightarrow \text{C}_6\text{H}_5\text{CH}_2\text{Br} + \text{HBr}$$

In order to minimize the possibility of further substitution taking place, the methylbenzene is present in excess.

### 30.3 REACTIONS OF HALOALKANES

Haloalkanes react readily with a variety of reagents undergoing nucleophilic substitution reactions (mechanism in Unit 38.6). Iodoalkanes react more rapidly than bromoalkanes which react more rapidly than chloroalkanes (Unit 38.6).

Some of the common reactions include:

### (i) Reaction with potassium hydroxide

Potassium hydroxide in aqueous solution reacts with haloalkanes on refluxing to produce an alcohol. These reactions are substitution reactions.

*E.g.*
$$\text{CH}_3\text{CH}_2\text{Br} + \text{OH}^- \rightarrow \underset{\text{ethanol}}{\text{CH}_3\text{CH}_2\text{OH}} + \text{Br}^-$$

A different reaction takes place if a haloalkane is refluxed with a solution of potassium hydroxide dissolved in ethanol. The reaction is an elimination reaction and produces an alkene.

*E.g.*
$$\text{CH}_3\text{CH}_2\text{Br} + \text{OH}^- \rightarrow \underset{\text{ethene}}{\text{CH}_2 = \text{CH}_2} + \text{Br}^- + \text{H}_2\text{O}$$

Potassium hydroxide is preferred to the cheaper sodium hydroxide in this reaction since it is more soluble in ethanol.

When a haloalkane is heated with moist silver(I) oxide, hydrolysis occurs similar to the reaction with aqueous potassium hydroxide solution.

*E.g.*
$$2\text{CH}_3\text{CH}_2\text{Br} + \text{Ag}_2\text{O} + \text{H}_2\text{O} \rightarrow 2\text{AgBr} + 2\text{CH}_3\text{CH}_2\text{OH}$$

### (ii) Reaction of a haloalkane with potassium cyanide

A haloalkane is converted to a nitrile (or cyanide) by refluxing the haloalkane with a solution of potassium cyanide in ethanol.

$$\text{CH}_3\text{CH}_2\text{Br} + \text{KCN} \rightarrow \underset{\text{propanenitrile}}{\text{CH}_3\text{CH}_2\text{CN}} + \text{KBr}$$

This reaction is an important step in increasing the number of carbon atoms in a molecule (called **ascending the homologous series** Unit 37.3)

### (iii) Reaction of a haloalkane with ammonia

When a solution of haloalkane dissolved in ethanol is heated with ammonia in a sealed vessel, an amine is produced, but the reaction does not stop there.

*E.g.*
$$\text{CH}_3\text{CH}_2\text{Br} + \text{NH}_3 \rightarrow \underset{\text{ethylamine}}{\text{CH}_3\text{CH}_2\text{NH}_2} + \text{HBr}$$

$$\text{CH}_3\text{CH}_2\text{NH}_2 + \text{CH}_3\text{CH}_2\text{Br} \rightarrow \underset{\text{diethylamine}}{(\text{CH}_3\text{CH}_2)_2\text{NH}} + \text{HBr}$$

$$(CH_3CH_2)_2NH + CH_3CH_2Br \rightarrow (CH_3CH_2)_3N + HBr$$
triethylamine

$$(CH_3CH_2)_3N + CH_3CH_2Br \rightarrow (CH_3CH_2)_4N^+Br^-$$
tetraethylammonium bromide.

Tetraethylammonium bromide is a quaternary ammonium salt. Quaternary ammonium salts are widely used as household fabric softeners.

The amines produced are bases and the reaction products of the above reaction will include salts.

*E.g.*

$$CH_3CH_2NH_2 + HBr \rightarrow CH_3CH_2NH_3^+Br^-$$
ethylammonium bromide

### (iv) Reaction of a haloalkane with a sodium salt of a carboxylic acid

Refluxing a haloalkane with the sodium salt of a carboxylic acid produces an ester.

$$CH_3CH_2Br + \quad CH_3COONa \rightarrow CH_3COOCH_2CH_3 + NaBr$$
bromoethane + sodium ethanoate → ethyl ethanoate + sodium bromide

A similar reaction can be carried out by heating a haloalkane with a silver salt of a carboxylic acid.

$$CH_3COOAg + CH_3CH_2Br \rightarrow CH_3COOCH_2CH_3 + AgBr$$

### (v) Reaction of a haloalkane with an alkoxide

Sodium ethoxide is an alkoxide produced by reacting sodium and ethanol.

$$2Na + 2CH_3CH_2OH \rightarrow 2CH_3CH_2O^-Na^+ + H_2$$
sodium + ethanol → sodium ethoxide + hydrogen

An ether is produced when this solution of sodium ethoxide in ethanol is refluxed with a haloalkane.

$$CH_3Br + CH_3CH_2O^-Na^+ \rightarrow CH_3CH_2OCH_3 + NaBr$$
bromomethane + sodium ethoxide → methoxyethane + sodium bromide

This method of producing an ether is called **Williamson's synthesis**.

### (vi) Formation of Grignard reagents

Haloalkanes can be converted into unstable compounds called **Grignard reagents**. These Grignard reagents are useful for converting one organic compound into another.

A Grignard reagent is prepared by adding magnesium turnings to a solution of haloalkane dissolved in dry ethoxyethane. The mixture is refluxed and all water is excluded.

$$CH_3CH_2Br + Mg \rightarrow CH_3CH_2MgBr$$

Grignard reagents can be used in a variety of reactions.

(*i*) Hydrolysis of Grignard reagent with a dilute acid.

$$CH_3CH_2MgBr + H^+ \rightarrow CH_3CH_3 + Mg^{2+} + Br^-$$

(*ii*) Reaction with methanal followed by refluxing with a dilute acid.

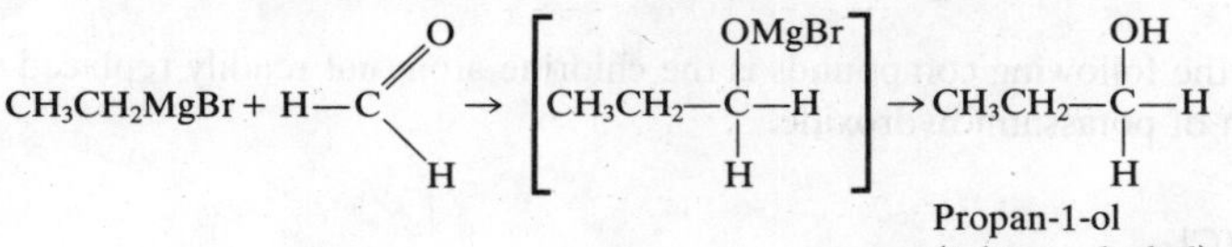

Propan-1-ol<br>(primary alcohol)

(*iii*) Reaction with an aldehyde (other than methanal) followed by refluxing with a dilute acid.

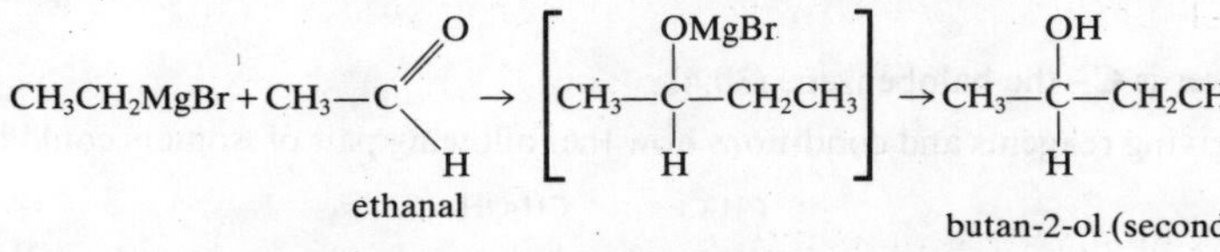

butan-2-ol (secondary alcohol)

(*iv*) Reaction with a ketone followed by refluxing with a dilute acid.

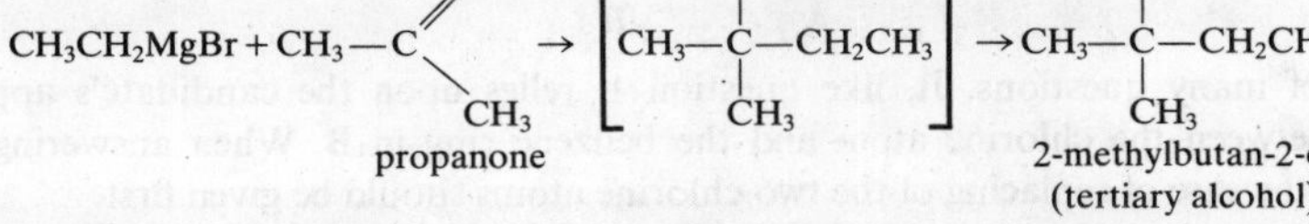

2-methylbutan-2-ol<br>(tertiary alcohol)

(*v*) Reaction with carbon dioxide followed by refluxing with a dilute acid.

$$CH_3CH_2MgBr + O{=}C{=}O \rightarrow \left[ CH_3CH_2{-}\overset{\displaystyle OMgBr}{\underset{\displaystyle O}{C}} \right] \rightarrow CH_3CH_2{-}\overset{\displaystyle OH}{\underset{\displaystyle O}{C}}$$

propanoic acid
(carboxylic acid)

### 30.4 REACTIONS OF ARYL HALIDES

Whereas nucleophilic substitution reactions occur readily with haloalkanes, aryl halides only undergo substitution reactions with the greatest difficulty.

*E.g.*

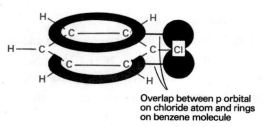

chlorobenzene + NaOH(aq) → phenol + NaCl

For this reaction to occur the reactants must be heated to 300°C under great pressure. For all practical purposes substitution of the halogen in the ring does not take place under laboratory conditions.

An explanation for this lack of reactivity is an overlap between a pair of electrons in a *p* orbital on the halogen atom and the overlapping *p* orbitals in the benzene ring system (called $\pi$ orbitals) (Fig. 30.1).

Overlap between p orbital
on chloride atom and rings
on benzene molecule

**Fig. 30.1** Structure of chlorobenzene

The halogen atom, chlorine in this case, is more strongly held as shown by the bond lengths.

C–Cl in chloroalkanes = 0.177nm
C–Cl in chlorobenzene = 0.169nm

### 30.5 REACTIONS OF (BROMOMETHYL)BENZENE

The bromine atom in (bromomethyl)benzene is not attached to a carbon atom of the benzene ring. There is no possibility of overlap between a *p* orbital and the $\pi$ orbital system of the benzene ring. The bromine atom is therefore more readily lost as a bromide ion, and (bromomethyl)-benzene undergoes reactions similar to those of haloalkanes, *e.g.* on refluxing with aqueous sodium hydroxide solution.

CH₂Br (benzene) + OH⁻(aq) → CH₂OH (benzene) + Br⁻

phenylmethanol

### QUESTIONS

**1** Multiple choice

In which one of the following compounds is the chlorine atom **not** readily replaced when refluxed with an aqueous solution of potassium hydroxide.

  A $C_2H_5Cl$
  B $(CH_3)_2CHCl$
  C $C_6H_5Cl$
  D $C_6H_5CH_2Cl$
  E $C_6H_5COCl$

The correct answer is C – the halobenzene (30.4).

**2** Explain clearly giving reagents and conditions how the following pair of isomers could be distinguished.

CH₂Cl (benzene) OH — **A**    CH₂OH (benzene) Cl — **B**

This is typical of many questions. It, like question 1, relies upon the candidate's appreciation of the stronger bond between the chlorine atom and the benzene ring in B. When answering this question a statement about the ease of replacing of the two chlorine atoms should be given first.

Then briefly explain how this might be shown. For example, these compounds could be added to separate test tubes and aqueous sodium hydroxide solution added to each tube. Both tubes are then heated. With A hydrolysis takes place producing free chloride ions.

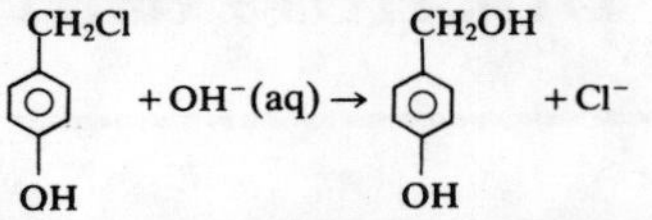

The contents of this tube are acidified with dilute nitric acid and silver nitrate solution added. A white precipitate of silver chloride is produced which turns slightly purple when exposed to strong light and which dissolves in ammonia solution. This is a test for free chloride ions.

$$Ag^+(aq) + Cl^-(aq) \rightarrow AgCl(s)$$

No hydrolysis takes place with B under similar conditions.

3 Hydrolysis of a compound A (formula $C_7H_6Cl_2$) was carried out by refluxing with excess potassium hydroxide solution. The resulting solution was acidified with dilute nitric acid and excess silver nitrate added to precipitate the chloride ions as silver chloride.

$\qquad$ 0.718g of silver chloride were formed from 0.805g of A.

(*i*) Calculate the number of chlorine atoms present in each molecule of A which are liberated by hydrolysis.

(*ii*) Suggest an explanation for the result obtained in (*i*).

(*iii*) Give a possible structure for A.

$(A_r(H) = 1, A_r(C) = 12, A_r(Cl) = 35.5, A_r(Ag) = 108)$

(*i*) **Relative molecular mass of A** $= (7 \times 12) + (6 \times 1) + (35.5 \times 2)$
$$= 161$$

**Number of moles of A used** $= \dfrac{0.805}{161} = 0.005$ moles

**Number of moles of silver chloride produced** $= \dfrac{0.718}{108 + 35.5}$
$$= 0.005 \text{ moles}$$

∴ 0.005 moles of silver chloride are liberated by 0.005 moles of A.

$\qquad$ Only one of the two chlorine atoms in the molecule is liberated during hydrolysis.

(*ii*) An explanation is that one chlorine atom is attached to the benzene ring while the other is attached to a side chain.

(*iii*) Possible structures for A are

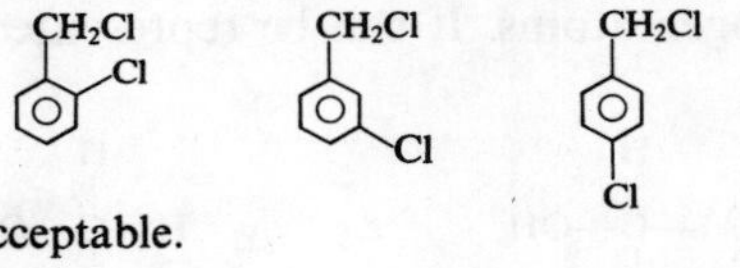

Any of these isomers would be acceptable.

4 Explain the following observations:
1–chloro, 1–bromo and 1–iodobutane were put into separate test tubes. A small volume of ethanol was added to each test tube followed by silver nitrate solution. The test tubes were put in a beaker of hot water. A precipitate forms in each test tube – the precipitate forming first in the tube with the 1–iodobutane and the last in the tube with the 1- chlorobutane.

The difference in the rate of formation of a silver halide precipitate indicates the difference in strength of the carbon–halogen bond. The differences in bond strengths are:

$\qquad\qquad$ C–I $\qquad$ 188 kJ mol$^{-1}$
$\qquad\qquad$ C–Br $\qquad$ 226 kJ mol$^{-1}$
$\qquad\qquad$ C–Cl $\qquad$ 278 kJ mol$^{-1}$

The carbon–iodine bond is weaker and the bond breaks easier. The precipitate therefore forms faster.
(Do not try to remember values such as these bond strengths).

# 31 Alcohols, phenols and ethers

## 31.1 INTRODUCTION

In this unit a comparison will be made between **alcohols** and **phenols**. Both contain a hydroxyl —OH group but in the alcohol the —OH group is attached to a chain derived from an alkane while a phenol has an —OH group attached directly to a benzene nucleus.

Examples of simple alcohols and phenols include:

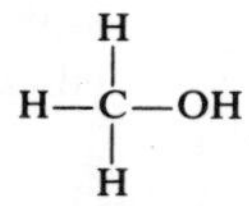

methanol 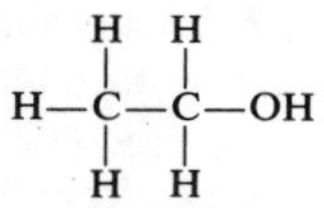    ethanol 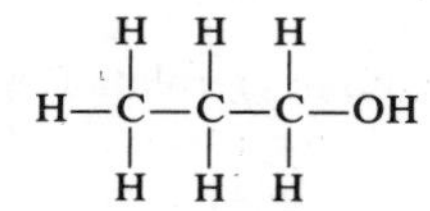    propanol 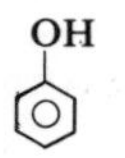    phenol

Ethers also contain a single oxygen atom and are structural isomers of alcohols. They do not however resemble alcohols in properties. Simple ethers include:

$CH_3OCH_3$    $CH_3CH_2OCH_2CH_3$    $CH_3OCH_2CH_3$
methoxymethane    ethoxyethane    methoxyethane

## 31.2 PRIMARY, SECONDARY AND TERTIARY ALCOHOLS

An understanding of the difference between primary, secondary and tertiary alcohols is most important. The classification depends upon the position of the —OH group in the molecule, unlike the classification of amines as primary secondary and tertiary (see Unit 35).

In a primary alcohol, the —OH group is attached to a carbon atom which is in turn attached directly to two or three hydrogen atoms. It can be represented as

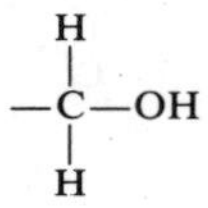

*e.g.* 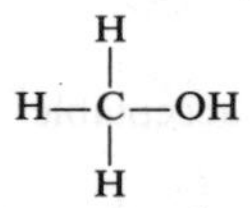    methanol

Only one hydrogen is attached to this carbon atom in a secondary alcohol:

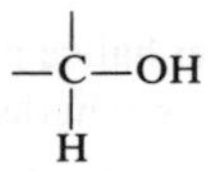

In a tertiary alcohol there are no hydrogen atoms attached to this carbon atom:

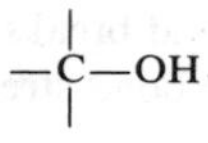

The four alcohols with a molecular formula $C_4H_9OH$ have the following structures.

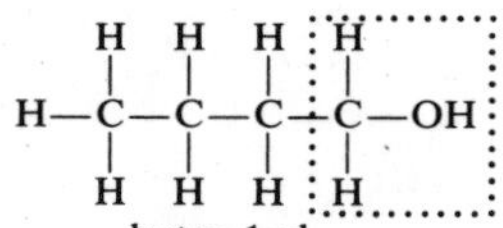

butan-1-ol
(primary alcohol)

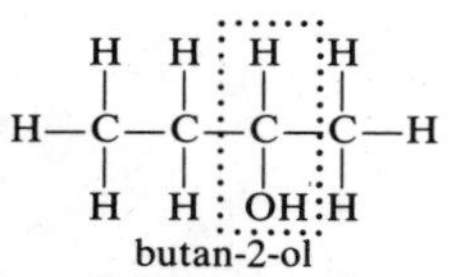

butan-2-ol
(secondary alcohol)

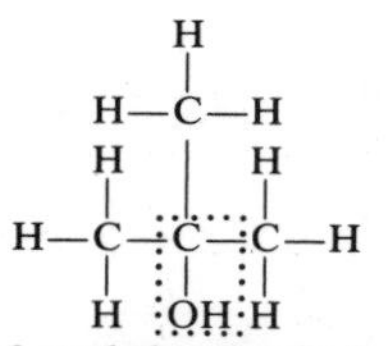

2-methylpropan-2-ol
(tertiary alcohol)

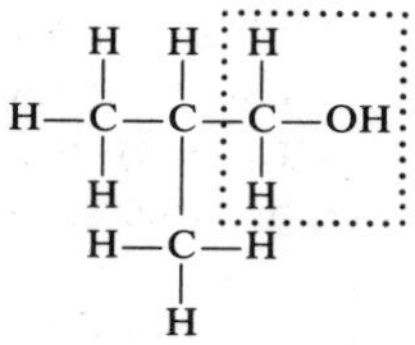

2-methylpropan-1-ol
(primary alcohol)

In Unit 31.4 the chemical tests necessary to differentiate between primary, secondary and tertiary alcohols are given.

### 31.3 METHODS OF PREPARATION OF ALCOHOLS

Ethanol, $C_2H_5OH$, can be prepared by fermentation of starch or sugar solution in the presence of enzymes in yeast, in *anaerobic* conditions (*i.e.* in absence of air) as in the wine making process.
*E.g.*
$$C_6H_{12}O_6 \xrightarrow{\text{zymase}} 2C_2H_5OH + 2CO_2$$

This process is carried out in slightly warm conditions (about $30-35°C$). Air must not be allowed to come into contact with the solution, or souring of the wine might take place. The souring involves the bacterial oxidation of the alcohol to a carboxylic acid:
$$C_2H_5OH + 2[O] \rightarrow CH_3COOH + H_2O$$

The solution resulting from anaerobic fermentation is a dilute solution of ethanol in water. This solution can be concentrated by fractional distillation (Unit 7.5). This does not produce 100% pure ethanol, however.

Methods of producing alcohols include

1  Hydrolysis of a haloalkane (Unit 30.3).
$$CH_3CH_2Br + OH^-(aq) \rightarrow CH_3CH_2OH + Br^-$$

2  Reduction of aldehydes, ketones or carboxylic acid (Unit 32.6 and 33.3).

(*i*) 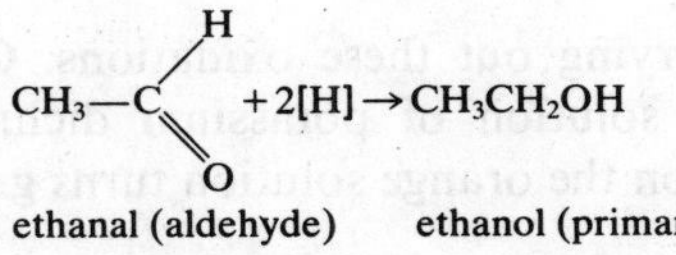  ethanal (aldehyde)      ethanol (primary alcohol)

(*ii*) 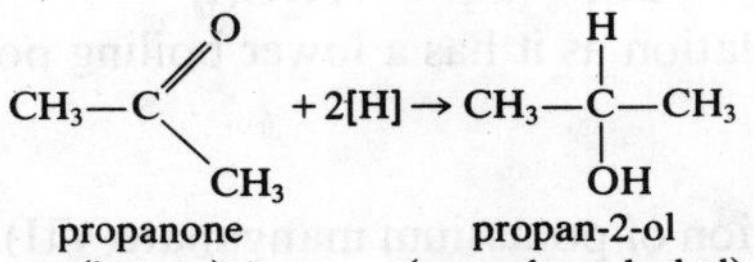  propanone (ketone)      propan-2-ol (secondary alcohol)

(*iii*) 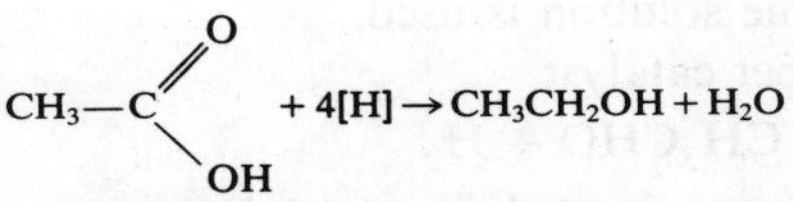  ethanoic acid (carboxylic acid)      ethanol (primary alcohol)

The reagents necessary for these reactions are never well known by examination candidates. For the examples (*i*) and (*ii*) any of the following reagents could be used: sodium and ethanol, hydrogen with platinum or nickel catalyst, lithium tetrahydridoaluminate(III) (lithium aluminium hydride), or sodium tetrahydridoborate(III) (sodium borohydride). Example (*iii*) requires lithium tetrahydridoaluminate(III); this reaction is carried out in a solution in dry ethoxyethane.

Alcohols can also be produced by the reduction of esters with lithium tetrahydridoaluminate(III) (Unit 34.4).

3  Action of nitrous acid (nitric(III) acid) on a primary amine (Unit 35.3).
*E.g.*
$$CH_3CH_2NH_2 + HONO \rightarrow CH_3CH_2OH + N_2 + H_2O$$

4  From the reaction of a Grignard reagent with an aldehyde or ketone (Unit 30.3).

Industrial methods for preparing alcohols are important. Vast quantities of the common alcohols are required by the chemical industry. Methanol is manufactured by heating carbon monoxide and hydrogen.
$$CO + 2H_2 \rightleftharpoons CH_3OH$$

The process is carried out at high pressures and between $350°C$ and $400°C$ in the presence of a chromium(III) oxide catalyst.

The other alcohols are obtained from alkenes. Alkenes are produced by cracking of higher fractions from petroleum refining. The alkene is reacted with concentrated sulphuric acid and the product is hydrolysed:

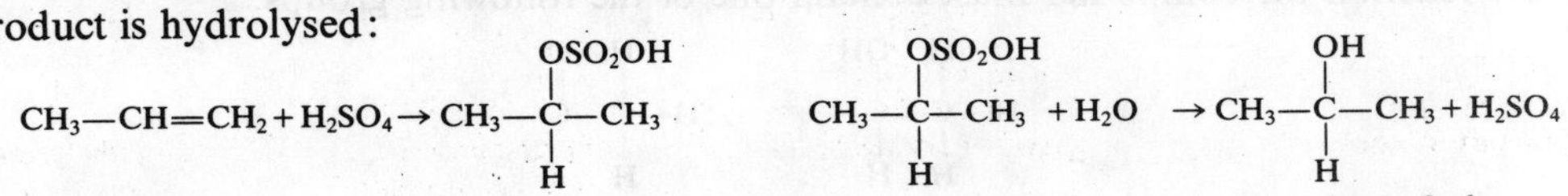

propene      propan-2-ol

For industrial purposes ethanol is made from ethene, rather than by fermentation, by hydration over a phosporic acid catalyst.

### 31.4    REACTIONS OF ALCOHOLS

#### (i) Oxidation of alcohols

Oxidation is used to distinguish primary, secondary and tertiary alcohols.

Primary alcohols are oxidized to produce aldehydes and then, on prolonged oxidation, carboxylic acids.

$$R-\underset{\underset{H}{|}}{\overset{\overset{H}{|}}{C}}-OH + [O] \rightarrow R-C\underset{O}{\overset{H}{\diagdown}} + H_2O \qquad R-C\underset{O}{\overset{H}{\diagdown}} + [O] \rightarrow R-C\underset{O}{\overset{OH}{\diagdown}}$$

aldehyde                          carboxylic acid

Secondary alcohols are oxidized to produce a ketone, and no further oxidation takes place under the stated conditions.

$$R-\underset{\underset{R'}{|}}{\overset{\overset{H}{|}}{C}}-OH + [O] \rightarrow R-C\underset{R'}{\overset{O}{\diagdown}}$$

ketone

Tertiary alcohols are only oxidized under very severe conditions, when the molecule is split. Under the conditions stated, no oxidation takes place.

There are a variety of methods available for carrying out these oxidations. One of the commonest methods is to heat the alcohol with a solution of potassium dichromate(VI) acidified with dilute sulphuric acid. During the oxidation the orange solution turns green due to the formation of $Cr^{3+}(aq)$ ions.

$$6e^- + Cr_2O_7^{2-}(aq) + 14H^+(aq) \rightarrow 2Cr^{3+}(aq) + 7H_2O(l)$$

The aldehyde usually distils off during the oxidation as it has a lower boiling point than the corresponding alcohol.

Other suitable methods of oxidation include:

*1* Heating with either an acidic or an alkaline solution of potassium manganate(VII). The purple colour is removed during the oxidation and a brown precipitate of manganese(IV) oxide (manganese dioxide) may be formed if an alkaline solution is used.

*2* Passing the alcohol vapour over a heated copper catalyst.

*E.g.*                    $CH_3CH_2OH \rightarrow CH_3CHO + H_2$

*3* Passing a mixture of alcohol vapour and air over a heated silver catalyst.

*E.g.*                    $2CH_3CH_2OH + O_2 \rightarrow 2CH_3CHO + 2H_2O$

*4* Warming with concentrated nitric acid.

#### (ii) Reactions of alcohols with concentrated sulphuric acid

The reaction of an alcohol with concentrated sulphuric acid can lead to the formation of an alkene or an ether depending upon the conditions (Unit 29.5).

#### (iii) Reaction with sodium

Sodium (and the other members of the alkali metal family) react with an alcohol to produce hydrogen. There are parallels between this reaction and the reaction of sodium with water, but the reaction is slower:

$$2CH_3CH_2OH + 2Na \rightarrow 2CH_3CH_2O^-Na^+ + H_2$$
ethanol                    sodium ethoxide

This reaction is used to dispose safely of waste scraps of alkali metals.

#### (iv) The triiodomethane (iodoform) reaction

This is a most important reaction for A level candidates. Many successful answers rely upon recognising that the triiodomethane reaction will take place, and on understanding the steps in the reaction.

Only certain alcohols (and certain aldehydes and ketones) undergo this reaction. In order to give the reaction the compound must contain one of the following groups.

$$H-\underset{\underset{H}{|}}{\overset{\overset{H}{|}}{C}}-\underset{\underset{H}{|}}{\overset{\overset{OH}{|}}{C}}- \qquad H-\underset{\underset{H}{|}}{\overset{\overset{H}{|}}{C}}-\overset{\overset{O}{\|}}{C}-$$

alcohols            ethanal or ketones

Methanol $CH_3OH$ does not undergo the triiodomethane reaction. The simplest alcohol to undergo this reaction is ethanol $CH_3CH_2OH$.

The reaction takes place when the compound is warmed with a solution of sodium hydroxide (to produce alkaline conditions) and iodine solution. Alternatively a solution of potassium iodide and sodium chlorate(I) (hypochlorite) can be used.

On warming, a yellow crystalline precipitate of triiodomethane is produced, which has an antiseptic smell.

The steps in this reaction using ethanol as the example are:

(*a*) Oxidation of alcohol with iodine:

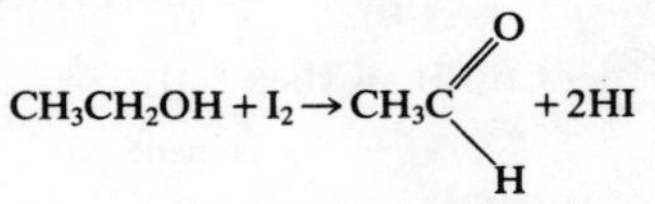

$$CH_3CH_2OH + I_2 \rightarrow CH_3C\overset{O}{\underset{H}{<}} + 2HI$$

ethanal

(*b*) Substitution of the iodine atoms into the aldehyde:

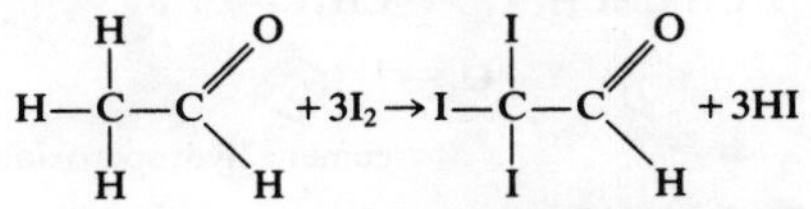

$$H-\overset{H}{\underset{H}{C}}-C\overset{O}{\underset{H}{<}} + 3I_2 \rightarrow I-\overset{I}{\underset{I}{C}}-C\overset{O}{\underset{H}{<}} + 3HI$$

triiodoethanal

(*c*) Hydrolysis in alkaline conditions

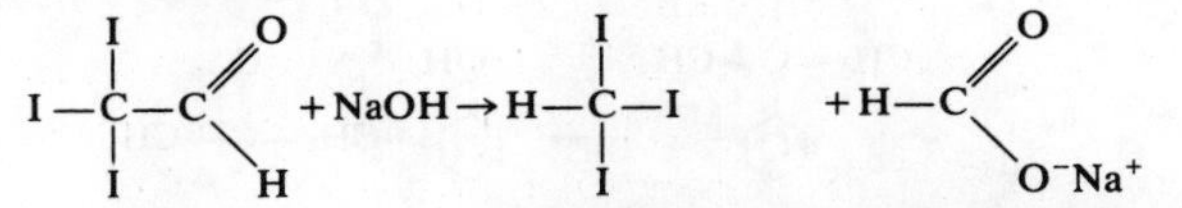

$$I-\overset{I}{\underset{I}{C}}-C\overset{O}{\underset{H}{<}} + NaOH \rightarrow H-\overset{I}{\underset{I}{C}}-I + H-C\overset{O}{\underset{O^-Na^+}{<}}$$

triiodomethane      sodium methanoate

If ethanal or a suitable ketone is used for this reaction, only steps (*b*) and (*c*) are necessary. An overall equation for this reaction is:

$$CH_3CH_2OH + 4I_2 + 6NaOH \rightarrow CHI_3 + HCOONa + 5NaI + 5H_2O$$

Examples of questions using this reaction will be found at the end of this unit.

### (v) Ester formation (see Unit 34.4)

### (vi) Reaction with phosphorus halides

See Unit 30.2.

## 31.5 PREPARATION OF PHENOL

In the laboratory, phenol can be prepared from benzenesulphonic acid and from benzenediazonium chloride.

### (i) From benzenesulphonic acid

Benzenesulphonic acid is prepared from benzene by sulphonation (Unit 29.12). The benzenesulphonic acid is neutralized with sodium hydroxide, and anhydrous sodium benzenesulphonate is fused with solid sodium hydroxide.

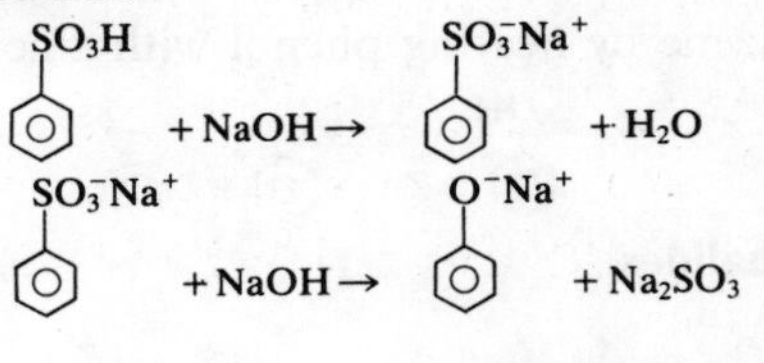

$$C_6H_5SO_3H + NaOH \rightarrow C_6H_5SO_3^-Na^+ + H_2O$$

$$C_6H_5SO_3^-Na^+ + NaOH \rightarrow C_6H_5O^-Na^+ + Na_2SO_3$$

sodium phenoxide
(sodium phenate)

Phenol is then produced by acidification.

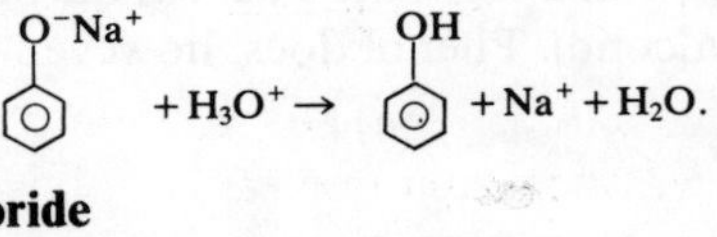

$$C_6H_5O^-Na^+ + H_3O^+ \rightarrow C_6H_5OH + Na^+ + H_2O.$$

### (ii) From benzenediazonium chloride

Benzenediazonium chloride is prepared by the action of nitrous acid (nitric(III) acid) on phenylamine below 5°C (Unit 35.4).

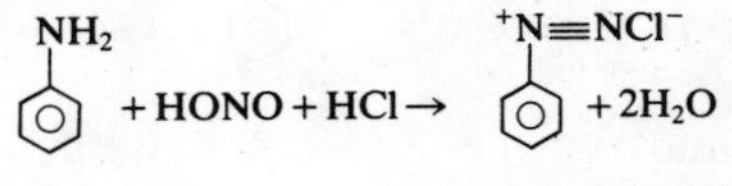

$$C_6H_5NH_2 + HONO + HCl \rightarrow C_6H_5^+N\equiv NCl^- + 2H_2O$$

benzenediazonium chloride

An aqueous solution of benzenediazonium chloride decomposes on heating to produce phenol:

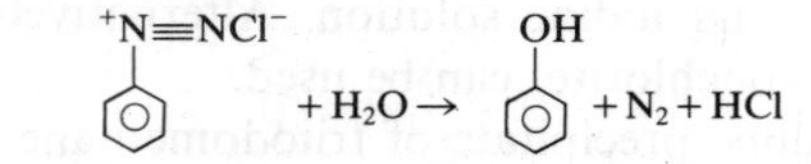

$$\text{(benzenediazonium chloride)} + H_2O \rightarrow \text{(phenol)} + N_2 + HCl$$

Industrially, phenol is produced from benzene. Propene and benzene vapours are mixed and passed over a heated catalyst to produce 2-phenylpropane (cumene).

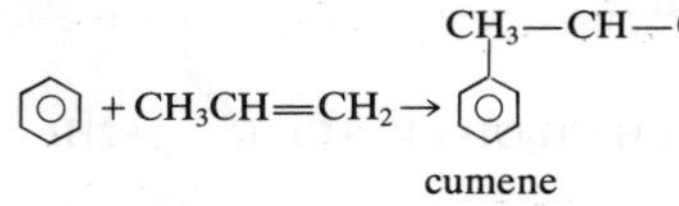

$$\text{(benzene)} + CH_3CH{=}CH_2 \rightarrow \text{(cumene)}$$

cumene

Cumene is then heated with air at 180°C.

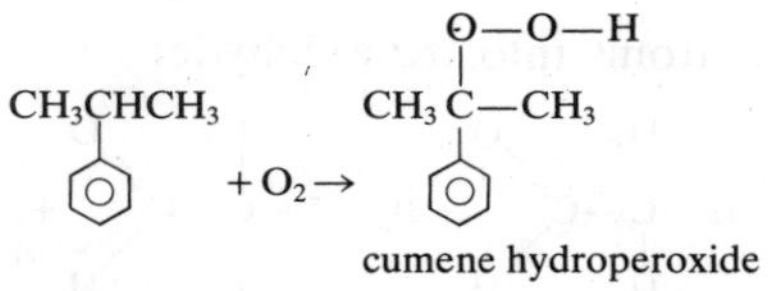

$$CH_3CHCH_3\text{(benzene)} + O_2 \rightarrow CH_3\overset{O-O-H}{\underset{}{C}}-CH_3\text{(benzene)}$$

cumene hydroperoxide

Phenol is produced by the hydrolysis of cumene hydroperoxide with dilute sulphuric acid.

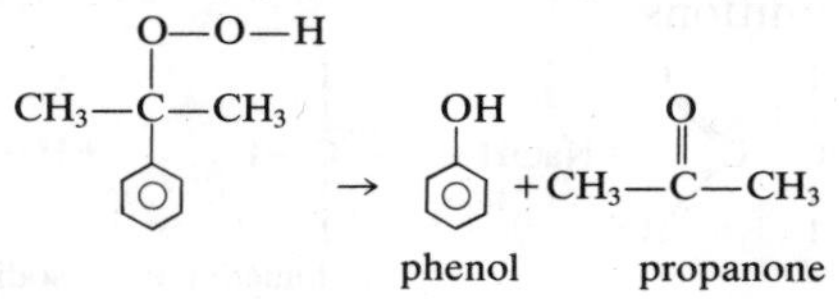

$$CH_3-\overset{O-O-H}{\underset{}{C}}-CH_3 \rightarrow \text{(phenol)} + CH_3-\overset{O}{\underset{}{C}}-CH_3$$

phenol    propanone

## 31.6  REACTIONS OF PHENOL

### (i) Acidic properties of phenol

Phenols are very slightly acidic. Phenol dissolves in sodium hydroxide solution to form sodium phenoxide. Alcohols do not carry out this reaction.

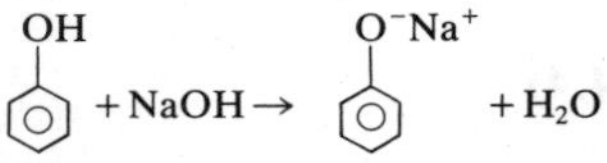

$$\text{(phenol)} + NaOH \rightarrow \text{(sodium phenoxide)} + H_2O$$

Phenol is not sufficiently acidic to liberate carbon dioxide from sodium hydrogencarbonate. This distinguishes a phenol from a carboxylic acid.

Phenol is recovered from sodium phenoxide by acidification with a dilute acid.

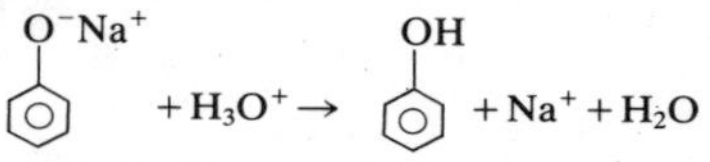

$$\text{(sodium phenoxide)} + H_3O^+ \rightarrow \text{(phenol)} + Na^+ + H_2O$$

The reasons for the acidity of phenol is discussed in Unit 39.

### (ii) Testing for phenol

Phenol contains the -enol group $\overset{}{\underset{}{>}}C{=}C\overset{OH}{\underset{}{<}}$  This can be tested for by adding a neutral solution of iron(III) chloride. A deep purple colouration is formed due to the formation of a complex.

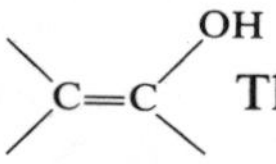

### (iii) Reduction of phenol

Phenol can be reduced to benzene by heating phenol with zinc dust at 400°C.

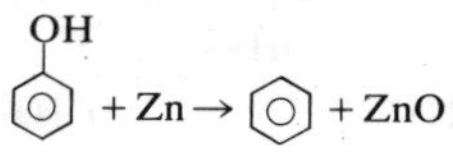

$$\text{(phenol)} + Zn \rightarrow \text{(benzene)} + ZnO$$

### (iv) Reaction with phosphorus halides

See Unit 30.2.

### (v) Ester formation (31.4)

Phenol does not produce an ester with a carboxylic acid in the presence of concentrated sulphuric acid. In this it differs from the alcohol. Phenol does, however, form an ester when treated with an acid chloride or anhydride.

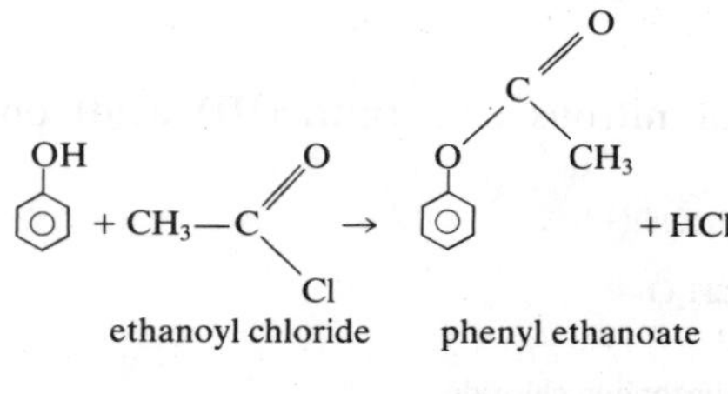

$$\text{(phenol)} + CH_3-\overset{O}{\underset{Cl}{C}} \rightarrow \text{(phenyl ethanoate)} + HCl$$

ethanoyl chloride    phenyl ethanoate

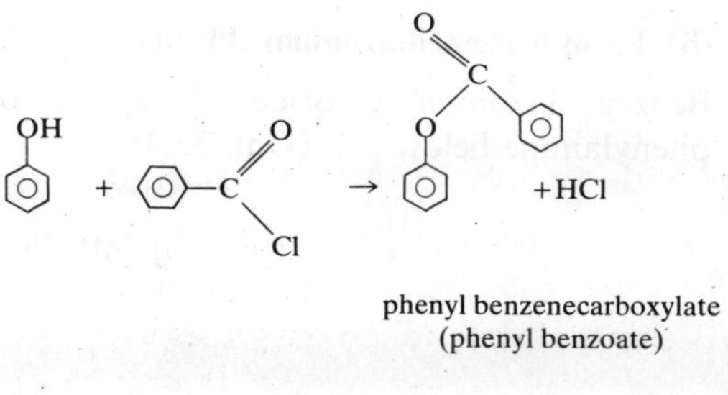

$$\text{(phenol)} + \text{(benzene)}-\overset{O}{\underset{Cl}{C}} \rightarrow \text{(phenyl benzenecarboxylate)} + HCl$$

phenyl benzenecarboxylate
(phenyl benzoate)

## *(vi) Substitution reactions

Phenol undergoes substitution reactions into the nucleus more readily than benzene. This is because of a tendency for the pair of non-bonding electrons of the —OH group to be drawn into the $\pi$ electron system. The positions 2,4 and 6 in the ring are particularly susceptible to electrophilic attack.

(a) Bromination of benzene requires bromine in the presence of a halogen carrier. With phenol, however, bromination takes place immediately when bromine water is added to phenol.

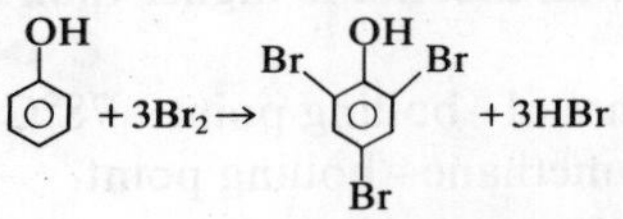

$$\text{OH} + 3Br_2 \rightarrow \text{2,4,6-tribromophenol} + 3HBr$$

2,4,6-tribromophenol
(white precipitate)

(b) Nitration of benzene requires a mixture of concentrated nitric and sulphuric acids. Phenol is, however, nitrated with dilute nitric acid.

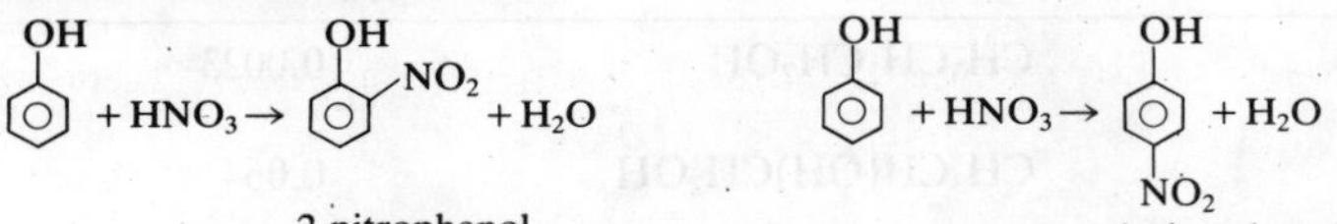

$$\text{OH} + HNO_3 \rightarrow \text{2-nitrophenol} + H_2O$$

2-nitrophenol

$$\text{OH} + HNO_3 \rightarrow \text{4-nitrophenol} + H_2O$$

4-nitrophenol

A mixture of 2-nitrophenol and 4-nitrophenol is produced. This mixture can be separated by steam distillation.

(c) With diazonium salts. The benzene diazonium ion $C_6H_5N_2^+$ is not sufficiently powerful an electrophile for substitution into benzene but it will react with an alkaline solution of phenol.

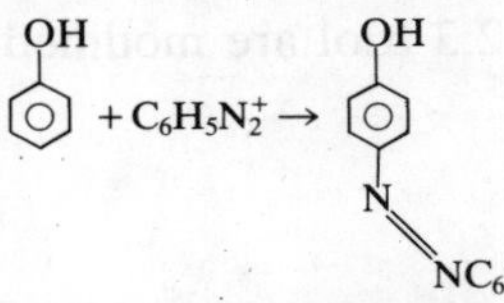

$$\text{OH} + C_6H_5N_2^+ \rightarrow \text{(4-hydroxyphenylazobenzene)}$$

(4-hydroxyphenylazobenzene)

An orange precipitate is formed in this coupling reaction.

## 31.7 TESTING FOR AN —OH GROUP IN ALCOHOLS

Phosphorus pentachloride can be used to show the presence of an —OH group in an alcohol. When solid phosphorus pentachloride is added to a carefully dried alcohol, fumes of hydrogen chloride are produced. If the alcohol is not dry, HCl will be generated by the water present.

$$CH_3CG_2OH + PCl_5 \rightarrow CH_3CG_2Cl + POCl_3 + HCl$$
chloroethane

Hydrogen chloride produces steamy fumes in moist air and dense white fumes with ammonia gas.

$$NH_3 + HCl \rightarrow NH_4Cl$$

Hydrogen chloride fumes are also produced if phosphorus pentachloride is added to a dry carboxylic acid.

## *31.8 ETHERS

Ethoxyethane can be prepared by the action of concentrated sulphuric acid on ethanol (Unit 29.4). Ethers can also be produced by the reactions of a haloalkane with an alkoxide (Unit 30.3). or by heating a haloalkane with dry silver oxide:

e.g. $$2CH_3CH_2Br + Ag_2O \rightarrow CH_3CH_2OCH_2CH_3 + 2AgBr$$

A phenoxide can be used to replace an alkoxide in the reaction with a haloalkane. This produces an ether containing an aromatic group.

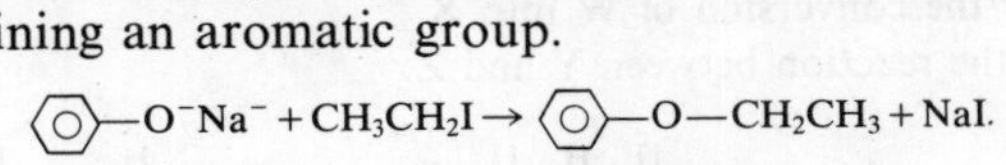

$$\text{O}^-Na^+ + CH_3CH_2I \rightarrow \text{O}-CH_2CH_3 + NaI$$

sodium phenoxide      ethoxybenzene

Chemically, ethers are much less reactive than alcohols. They do not contain a hydrogen attached directly to the oxygen atom.

### 31.9 HYDROGEN BONDING (see also 4.8)

Hydrogen bonding takes place between alcohol molecules but it is not possible with ethers.

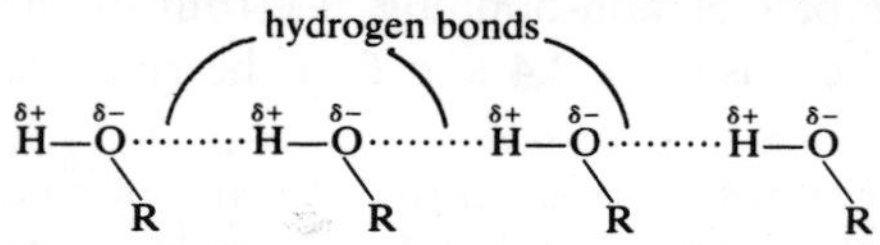

hydrogen bonds

$$\overset{\delta+}{H}-\overset{\delta-}{O}\cdots\overset{\delta+}{H}-\overset{\delta-}{O}\cdots\overset{\delta+}{H}-\overset{\delta-}{O}\cdots\overset{\delta+}{H}-\overset{\delta-}{O}$$
$$\qquad R \qquad\quad R \qquad\quad R \qquad\quad R$$

As a result, the boiling point of an alcohol is higher than the boiling point of the isomeric ether.

*E.g.*
ethanol – boiling point    78°C
methoxymethane – boiling point    −24°C

Table 31.1 gives the formula and viscosity at 20°C of a number of alcohols.

| Alcohol | Formula | Viscosity N s $m^{-2}$ |
|---|---|---|
| Propan-1-ol | $CH_3CH_2CH_2OH$ | 0.0023 |
| Propane-1,2-diol | $CH_3CH(OH)CH_2OH$ | 0.064 |
| Propane-1,2,3-triol | $CH_2(OH)CH(OH)CH_2OH$ | 1.07 |

The viscosity increases as the number of hydroxyl groups in the molecule increases. This is due to the increased amount of hydrogen bonding which is possible with the presence of the extra —OH groups.

If the —OH groups in propane-1,2,3-triol are modified by ester formation, the viscosity should decrease.

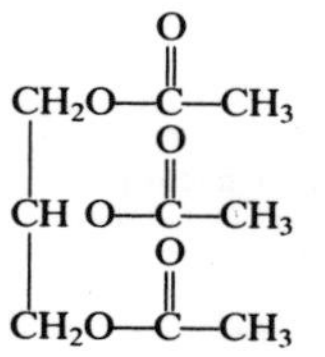

$$\begin{array}{l} CH_2O-\overset{\overset{\displaystyle O}{\|}}{C}-CH_3 \\ CH\ O-\overset{\overset{\displaystyle O}{\|}}{C}-CH_3 \\ CH_2O-\overset{\overset{\displaystyle O}{\|}}{C}-CH_3 \end{array}$$

viscosity at 20°C = 0.023 N s m$^{-2}$

### QUESTIONS

1 Classify the following alcohols as primary, secondary or tertiary.
   (*i*) $(CH_3)_2C(OH)C_2H_5$
   (*ii*) $CH_3CH(OH)CH_2CH_3$
   (*iii*) $C_6H_5CH_2OH$
   (*iv*) $(CH_3)_2CHOH$
   (*v*) $(CH_3CH_2)_2CH_2OH$

(*i*) tertiary (*ii*) secondary (*iii*) primary (*iv*) secondary (*v*) primary.
If you have any doubt, draw out the structural formulae in full.

2 The following diagram shows the relationships between the molecular formulae of some compounds

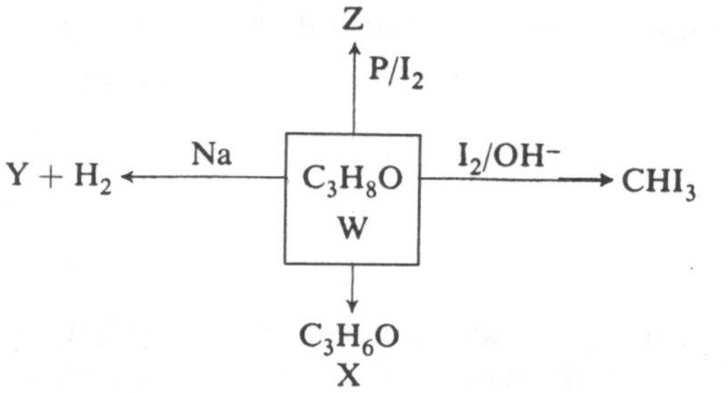

$$Z$$
$$\uparrow P/I_2$$
$$Y + H_2 \xleftarrow{\text{Na}} \boxed{\begin{array}{c} C_3H_8O \\ W \end{array}} \xrightarrow{I_2/OH^-} CHI_3$$
$$\downarrow$$
$$C_3H_6O$$
$$X$$

(*a*) Name and write out structural formulae for three isomers of W.
(*b*) Which one of these isomers corresponds to the compound W? Give a reason for your answer.
(*c*) Identify the compounds X, Y and Z.
(*d*) Give the conditions for the conversion of W into X.
(*e*) Write the equation for the reaction between Y and Z.

(*a*)

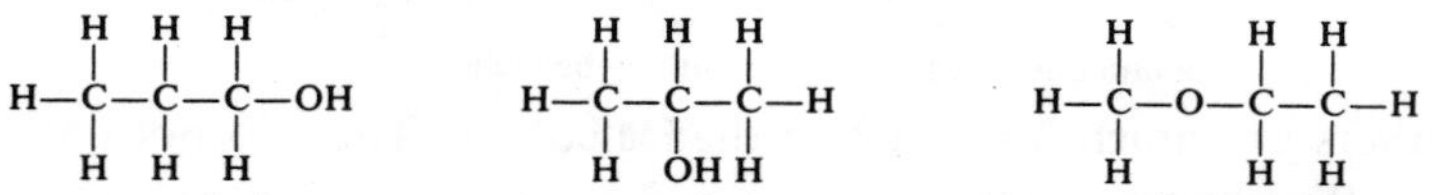

$$\begin{array}{c} \overset{H}{\underset{H}{|}}\ \overset{H}{\underset{H}{|}}\ \overset{H}{\underset{H}{|}} \\ H-C-C-C-OH \\ \end{array} \qquad \begin{array}{c} H\ H\ H \\ H-C-C-C-H \\ H\ OH\ H \end{array} \qquad \begin{array}{c} H\qquad H\ H \\ H-C-O-C-C-H \\ H\qquad H\ H \end{array}$$
propan-1-ol            propan-2-ol            methoxyethane

Don't forget that alcohols and ethers are isomeric.

(b) Here the knowledge of the triiodomethane (iodoform) reaction (31.4) is essential.

Only propan-2-ol of the possible compounds undergoes the triiodomethane reaction because it contains

the $\begin{array}{c} H \ \ H \\ | \ \ \ | \\ -C-C-H \\ | \ \ \ | \\ OH \ H \end{array}$ group.

Therefore **W** is the secondary alcohol propan-2-ol.

(c)         **X** is the ketone $CH_3-\overset{\overset{\displaystyle O}{\|}}{C}-CH_3$

     **Y** is $CH_3-\overset{\overset{\displaystyle H}{|}}{\underset{\underset{\displaystyle O^-Na^+}{|}}{C}}-CH_3$

     **Z** is $CH_3-\overset{\overset{\displaystyle H}{|}}{\underset{\underset{\displaystyle I}{|}}{C}}-CH_3$

(d) The oxidation of **W** into **X** can be achieved by heating with acidified potassium dichromate(VI) solution.

(e)

$$CH_3-\overset{\overset{\displaystyle H}{|}}{\underset{\underset{\displaystyle O^-Na^+}{|}}{C}}-CH_3 + CH_3-\overset{\overset{\displaystyle H}{|}}{\underset{\underset{\displaystyle I}{|}}{C}}-CH_3 \rightarrow CH_3-\overset{\overset{\displaystyle H}{|}}{\underset{\underset{\displaystyle CH_3}{|}}{C}}-O-\overset{\overset{\displaystyle H}{|}}{\underset{\underset{\displaystyle CH_3}{|}}{C}}-CH_3$$

The product is an ether. You are not asked to name this ether. (The old name is isopropyl ether; its modern name is 2-methylethoxy-2-methylethane.)

**3** The following table lists the relative molecular masses and boiling points of some organic compounds.

| Compound | Relative molecular mass | Boiling point /K |
|---|---|---|
| benzene | 78 | 353 |
| methylbenzene | 92 | 383 |
| phenol | 94 | 454 |
| nitrobenzene | 123 | 483 |
| 2-nitrophenol | 139 | 494 |
| 4-nitrophenol | 139 | 518 |

(a) What reason(s) can you suggest for the *general* rise in boiling point with increasing relative molecular mass of these compounds?

(b) Why does water not fit into the pattern of the table?

(c)  (i) Draw the structural formulae of 2-nitrophenol and 4-nitrophenol.

    (ii) Suggest a reason why the boiling point of 4-nitrophenol is higher than that of 2-nitrophenol.

    (iii) Which of the two isomers is likely to be more soluble in water? Give a reason.     (*Nuffield*)

(a) **Generally, the boiling point rises as molecules get larger. More energy is required to move larger molecules.**

(b) **Hydrogen bonding between water molecules.**

(c)  (i)

2-nitrophenol             4-nitrophenol

    (ii) **4-nitrophenol. Hydrogen bonding between molecules.**
          **2-nitrophenol. Hydrogen bonding within molecules. Isolated molecules.**

    (iii) **4-nitrophenol. More possibility of hydrogen bonding with water molecules.**

**4** (a) Two isomeric compounds A and B, with the molecular formula, $C_3H_8O$, can be oxidised to C and D respectively. C reacts with Fehling's solution to produce a red-brown precipitate E and another compound F. D has no reaction with Fehling's solution, but gives a yellow crystalline product G when treated with 2,4-dinitrophenylhydrazine solution.

  (i) Give the names and structural formulae of compounds A, B, C, D and F.

  (ii) Name the compound E.

(*iii*) Draw the structural formula of G.
(*b*) Describe what you would observe and name all the products formed when C is warmed with sodium dichromate(VI) solution acidified with dilute sulphuric acid. Write an equation (or equations) for the reactions taking place.
(*c*) Describe how D is obtained industrially from crude petroleum.

(*Southern Universities Joint Board*)

**This question uses information from Unit 32 but illustrates the distinguishing of a primary and a secondary alcohol.**

**(*a*) (*i*) Since these isomers are easily oxidized, they cannot be ethers.**

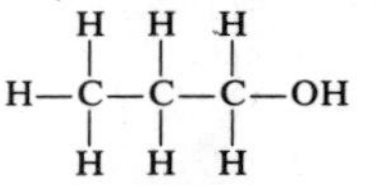
propan-1-ol

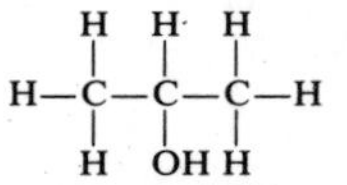
propan-2-ol

**C reacts with Fehling's solution and must, therefore, be the aldehyde. D must be the ketone as it gives a reaction with 2,4-dinitrophenylhydrazine but not with Fehling's solution.**

**A must be propan-1-ol (primary alcohol) which oxidizes to give the aldehyde C.**

**B must be propan-2-ol (secondary alcohol) which oxidizes to give the ketone D.**

C is 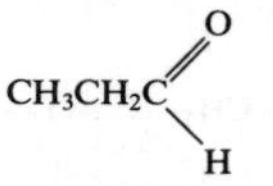
propanal

D is 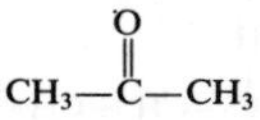
propanone

F is 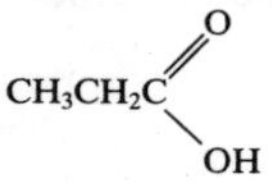

**(*ii*) E is copper(I) oxide, $Cu_2O$.**

**(*iii*)**

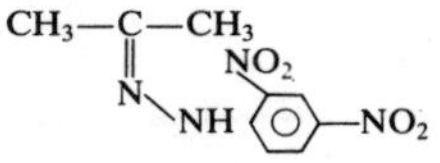

**(*b*) During the reaction the orange solution would turn green.**

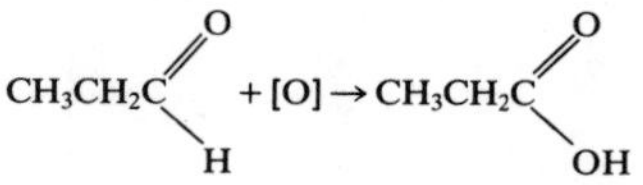
propanoic acid

$$Cr_2O_7^{2-} + 14H^+ + 6e^- \rightarrow 2Cr^{3+} + 7H_2O$$

**The products are propanoic acid, chromium(III) sulphate and water.**

**(*c*) Propanone is produced from crude petroleum by fractional distillation followed by cracking to produce propene. The industrial method of obtaining phenol (31.5) produces propanone as a by-product.**

# 32 Aldehydes and ketones

## 32.1 INTRODUCTION

In this Unit aldehydes and ketones will be considered. They both contain the carbonyl group. It is important to know the similarities and differences between aldehydes and ketones.
Some common aldehydes and ketones are:

*Aldehydes*

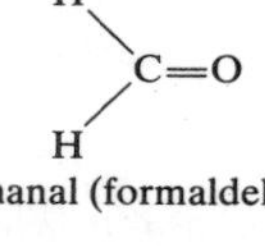
methanal (formaldehyde)

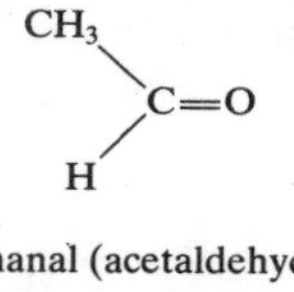
ethanal (acetaldehyde)

*Ketones*

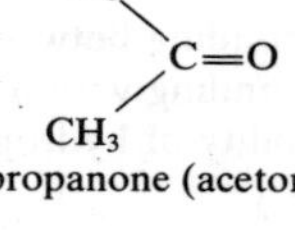
propanone (acetone)

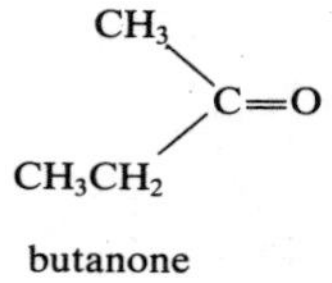
butanone

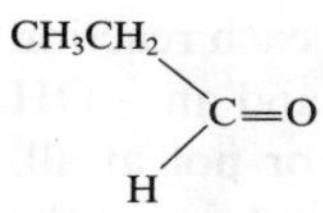

CH₃CH₂
\|
C=O
\|
H
propanal

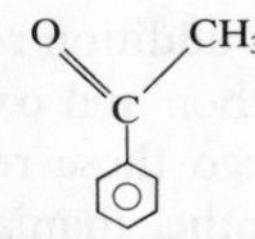

$$\text{CH}_3\text{CH}_2 - \underset{\underset{H}{|}}{C} = O$$
propanal

$$\text{CH}_3 - \underset{\underset{}{||}}{\overset{O}{C}}$$
phenylethanone

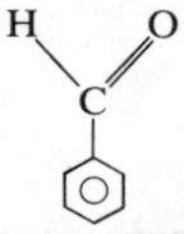

benzenecarbaldehyde (benzaldehyde)

Both aldehydes and ketones contain the C=O (carbonyl) group. In an aldehyde there is a hydrogen atom attached to the carbonyl group. Aldehydes and ketones can be represented as

$$\underset{R}{\overset{H}{\phantom{|}}}C=O \qquad \underset{R}{\overset{R'}{\phantom{|}}}C=O$$

aldehyde       ketone

Methanal is a gas at room temperature and pressure but is used in solution in water. The other simple aldehydes are liquids under normal laboratory conditions.

## 32.2 PREPARATION OF ALDEHYDES AND KETONES

### (i) Oxidation of alcohols

Aldehydes and ketones are usually prepared by the oxidation of alcohols using acidified potassium dichromate(VI) solution. Aldehydes are produced by oxidation of primary alcohols, and ketones by the oxidation of secondary alcohols.

*E.g.*      $\text{CH}_3\text{CH}_2\text{OH} + [\text{O}] \rightarrow \text{CH}_3\text{CHO} + \text{H}_2\text{O}$
                            ethanal

$\text{CH}_3\text{CH(OH)CH}_3 + [\text{O}] \rightarrow \text{CH}_3\text{COCH}_3 + \text{H}_2\text{O}$

Prolonged oxidation of primary alcohols leads to the formation of carboxylic acids.

### (ii) Hydrolysis of certain dihalides

Hydrolysis of compounds with two halogen atoms attached to the same carbon atom by boiling with dilute acid can produce aldehydes or ketones. If the two halogen atoms are attached to a terminal (or end) carbon atom an aldehyde is produced, and if attached to a middle carbon atom a ketone is produced.

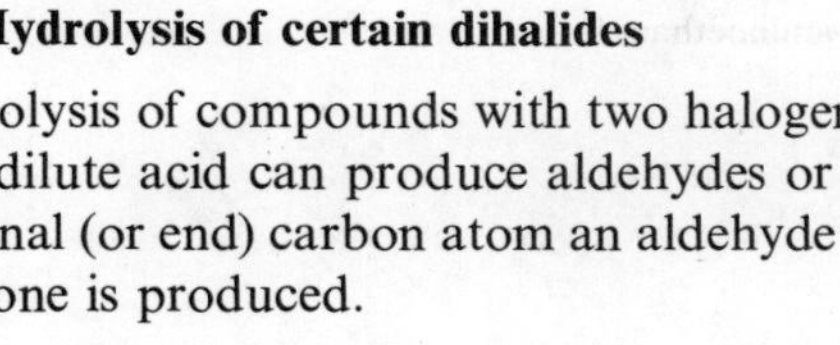

1,1-dichloroethane     ethanal 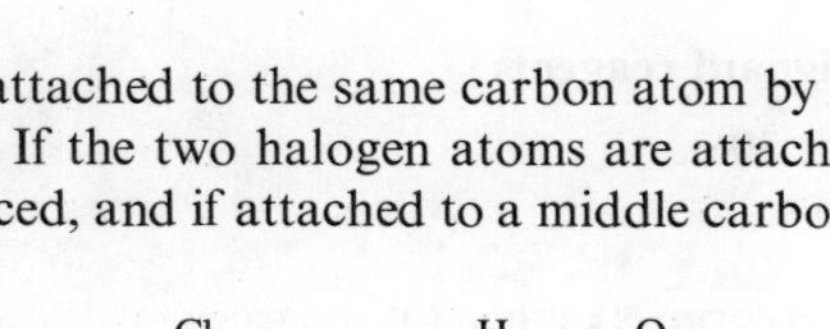          (dichloromethyl) benzene    benzenecarbaldehyde

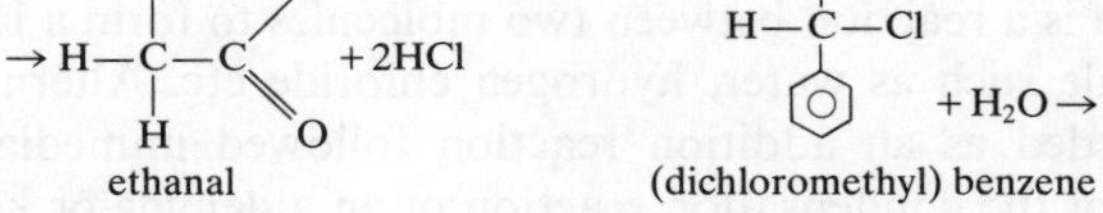

2,2-dichloropropane     propanone.

There are other methods for preparing specific aldehydes and ketones. These include:
(*i*) Friedel Crafts reaction to produce phenylethanone (Unit 29.11)
(*ii*) Hydration to ethyne to produce ethanal (Unit 29.9)
(*iii*) Action of heat on calcium ethanoate produces propanone (Unit 33.5)
(*iv*) Propanone obtained as a byproduct of the manufacture of phenol (Unit 31.5)

## 32.3 STRUCTURE OF THE CARBONYL GROUP

There are four electrons between the carbon and oxygen atoms in the double bond. Because oxygen is more electronegative than carbon, there is a slight electron shift towards the oxygen.

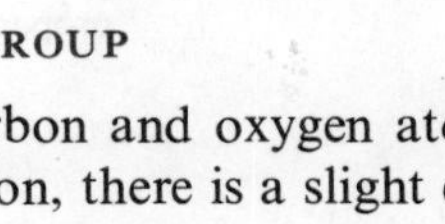

$$\overset{\delta+}{C}\!=\!\overset{\delta-}{O}$$

This makes the carbon atom more susceptible to nucleophilic attack (Unit 38.5).

## 32.4 ADDITION REACTIONS

There are a number of important addition reactions of aldehydes and ketones. In each reaction the double bond between the carbon and oxygen is converted into a single bond and an $-OH$ group is formed. Ketones undergo these reactions less readily than aldehydes, or not at all, especially if the ketone contains other than a methyl group. This is due to steric hindrance – the attacking nucleophile is unable to reach the carbonyl group. The addition reactions of aldehydes and ketones include:

### *(i) Addition of sodium hydrogensulphite

When a saturated solution of sodium hydrogensulphite is added to an aldehyde, a white crystalline addition product is formed.

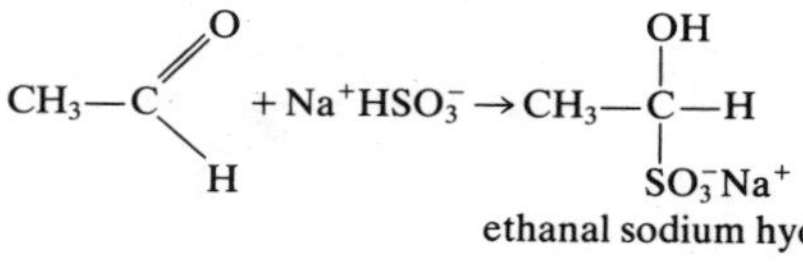

$$CH_3-C{\overset{O}{\underset{H}{}}} + Na^+HSO_3^- \rightarrow CH_3-\underset{\underset{SO_3^-Na^+}{|}}{\overset{\overset{OH}{|}}{C}}-H$$

ethanal sodium hydrogensulphite

This addition product decomposes on warming with a dilute acid to produce the original carbonyl compound. This provides a good method of purification of aldehydes.

### (ii) Addition of hydrogen cyanide

Addition of hydrogen cyanide to an aldehyde or ketone produces a cyanohydrin.

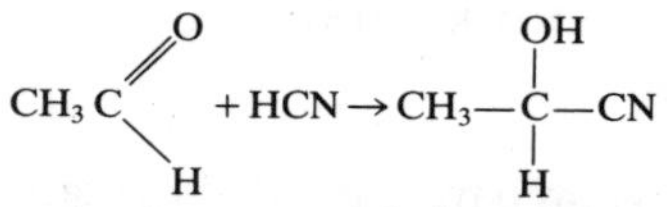

$$CH_3C{\overset{O}{\underset{H}{}}} + HCN \rightarrow CH_3-\underset{\underset{H}{|}}{\overset{\overset{OH}{|}}{C}}-CN$$

2-hydroxypropanenitrile
(ethanal cyanohydrin)

### (iii) Addition of ammonia

Ammonia forms a solid addition product with aldehydes (but not with methanal). Similar compounds using ketones in place of aldehydes are unstable.

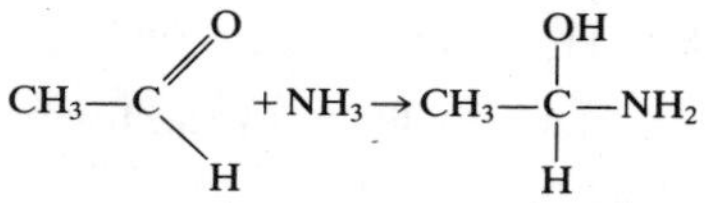

$$CH_3-C{\overset{O}{\underset{H}{}}} + NH_3 \rightarrow CH_3-\underset{\underset{H}{|}}{\overset{\overset{OH}{|}}{C}}-NH_2$$

1-aminoethanol

### (iv) Grignard reagents

See Unit 30.3.

## 32.5 CONDENSATION REACTIONS

A **condensation reaction** is a reaction between two molecules to form a larger molecule, with the loss of a small molecule such as water, hydrogen chloride etc. Alternatively, a condensation reaction may be regarded as an addition reaction followed immediately by an elimination reaction. The product of the condensation reaction of an aldehyde or ketone always contains a double bond.

### (i) Reaction with hydrazine, $NH_2NH_2$

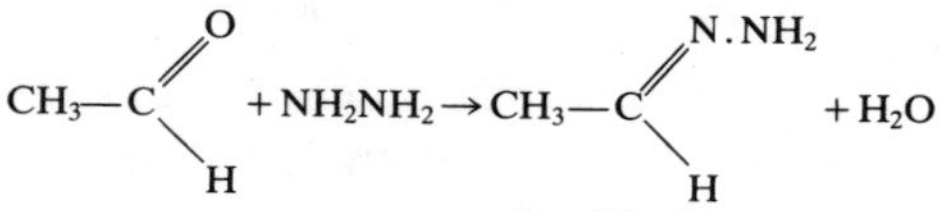

$$CH_3-C{\overset{O}{\underset{H}{}}} + NH_2NH_2 \rightarrow CH_3-C{\overset{N.NH_2}{\underset{H}{}}} + H_2O$$

ethanal hydrazone

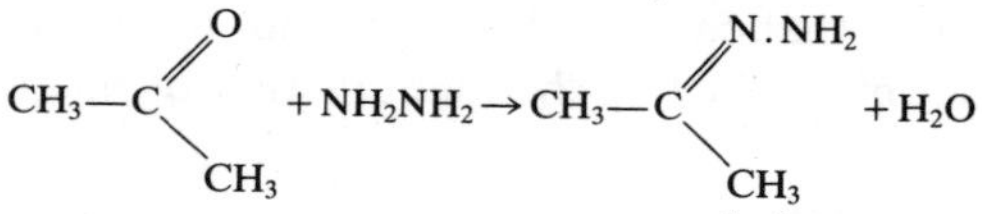

$$CH_3-C{\overset{O}{\underset{CH_3}{}}} + NH_2NH_2 \rightarrow CH_3-C{\overset{N.NH_2}{\underset{CH_3}{}}} + H_2O$$

propanone hydrazone

### (ii) Reaction with phenyhydrazine

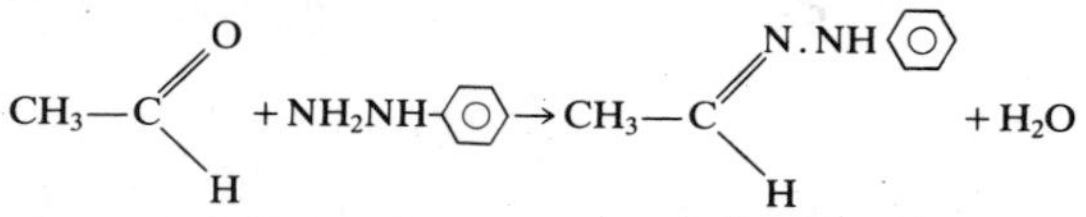

$$CH_3-C{\overset{O}{\underset{H}{}}} + NH_2NH-\bigcirc \rightarrow CH_3-C{\overset{N.NH\bigcirc}{\underset{H}{}}} + H_2O$$

ethanal phenylhydrazone

### (iii) Reaction with 2,4-dinitrophenylhydrazine

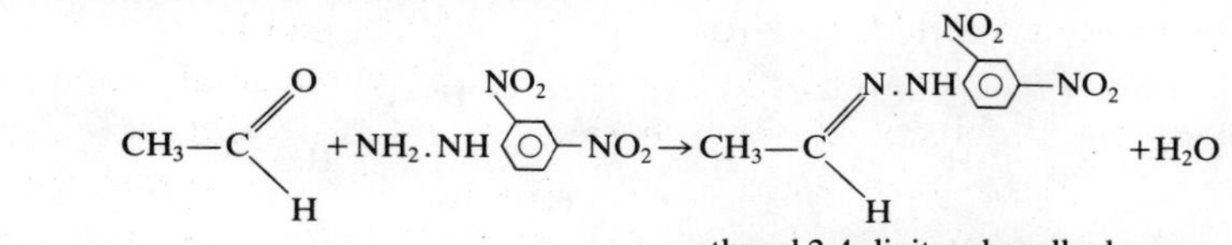

ethanal 2,4-dinitrophenylhydrazone

### *(iv) Reaction with hydroxylamine, NH$_2$OH

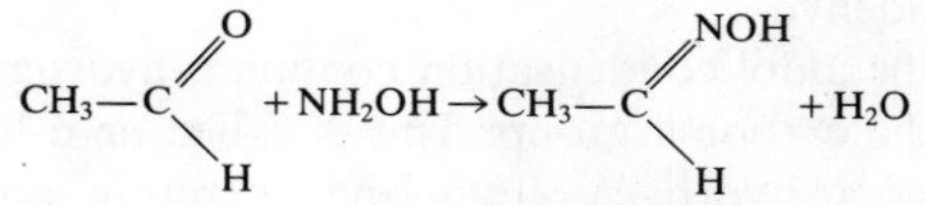

ethanal oxime

The products of these condensation reactions are obtained as solids by mixing a solution of the reagent with the aldehyde or ketone. These derivatives, particularly the phenylhydrazones and the 2,4-dinitrophenylhydrazones which are less soluble and more easily precipitated, are used to identify a particular aldehyde as the derivatives have characteristic melting points.

The mechanism for these reactions is explained in Unit 38.

### 32.6 OXIDATION AND REDUCTION REACTIONS

Aldehydes and ketones can be reduced to the appropriate alcohol by using lithium tetrahydrido-aluminate(III) (lithium aluminium hydride) or sodium and ethanol or sodium tetrahydrido-borate(III) (sodium borohydride).

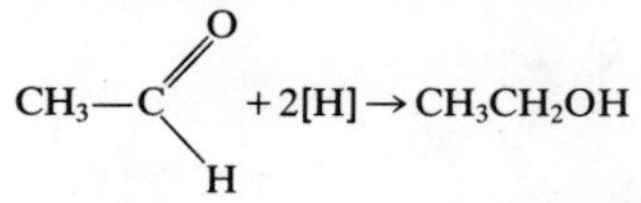

ethanol (primary alcohol) 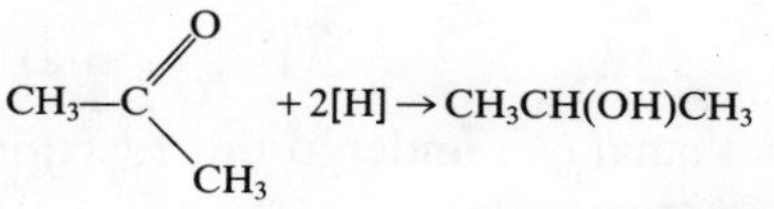     propan-2-ol (secondary alcohol)

Lithium tetrahydridoaluminate(III) is used in solution with ethoxyethane as solvent. Sodium tetrahydridoborate(III) is less reactive and is used in solution in water or ethanol.

Aldehydes can be oxidized to acids by:

(*i*) passing the aldehyde vapour into a solution of dilute sulphuric acid containing a manganese(II) salt as catalyst. The solution should be heated to about 50°C.

(*ii*) Refluxing the aldehyde with an excess of acidified sodium dichromate(VI) solution.

Aldehydes and ketones can be distinguished by oxidation. Aldehydes are readily oxidized to carboxylic acids but ketones cannot readily be oxidized.

(*i*) Aldehydes readily restore the colour to Schiff's reagent. Schiff's reagent consists of a pinkish-purple solution of magenta decolourized by passing sulphur dioxide through the solution. The pinkish-purple colour is restored when an aldehyde is added.

(*iii*) Tollen's reagent is essentially a complexed solution of silver(I) ions produced by dissolving silver(I) oxide in excess ammonia. When an aldehyde is mixed with Tollens' solution and the mixture heated, a silver mirror forms on the inside of the test tube. No silver mirror is produced with a ketone

$$Ag_2O + CH_3CHO \rightarrow CH_3COOH + 2Ag$$

(*iii*) Fehling's and Benedict's solutions are two very similar tests for reducing sugars. Either can be employed to distinguish between an aldehyde and a ketone.

Both reagents consist of a complexed solution of copper(II) ions. When one of these solutions is mixed with an aldehyde and the mixture warmed, a precipitate of copper(I) oxide is produced. The precipitate can vary considerably in colour from red to orange or even yellow. No precipitate is obtained with ketones.

$$RCHO + 2CuO \rightarrow RCOOH + Cu_2O$$

### 32.7 REACTIONS WITH ALKALIS

The reactions of aldehydes with alkalis are very complicated and depend upon the particular aldehyde, the concentration of the alkali and the reaction conditions.

Aliphatic aldehydes except methanal react with cold, dilute sodium hydroxide solution (or aqueous potassium carbonate solution.) For example ethanal reacts to produce 3-hydroxybutanal (aldol). This reaction is called the **aldol condensation**.

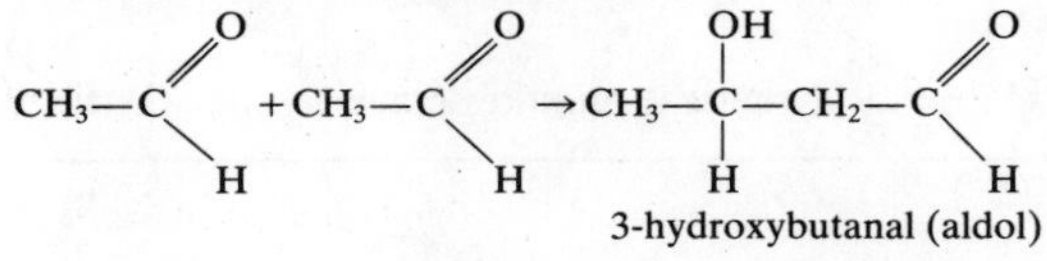

3-hydroxybutanal (aldol)

The aldol is inclined to lose water.

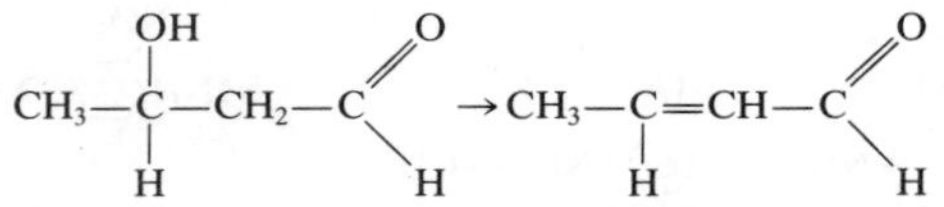

$$CH_3-\underset{\underset{H}{|}}{\overset{\overset{OH}{|}}{C}}-CH_2-\overset{\overset{O}{\parallel}}{C}\diagdown_{H} \rightarrow CH_3-\underset{\underset{H}{|}}{C}=CH-\overset{\overset{O}{\parallel}}{C}\diagdown_{H}$$

but-2-enal

Ketones do not undergo this type of condensation.

With hot, concentrated sodium hydroxide solution a complex yellow-brown resin is produced from polymerization of the aldehyde.

Aldehydes that undergo the aldol condensation contain a hydrogen atom attached directly to the carbon atom next to the carbonyl group. This is called an α-hydrogen atom. Methanal and benzaldehyde possess no α-hydrogen atom and therefore do not undergo the aldol condensation. These aldehydes, when shaken with sodium hydroxide solution, undergo disproportionation. The products are the corresponding alcohol and the sodium salt of the corresponding acid. This is **Cannizzarro's reaction**.

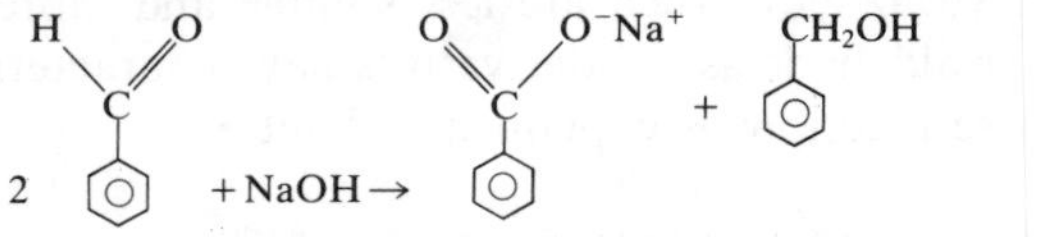

$$2 \;\; C_6H_5CHO + NaOH \rightarrow C_6H_5COO^-Na^+ + C_6H_5CH_2OH$$

sodium benzenecarboxylate     phenylmethanol

Benzenecarboxylic acid can be obtained by acidification with a dilute acid.

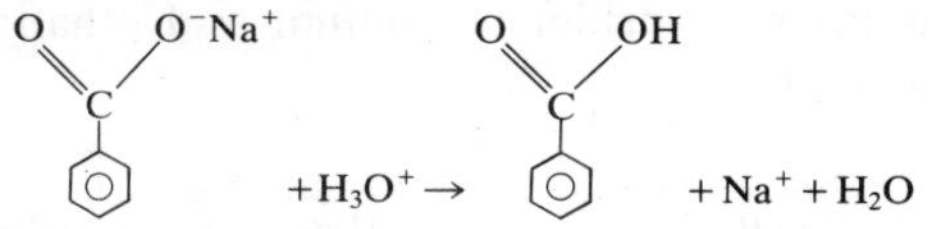

$$C_6H_5COO^-Na^+ + H_3O^+ \rightarrow C_6H_5COOH + Na^+ + H_2O$$

Similarly methanal can undergo this reaction

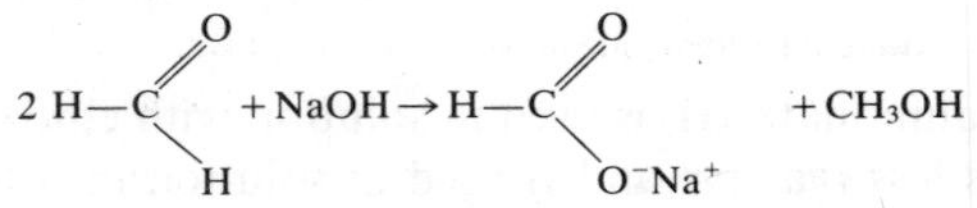

$$2 \; H-\overset{\overset{O}{\parallel}}{C}\diagdown_{H} + NaOH \rightarrow H-\overset{\overset{O}{\parallel}}{C}\diagdown_{O^-Na^+} + CH_3OH$$

sodium methanoate   methanol

## 32.8   COMPARISON OF ALDEHYDES AND KETONES

Table 32.1 compares the properties of a typical aldehyde and ketone.

**Table 32.1  Comparison of aldehydes and ketones**

|  | Aldehydes | Ketones |
| --- | --- | --- |
| Method of preparation | Oxidation of primary alcohol | Oxidation of secondary alcohol |
| Reduction with lithium tetrahydridoaluminate(III) | Primary alcohol formed | Secondary alcohol formed |
| Reaction with sodium hydrogensulphite | Addition product formed | Addition product formed |
| Reaction with hydrogen cyanide | Addition product formed | Addition product formed |
| Reaction with ammonia | Addition product formed (not with methanal) | No addition product formed |
| Reaction with hydrazine, phenylhydrazine, 2,4-dinitrophenylhydrazine | Precipitate formed  Condensation reaction | Precipitate formed  Condensation reaction |
| Fehling's solution | Red/brown precipitate | No reaction |
| Tollens' reagent | Silver mirror | No reaction |
| Schiff's reagent | Pinkish-purple colour restored | No reaction |
| Triiodomethane (iodoform) reaction | Yellow precipitate only formed with ethanal | Yellow precipitate formed with any ketone containing $CH_3-\overset{\overset{}{\underset{\underset{O}{\parallel}}{C}}}-$ group |
| Polymerization | Polymerization takes place | Little tendency to polymerize |

QUESTIONS

**1** The organic compound X, which contains carbon, hydrogen and oxygen only, was found to have a relative molecular mass of about 85.

When 0.43g of X is burnt in excess oxygen, 1.10g of carbon dioxide and 0.45g of water are formed.

$$(A_r(H) = 1, A_r(C) = 12, A_r(O) = 16)$$

(a) What is the empirical formula of X?

(b) What is the molecular formula of X?

(c) Write an equation for the complete combustion of X.

(d) X undergoes a condensation reaction with hydroxylamine and also with 2,4-dinitrophenylhydrazine. Write the structural formulae of **five** possible non-cyclic compounds which X could be, and give the systematic names for each of the five formulae.

(e) What is meant by a condensation reaction?

(f) Write an equation for the reaction of **one** of the five possible compounds for X with hydroxylamine.

(g) Look closely at the five structures you have written for X.

(i) Which will give a yellow precipitate on warming with a mixture of iodine and alkali?

(ii) Which will reduce diamminosilver(I) ions?

(a)

44g of carbon dioxide contains 12g of carbon

1.10g of carbon dioxide contains $\frac{12}{44} \times 1.1$g of carbon

$$= 0.30\text{g of carbon}$$

18g of water contains 2g of hydrogen

0.45g of water contains $\frac{2}{18} \times 0.45$g of hydrogen

$$= 0.05\text{g of hydrogen}$$

0.43g of X contains 0.30g of carbon and 0.05g of hydrogen
The remainder must be oxygen = 0.08g of oxygen

Divide masses of each element by the relative atomic mass of each element.

| C | H | O |
|---|---|---|
| 0.025 | 0.05 | 0.005 |

Empirical formula $= C_5H_{10}O$

(b) Adding up the relative atomic masses of atoms in the empirical formula

$$= (12 \times 5) + (10 \times 1) + (1 \times 16)$$
$$= 86$$

Molecular formula $=$ empirical formula $= C_5H_{10}O$

(c) Complete combustion of the compound would produce carbon dioxide and water.

$$C_5H_{10}O + 7O_2 \rightarrow 5CO_2 + 5H_2O$$

(d) Since X undergoes a condensation reaction with hydroxylamine and forms a 2,4-dinitrophenyl-hydrazone, it is an aldehyde or ketone.

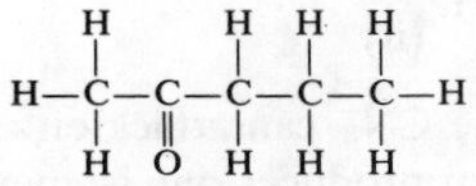

pentan-2-one

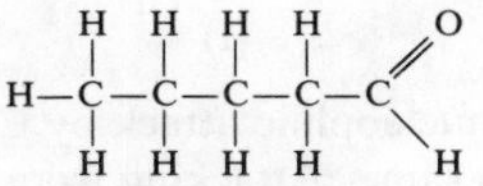

pentanal

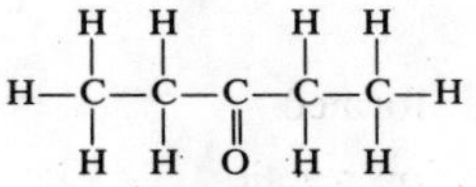

pentan-3-one

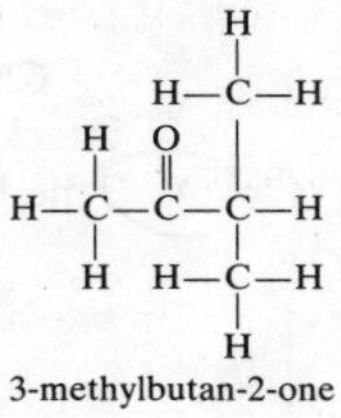

3-methylbutan-2-one

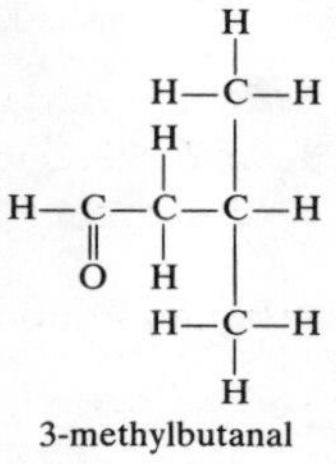

3-methylbutanal

(e) A condensation reaction occurs between two molecules to form a larger molecule, with the loss of water or a similar small molecule.

*E.g.*

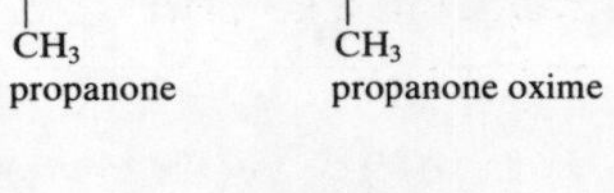

propanone          propanone oxime

It can also be defined as an addition reaction taking place immediately followed by an elimination reaction.
NB It is a good idea to give an example when defining a term like this.

(*f*)

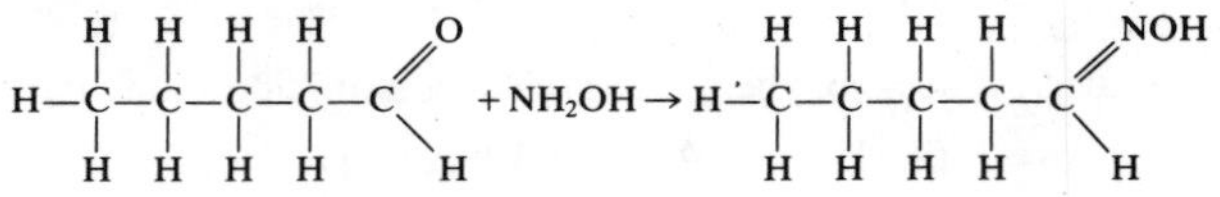

$$
H-\overset{\overset{\displaystyle H}{|}}{\underset{\underset{\displaystyle H}{|}}{C}}-\overset{\overset{\displaystyle H}{|}}{\underset{\underset{\displaystyle H}{|}}{C}}-\overset{\overset{\displaystyle H}{|}}{\underset{\underset{\displaystyle H}{|}}{C}}-\overset{\overset{\displaystyle H}{|}}{\underset{\underset{\displaystyle H}{|}}{C}}-C\!\!\overset{O}{\underset{H}{\diagdown}} \;+\;NH_2OH \;\rightarrow\; H-\overset{\overset{\displaystyle H}{|}}{\underset{\underset{\displaystyle H}{|}}{C}}-\overset{\overset{\displaystyle H}{|}}{\underset{\underset{\displaystyle H}{|}}{C}}-\overset{\overset{\displaystyle H}{|}}{\underset{\underset{\displaystyle H}{|}}{C}}-\overset{\overset{\displaystyle H}{|}}{\underset{\underset{\displaystyle H}{|}}{C}}-C\!\!\overset{NOH}{\underset{H}{\diagdown}}
$$

(*g*) (*i*) Triiodomethane (iodoform) reaction again.
Pentan–2–one, 3–methylbutan–2–one will give a positive result as they contain the $CH_3CO-$ group.

(*ii*) Diamminosilver(I) ion is the active component of Tollens' reagent. Aldehydes will reduce this reagent to form silver.
Pentanal and 3-methylbutanal.

2 Glycerol ($CH_2OHCH(OH)CH_2OH$), on heating with anhydrous magnesium sulphate, yields an acrid-smelling distillate of molecular formula $C_3H_4O$, which immediately decolourizes bromine water and restores the colour to Schiff's reagent. Indicate a possible structure for the compound and suggest its mode of formation from glycerol.
The product must contain the –CHO aldehyde group as it restores the colour to Schiff's reagent. The product also contains a double or triple bond as it decolourizes bromine water.
The product is therefore

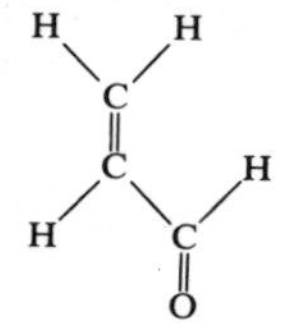

This compound is formed by the dehydration of glycerol. Anhydrous magnesium sulphate acts as the dehydrating agent.

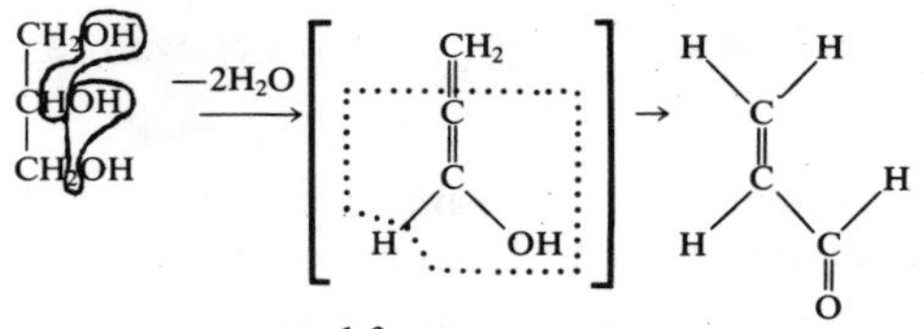

enol form rearranges

3 The reaction between ethanal and hydrogen cyanide can be represented by the equation:

$$CH_3CHO + HCN \rightarrow CH_3CH\,(OH)\,CN$$

(*a*) Explain how it is possible for the product to be a mixture of two isomers. (Hint: See page 51)
(*b*) Draw spatial formulae showing these two isomers.
(*c*) Why are the isomers present in equal quantities?

(*a*) The product, 2-hydroxypropanenitrile, contains a chiral carbon atom. It can, therefore, exist in two optical isomers.
(*b*)

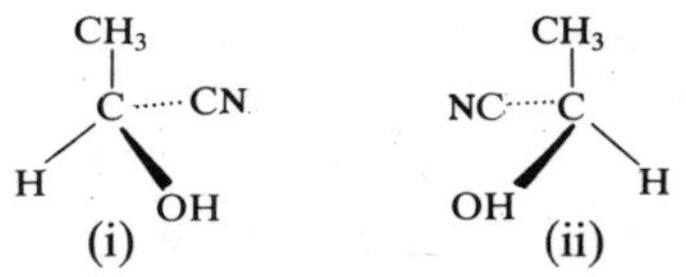

(*c*) The reaction involves nucleoplilic attack by $CN^-$. The $CN^-$ can attack either from the back or the front and the probability is same. Attacking from the front produces one isomer and from the back the other.

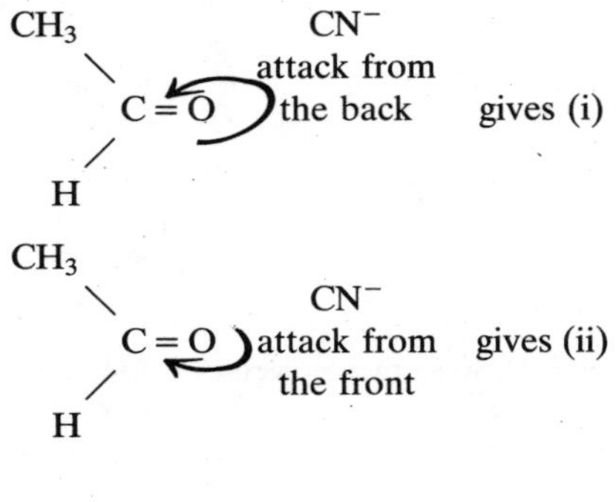

# 33 Carboxylic acids and their salts

## 33.1 INTRODUCTION

This unit contains information about carboxylic acids and the salts formed by these acids. Unit 39 compares the acidity of various carboxylic acids.

Carboxylic acids contain the carboxyl group, which, as its name suggests, is composed of a carbonyl and a hydroxyl group. The carboxyl group is

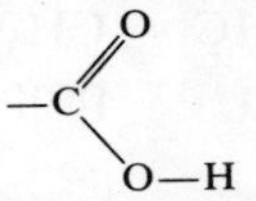

Despite having a carbonyl group within the structure, carboxylic acids do not undergo the typical addition and condensation reactions of aldehydes and ketones. This is because of the effect of the adjacent oxygen supplying electrons making the carbon of the carbonyl group less electron deficient.

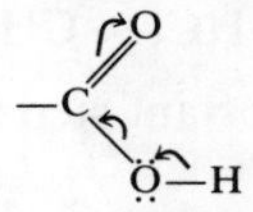

The movement of electrons away from the hydrogen atom weakens the bond between the hydrogen and oxygen atoms. The breaking away of this hydrogen to form a H$^+$ ion accounts for the acidity of the acid.

Common acids include:

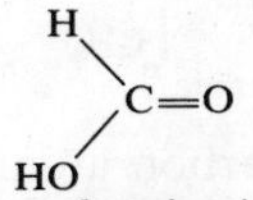

methanoic acid
(formic acid)

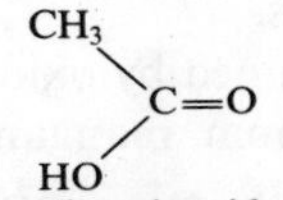

ethanoic acid
(acetic acid)

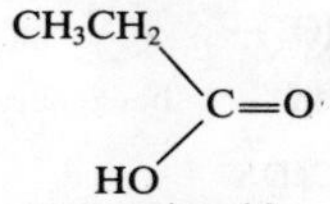

propanoic acid

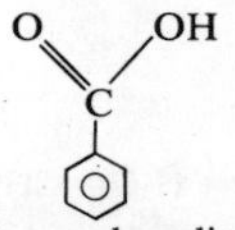

benzenecarboxylic acid
(benzoic acid)

All of these acids are monobasic, *i.e.* they contain only one hydrogen which can be lost as an H$^+$ ion.

Acids are named from the corresponding hydrocarbon by removing the –e and adding –oic acid.

Hydrogen bonding has a noticeable effect on the properties of the lower acids. Because of hydrogen bonding, methanoic and ethanoic acids are more soluble in water and have higher boiling points than expected. Higher members of the homologous series, with longer carbon chains, are less influenced by hydrogen bonding and are less soluble in water.

Benzenecarboxylic acid (benzoic acid) is only sparingly soluble in cold water but is more soluble in hot water.

## 33.2 PREPARATION OF ALIPHATIC AND AROMATIC CARBOXYLIC ACIDS

Aliphatic carboxylic acids can be prepared by the following methods.

(*i*) Prolonged oxidation of primary alcohols or aldehydes (Unit 31.4)

$E.g.$

$$CH_3CH_2OH + 2[O] \rightarrow CH_3COOH + H_2O$$
$$\text{ethanoic acid}$$

$$CH_3CH_2CHO + [O] \rightarrow CH_3CH_2COOH$$
$$\text{propanoic acid}$$

Heating with excess acidified potassium dichromate(VI) solution is one of the methods which could be used for this purpose.

(*ii*) Hydrolysis of nitriles (Unit 34.6). Nitriles are hydrolyzed by boiling with water but are hydrolyzed more quickly if the hydrolysis is carried out with a dilute acid or a dilute alkali.

*Acid hydrolysis*

$$CH_3CH_2CN + 2H_2O + HCl \rightarrow CH_3CH_2COOH + NH_4Cl$$
$$\text{propanenitrile} \qquad\qquad\qquad \text{propanoic acid}$$

*Alkaline hydrolysis*

$$CH_3CH_2CN + NaOH + H_2O \rightarrow CH_3CH_2COO^-Na^+ + NH_3$$
$$\text{sodium propanoate}$$

The free acid is obtained by acidification of the product.

$$CH_3CH_2COO^-Na^+ + HCl \rightarrow CH_3CH_2COOH + NaCl$$

Hydrolysis of nitriles can form an important step in ascending the homologous series (Unit 37.3)

(*iii*) From Grignard reagents (Unit 30.3). The Grignard reagent is poured onto solid carbon dioxide and the product is hydrolyzed by refluxing with a dilute acid

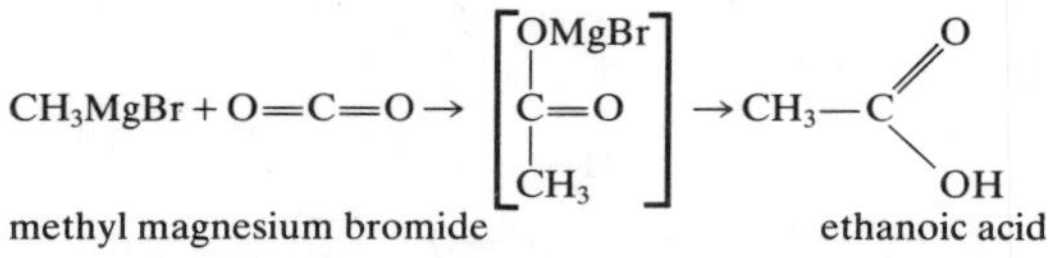

methyl magnesium bromide                    ethanoic acid

NB The product contains one more carbon atom than the Grignard reagent used. This is another method, therefore, of ascending the homologous series (Unit 37.3).

(*iv*) By hydrolysis of esters (Unit 34.4). Naturally occurring esters present in fats and oils can be used as sources of certain carboxylic acids.

Aromatic carboxylic acids can be obtained by oxidation of side-chains with strong oxidizing agents, *e.g.* alkaline or acidified potassium manganate(VII) or acidified potassium dichromate(VI).

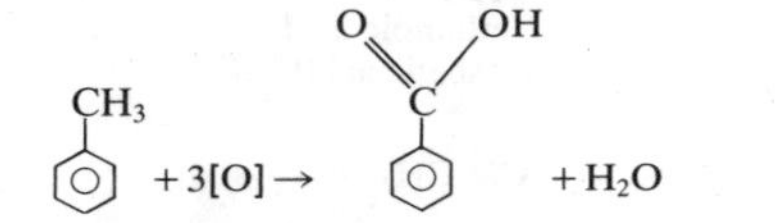

methylbenzene        benzenecarboxylic acid

### 33.3   REACTIONS OF CARBOXYLIC ACIDS

**(i) Acidic properties**

Carboxylic acids are weak acids

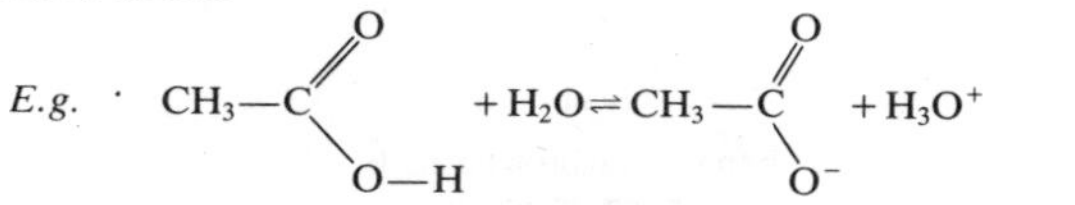

Carboxylic acids liberate carbon dioxide from a carbonate,

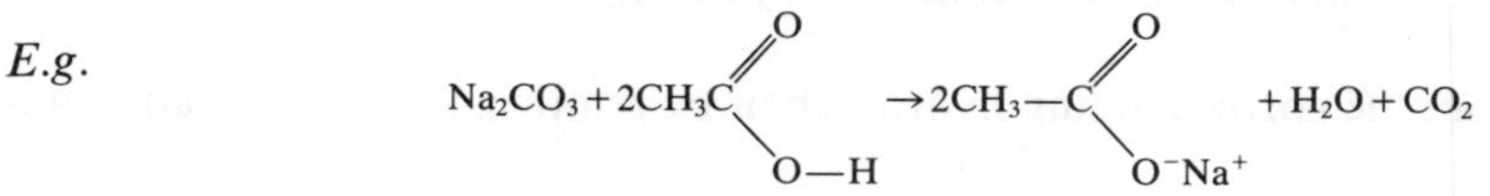

$E.g.$

sodium ethanaoate

**(ii) Reduction**

Carboxylic acids are reduced by lithium tetrahydridoaluminate(III) (lithium aluminium hydride) in dry ethoxyethane to produce a primary alcohol.

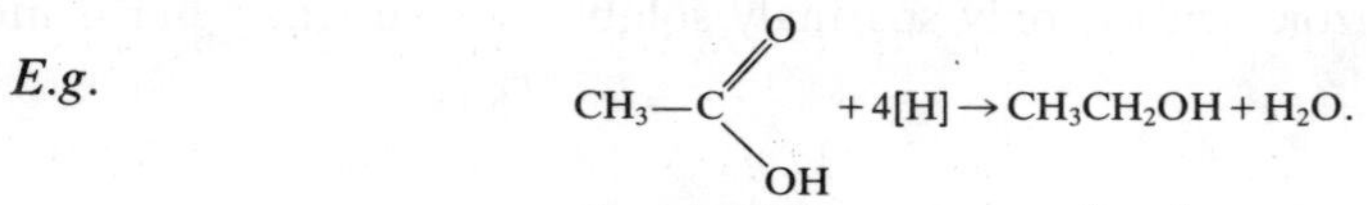

$E.g.$

ethanoic acid                    ethanol

It is not possible to reduce carboxylic acids and produce an aldehyde directly.

**(iii) Reaction with phosphorus pentachloride and sulphur dichloride oxide (thionyl chloride)**

The products of these reactions include an acid chloride and hydrogen chloride:

*E.g.*

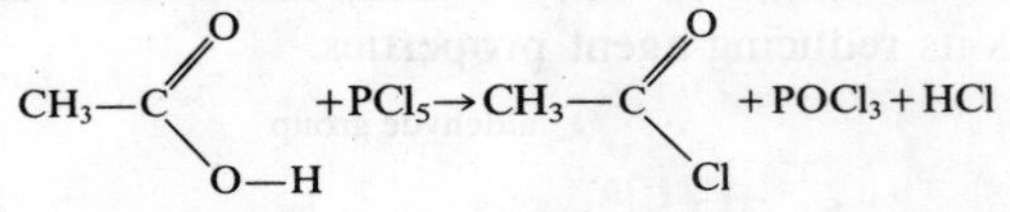

$$CH_3-C\underset{O-H}{\overset{O}{\big\langle}} + PCl_5 \rightarrow CH_3-C\underset{Cl}{\overset{O}{\big\langle}} + POCl_3 + HCl$$

ethanoic acid + phosphorus pentachloride → ethanoyl chloride + phosphorus trichloride oxide
+ hydrogen chloride

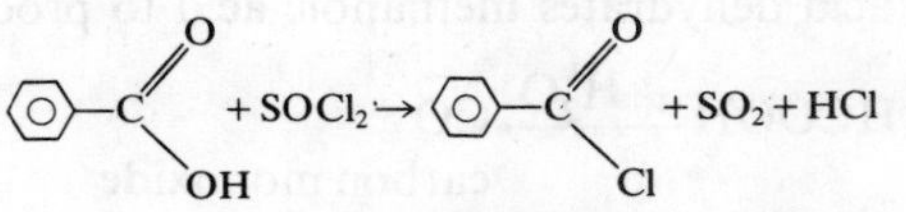

$$\text{⬡}-C\underset{OH}{\overset{O}{\big\langle}} + SOCl_2 \rightarrow \text{⬡}-C\underset{Cl}{\overset{O}{\big\langle}} + SO_2 + HCl$$

benzenecarbonyl chloride
(benzoyl chloride)

The formation of hydrogen chloride is used to test for the presence of an –OH group. Hydrogen chloride is produced with alcohols (Unit 29.2) and carboxylic acids.

**(iv) Halogenation of the side chain**

Chlorine or bromine substitute in alkyl group to form chloro- or bromo-substituted acids. The reaction takes place when the acid and halogen are mixed and subjected to strong light or a catalyst of iodine or red phosphorus:

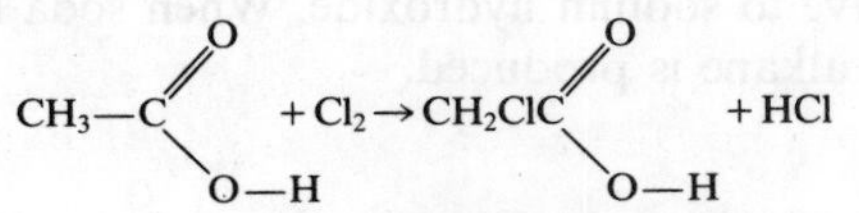

$$CH_3-C\underset{O-H}{\overset{O}{\big\langle}} + Cl_2 \rightarrow CH_2ClC\underset{O-H}{\overset{O}{\big\langle}} + HCl$$

chloroethanoic acid

$$CH_2Cl\,C\underset{O-H}{\overset{O}{\big\langle}} + Cl_2 \rightarrow CHCl_2C\underset{O-H}{\overset{O}{\big\langle}} + HCl$$

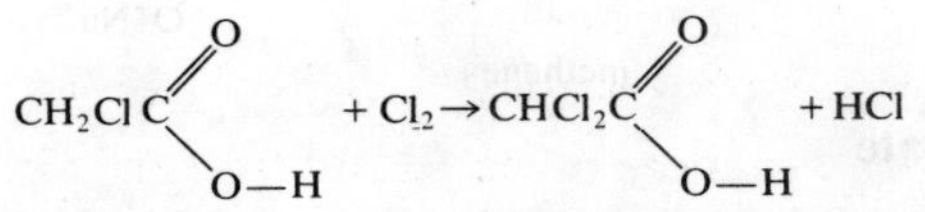

dichloroethanoic acid

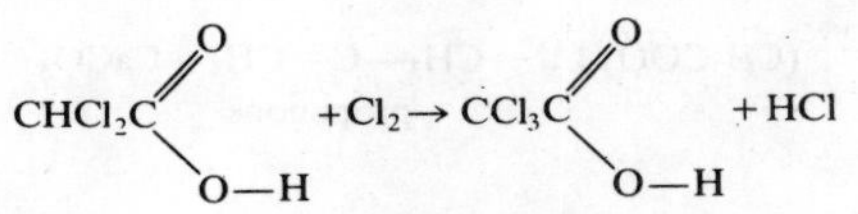

$$CHCl_2C\underset{O-H}{\overset{O}{\big\langle}} + Cl_2 \rightarrow CCl_3C\underset{O-H}{\overset{O}{\big\langle}} + HCl$$

trichloroethanoic acid

NB Unlike the chlorination of alkanes, this is a *controllable* reaction.

Where the carboxylic acid with a longer carbon chain is used, the substitution occurs at the α carbon atom.

$$CH_3CH_2CH_2COOH + Cl_2 \rightarrow CH_3CH_2CHClCOOH$$
2-chlorobutanoic acid

**(v) Ester formation**

When a mixture of carboxylic acid and alcohol are refluxed in the presence of concentrated sulphuric acid, an ester is formed.

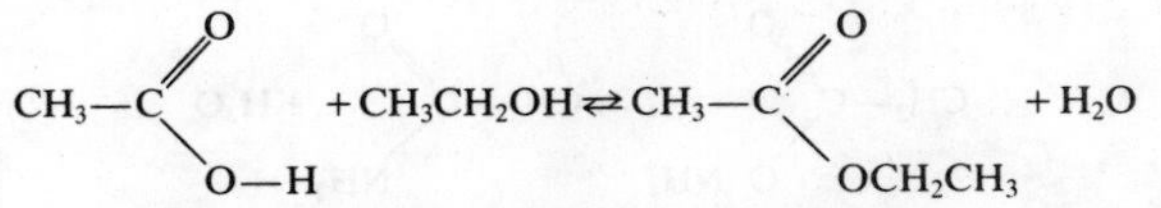

$$CH_3-C\underset{O-H}{\overset{O}{\big\langle}} + CH_3CH_2OH \rightleftharpoons CH_3-C\underset{OCH_2CH_3}{\overset{O}{\big\langle}} + H_2O$$

Acid    +    Alcohol   ⇌    Ester    +    Water

ethanoic acid + ethanol ⇌ ethyl ethanoate + water

**33.4 METHANOIC ACID AS A REDUCING AGENT**

The first member of any homologous series often shows atypical properties, *i.e.* shows some properties that are not carried out by other members of the series. This is certainly true with methanoic acid.

Methanoic acid, unlike other carboxylic acids, is easily oxidized by refluxing with acidified potassium manganate(VII) solution.

$$HCOOH + [O] \rightarrow H_2O + CO_2$$

Methanoic acid produces a silver mirror with Tollens' reagent and a brown precipitate with Fehling's solution.

Methanoic acid, unlike any other carboxylic acid, contains an aldehyde group within its structure and this explains its reducing agent properties.

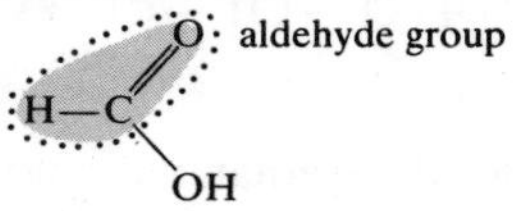

Concentrated sulphuric acid dehydrates methanoic acid to produce carbon monoxide.

$$HCOOH \xrightarrow{-H_2O} CO$$
carbon monoxide

## *33.5   SALTS OF CARBOXYLIC ACIDS

There are several reactions of salts of carboxylic acids that are important. Salts are ionized and contain the carboxylate ion.

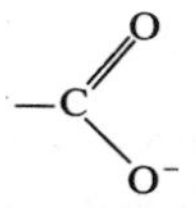

### (i) Heating the sodium salt of a carboxylic acid with soda lime

Soda lime is a safer alternative to sodium hydroxide. When soda lime is heated with a sodium salt of a carboxylic acid, an alkane is produced.

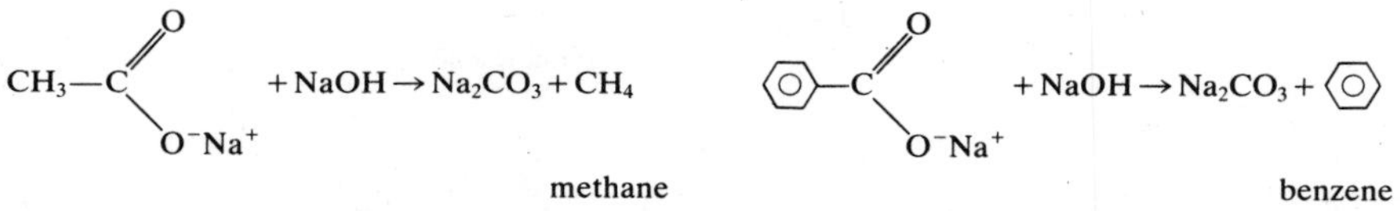

methane                                                benzene

### (ii) Heating calcium ethanoate

When dry calcium ethanoate is heated strongly, a poor yield of propanone is produced:

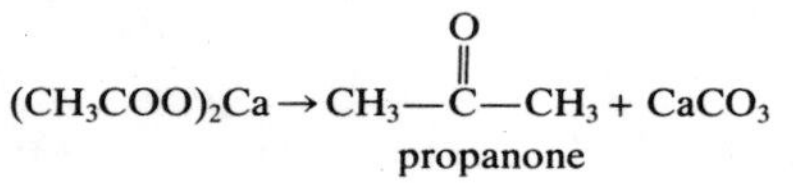

$$(CH_3COO)_2Ca \rightarrow CH_3-\overset{O}{\overset{\|}{C}}-CH_3 + CaCO_3$$
propanone

Despite the poor yield, this reaction figures frequently in A level questions.

### (iii) Heating a mixture of calcium methanoate and the calcium salt of another carboxylic acid

When the mixture of two calcium salts is heated, an aldehyde is produced in very poor yield:

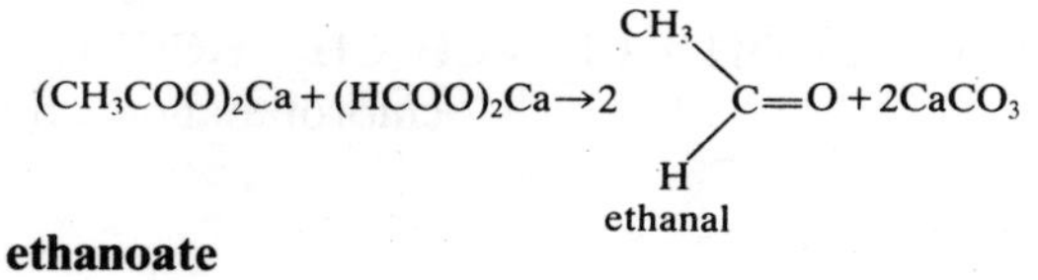

$$(CH_3COO)_2Ca + (HCOO)_2Ca \rightarrow 2 \underset{\underset{ethanal}{H}}{\overset{CH_3}{}}C=O + 2CaCO_3$$

### (iv) Heating ammonium ethanoate

Ammonium ethanoate is prepared by mixing solutions of ethanoic acid and ammonia. On heating the solid ethanoate, ethanamide is produced.

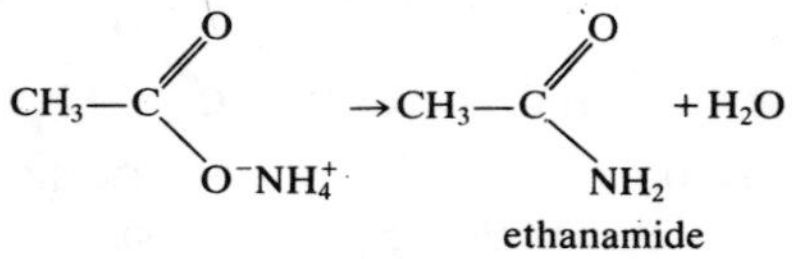

ethanamide

## *33.6   COMMON DIBASIC ACIDS

Ethanedioic acid (oxalic acid) is a dibasic acid containing two −COOH groups

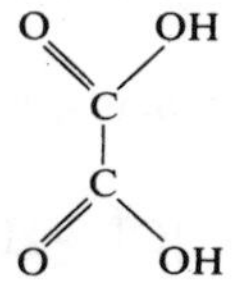

Two hydrogen atoms can be lost from each molecule as oxonium ions

$$\begin{array}{l}\text{COOH}\\|\\\text{COOH}\end{array} + 2H_2O \rightarrow \left[\begin{array}{l}\text{COO}\\|\\\text{COO}\end{array}\right]^{2-} + 2H_3O^+$$

ethanedioate ions

Ethanedioic acid, like methanoic acid, can be readily oxidised with acidified potassium manganate(VII).

$$\left[\begin{array}{l}\text{COO}\\|\\\text{COO}\end{array}\right]^{2-} \rightarrow 2CO_2 + 2e^-$$

Ethanedioic acid is also dehydrated by concentrated sulphuric acid on heating.

$$\begin{array}{l}\text{COOH}\\|\\\text{COOH}\end{array} \xrightarrow{-H_2O} CO_2 + CO$$

Hexane-1,6-dioic acid (adipic acid) is used in the manufacture of nylon.

$$HOOCCH_2CH_2CH_2CH_2COOH$$

It is manufactured by oxidation of cyclohexene.

$$\bigcirc \xrightarrow{[O]} HOOC(CH_2)_4COOH$$

There are three isomeric benzenedicarboxylic acids (phthalic acids)

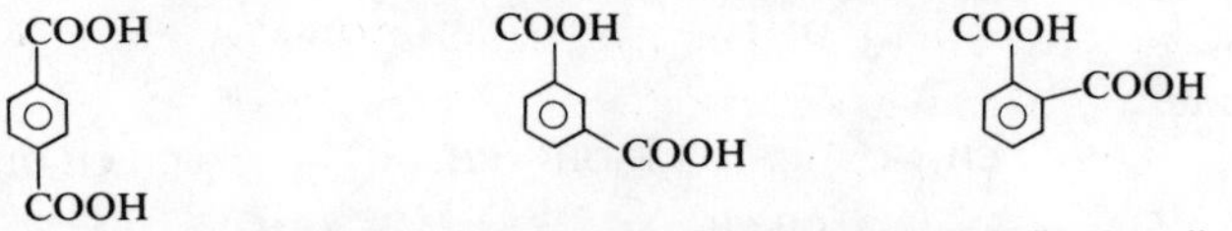

benzene-1,4-dicarboxylic acid (terephthalic acid)    benzene-1,3-dicarboxylic acid    benzene-1,2-dicarboxylic acid (phthalic acid)

Benzene-1,4-dicarboxylic acid is used in the manufacture of 'Terylene'. It is made by the oxidation of 1,4-dimethylbenzene with nitric acid.

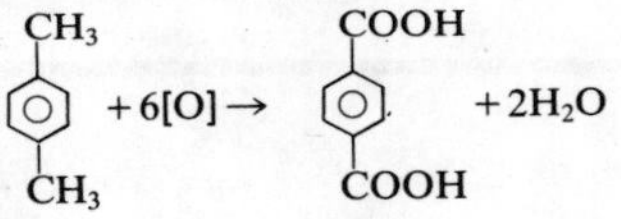

$$\bigcirc\text{(CH}_3)_2 + 6[O] \rightarrow \bigcirc\text{(COOH)}_2 + 2H_2O$$

## Questions

1 Describe how aminoethanoic acid could be prepared from ethanoic acid.

Monochloroethanoic acid is prepared by passing chlorine into warm ethanoic acid in strong ultraviolet light. Only a limited amount of chlorine is used to prevent further substitution.

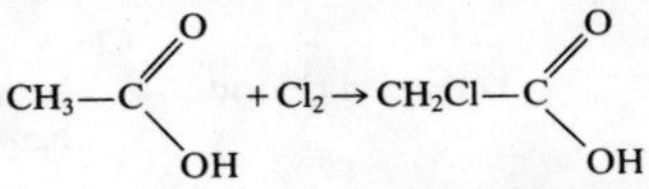

$$CH_3-C\overset{O}{\underset{OH}{<}} + Cl_2 \rightarrow CH_2Cl-C\overset{O}{\underset{OH}{<}}$$

monochloroethanoic acid

Monochloroethanoic acid reacts with concentrated ammonia solution at room temperature to form aminoethanoic acid.

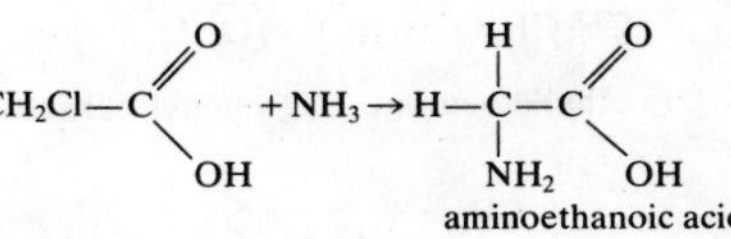

$$CH_2Cl-C\overset{O}{\underset{OH}{<}} + NH_3 \rightarrow H-\underset{NH_2}{\overset{H}{C}}-C\overset{O}{\underset{OH}{<}}$$

aminoethanoic acid

2 Identify the following substances. Explain fully the changes which have been recorded and give equations where possible.

(*i*) A is a white solid which, on acidification with dilute hydrochloric acid, produces a substance B which has a molecular formula $C_2H_4O_2$. A solution of B has a pH less than 7.

When A is heated dry in a test tube, a vapour is produced which, on cooling, produces a liquid C with a molecular formula $C_3H_6O$. C forms a yellow precipitate with 2,4-dinitrophenylhydrazine but does not form a silver mirror when warmed with Tollen's reagent.

(*ii*) D is a sweet-smelling liquid which reacts with aqueous sodium hydroxide solution to produce two compounds E and F.

When E is in the anhydrous state and is heated with soda lime, a hydrocarbon with relative molecular mass of 16 is given off.

F on oxidation with acidified potassium dichromate(VI) produces a compound G which on reaction with sodium hydroxide again produces E.

(*i*) B is an acid (pH less than 7)

B is ethanoic acid $CH_3COOH$

A is a salt of ethanoic acid.

C forms a yellow precipitate with 2,4-dinitrophenylhydrazine and must be an aldehyde or ketone (Unit 32.5). As it does not produce a silver mirror with Tollens' reagent, it must be a ketone.

C must be propanone

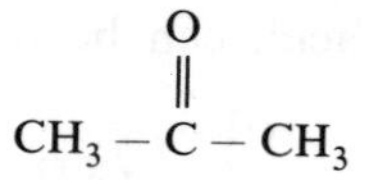

$$CH_3 - \overset{\overset{\displaystyle O}{\|}}{C} - CH_3$$

A is calcium ethanoate

$$(CH_3COO)_2Ca \rightarrow CaCO_3 + CH_3COCH_3$$

Acidification of calcium ethanoate liberates ethanoic acid.

$$(CH_3COO)_2Ca + 2H_3O^+ \rightarrow 2CH_3COOH + Ca^{2+} + 2H_2O$$

(*ii*) Since D is a sweet-smelling liquid, this suggests that it is an ester. E is sodium ethanoate. When heated with soda lime, methane is produced (relative molecular mass 16)

$$CH_3COONa + NaOH \rightarrow CH_4 + Na_2CO_3$$
$$\text{methane}$$

F is ethanol $CH_3CH_2OH$

Oxidation of ethanol produces ethanoic acid which on neutralization produces sodium ethanoate.

$$CH_3CH_2OH + 2[O] \rightarrow CH_3COOH + H_2O$$

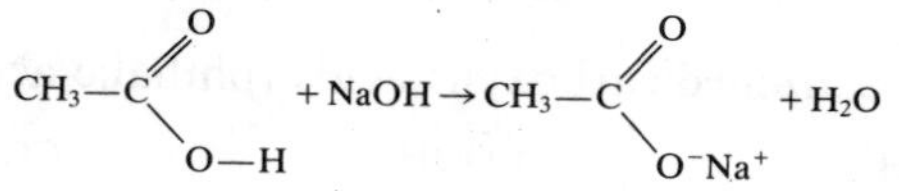

D is ethyl ethanoate

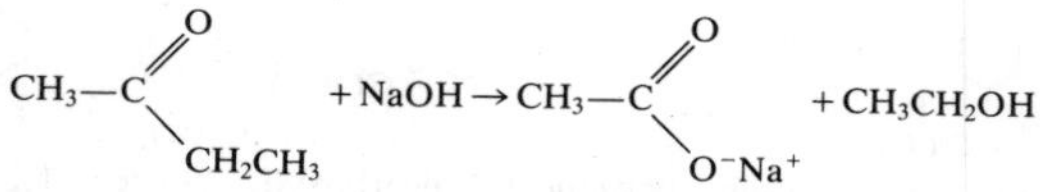

$$CH_3-\overset{\overset{\displaystyle O}{\|}}{\underset{O-H}{C}} + NaOH \rightarrow CH_3-\overset{\overset{\displaystyle O}{\|}}{\underset{O^-Na^+}{C}} + H_2O$$

$$CH_3-\overset{\overset{\displaystyle O}{\|}}{\underset{CH_2CH_3}{C}} + NaOH \rightarrow CH_3-\overset{\overset{\displaystyle O}{\|}}{\underset{O^-Na^+}{C}} + CH_3CH_2OH$$

# 34 Acid derivatives

## 34.1 INTRODUCTION

In this unit various compounds closely related to carboxylic acids will be discussed.

These derivatives are all based upon carboxylic acids but with the —OH group of the carboxyl group replaced.

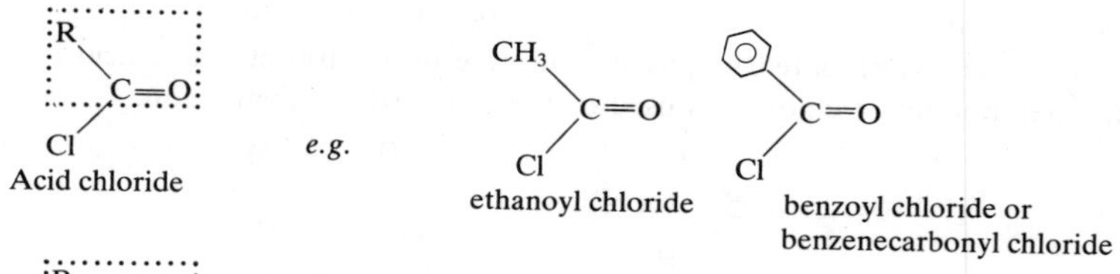

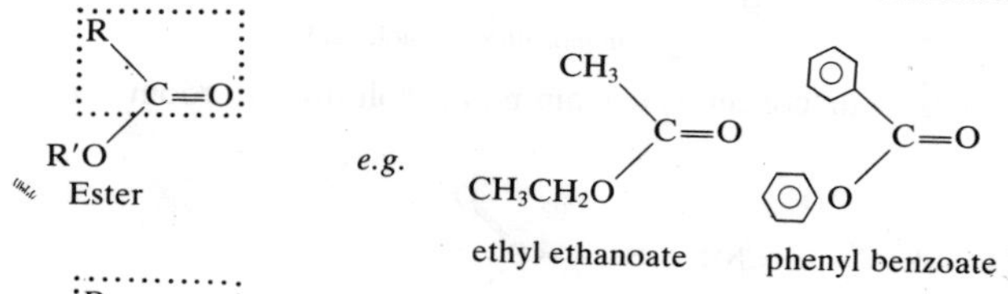

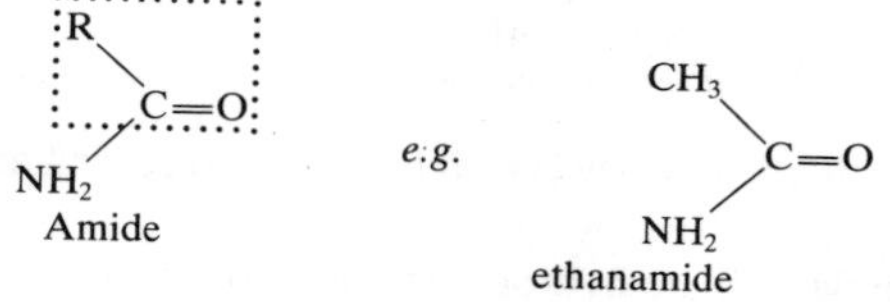

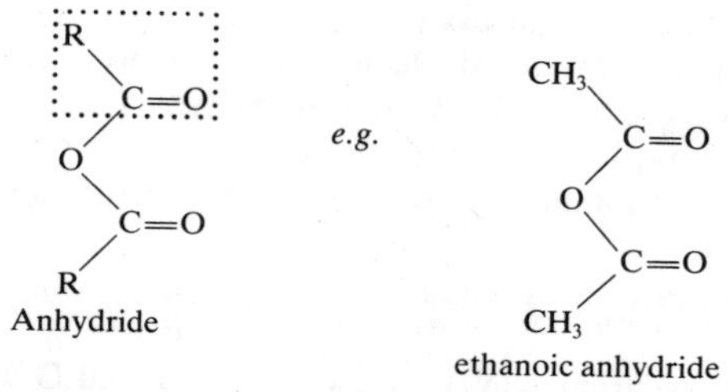

In addition nitriles will be included as they are closely related to carboxylic acids.

**34.2** ACID CHLORIDES

Acid chlorides are prepared by warming the corresponding carboxylic acid with sulphur dichloride oxide (thionyl chloride) or phosphorus chlorides.

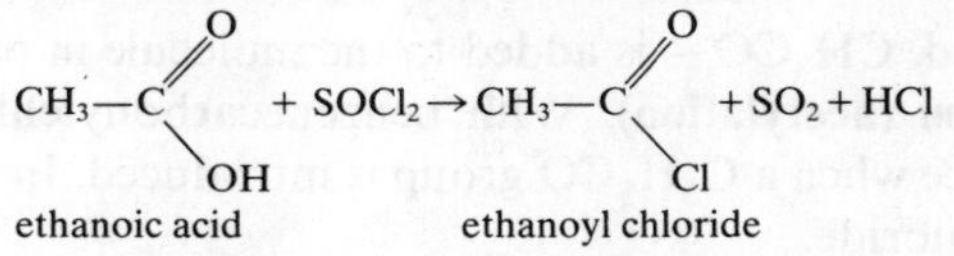

$$CH_3-\overset{O}{\underset{OH}{C}} + SOCl_2 \rightarrow CH_3-\overset{O}{\underset{Cl}{C}} + SO_2 + HCl$$

ethanoic acid   ethanoyl chloride

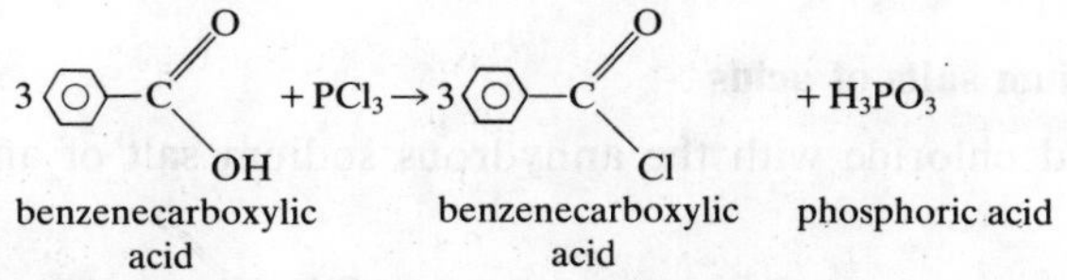

$$3\ \langle O \rangle-\overset{O}{\underset{OH}{C}} + PCl_3 \rightarrow 3\ \langle O \rangle-\overset{O}{\underset{Cl}{C}} + H_3PO_3$$

benzenecarboxylic   benzenecarboxylic   phosphoric acid
acid    acid

  Acid chlorides are colourless liquids which fume in moist air. They can be used to prepare a wide range of compounds
  Reactions of acid chlorides include:

**(i) Reaction with water**

Aliphatic acid chlorides are **hydrolyzed** rapidly with cold water.

$$E.g. \quad CH_3-\overset{O}{\underset{Cl}{C}} + H_2O \rightarrow CH_3-\overset{O}{\underset{OH}{C}} + HCl$$

ethanoic acid

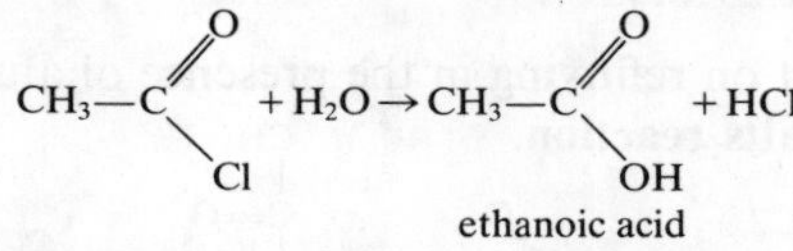

Aromatic acid chlorides are hydrolyzed more slowly than aliphatic.

$$E.g. \quad \langle O \rangle-\overset{O}{\underset{Cl}{C}} + H_2O \rightarrow \langle O \rangle-\overset{O}{\underset{OH}{C}} + HCl$$

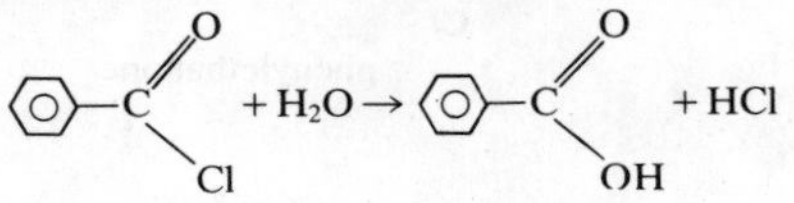

**(ii) Reactions with alcohols**

Acid chlorides react with alcohols and phenols to produce **esters** (Unit 34.4). These reactions take place on mixing the acid chloride and the alcohol without heating.

$$E.g. \quad CH_3-\overset{O}{\underset{Cl}{C}} + CH_3CH_2OH \rightarrow CH_3-\overset{O}{\underset{OCH_2CH_3}{C}} + HCl$$

ethyl ethanoate

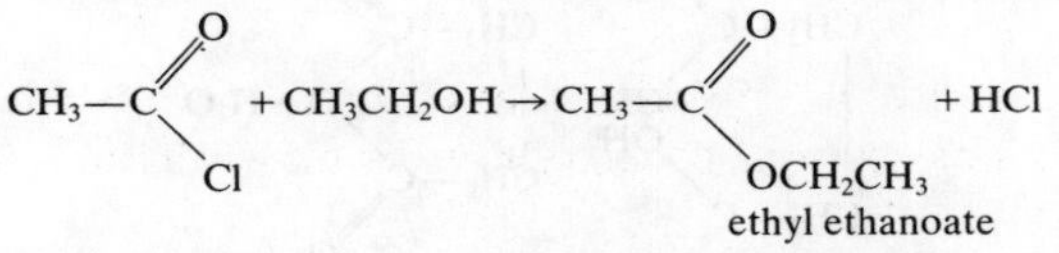

  With aromatic acid chlorides, sodium hydroxide solution is added to speed up the reaction.

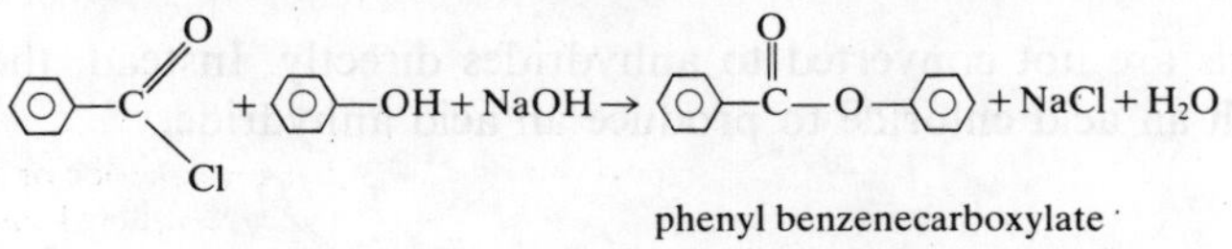

$$\langle O \rangle-\overset{O}{\underset{Cl}{C}} + \langle O \rangle-OH + NaOH \rightarrow \langle O \rangle-\overset{O}{C}-O-\langle O \rangle + NaCl + H_2O$$

phenyl benzenecarboxylate

**(iii) Reaction with ammonia**

Acid chlorides react with a concentrated aqueous solution of ammonia to produce **amides**. These reactions take place on mixing the reactants and without heating.

$$E.g. \quad CH_3-\overset{O}{\underset{Cl}{C}} + NH_3 \rightarrow CH_3-\overset{O}{\underset{NH_2}{C}} + HCl$$

ethanoyl chloride   ethanamide

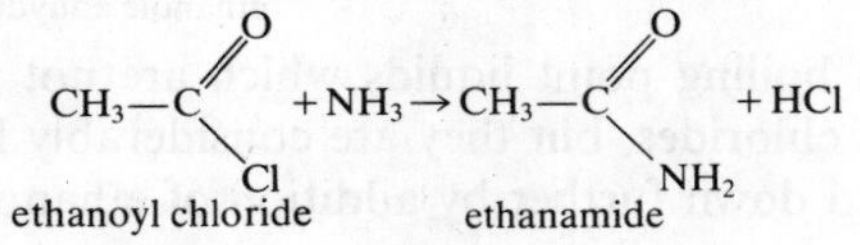

**(iv) Reactions with amines**

Acid chlorides react with primary and secondary amines to produce **substituted amides**.

$$E.g. \quad CH_3-\overset{O}{\underset{Cl}{C}} + \langle O \rangle-NH_2 \rightarrow CH_3-\overset{O}{\underset{\underset{\langle O \rangle}{NH}}{C}} + HCl$$

ethanoyl    phenylamine   *N*-phenylethanamide
chloride        (acetanilide)

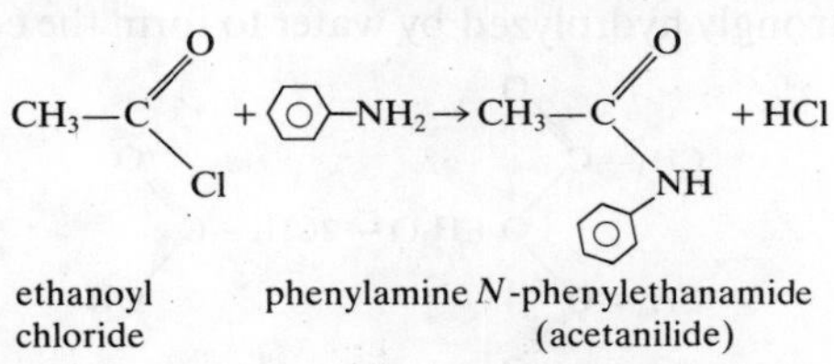

(In the naming **N-phenyl** means that a phenyl group is attached to the nitrogen atom of the amide.)

NB In all of these reactions, a hydrogen atom attached to an oxygen or a nitrogen atom is replaced by a $\underset{/}{\overset{R}{\diagdown}}\text{C}{=}\text{O}$ group. This process is called **acylation**.

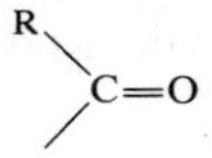

If ethanoyl chloride is used, $CH_3CO-$ is added to the molecule in place of a hydrogen. This process is called **ethanoylation (acetylation)**. With benzenecarbonylchloride, **benzenecarbonylation (benzoylation)** takes place when a $C_6H_5CO$ group is introduced. In all of these reactions the other product is hydrogen chloride.

### (v) Reaction with sodium salts of acids

Distillation of an acid chloride with the anhydrous sodium salt of an acid produces an **acid anhydride**.

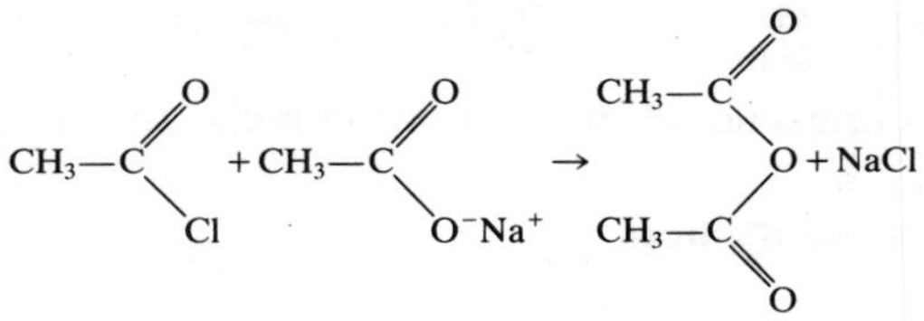

| ethanoyl chloride | sodium ethanoate | ethanoic anhydride |

### (vi) Reaction with benzene (Unit 29.11)

Benzene and acid chlorides react on refluxing in the presence of aluminium chloride to produce a **ketone**. This is the **Friedel-Crafts reaction**.

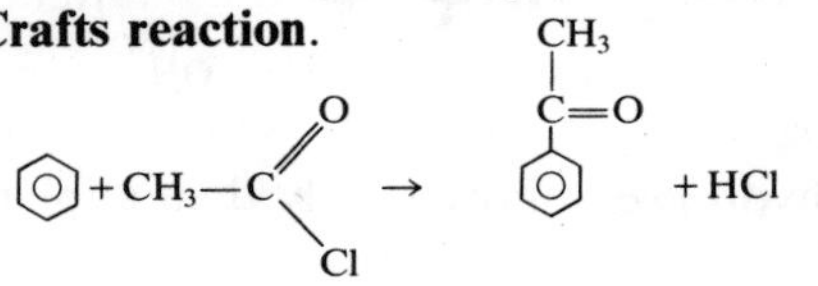

phenylethanone

## 34.3 ACID ANHYDRIDES

An **acid anhydride** is the product obtained by joining two carboxylic acid molecules together with the elimination of one molecule of water. Methanoic acid does not form an anhydride.

Some anhydrides can be made simply by distilling dibasic acids.

*E.g.*

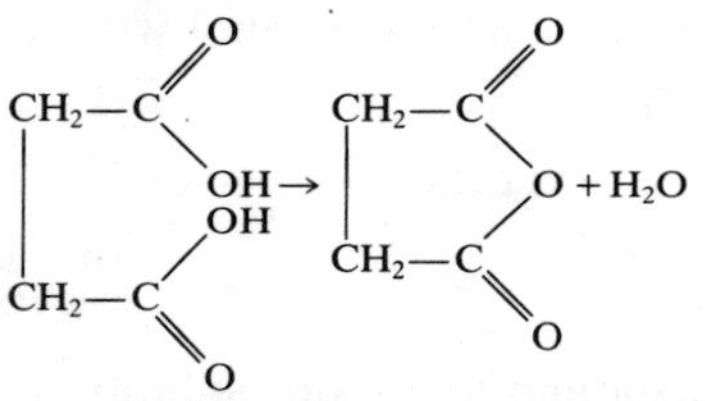

butane-1,4-dioic acid     butane-1,4-dioic anhydride

Monobasic acids are not converted to anhydrides directly. Instead, the sodium salt of an acid is distilled with an acid chloride to produce an acid anhydride.

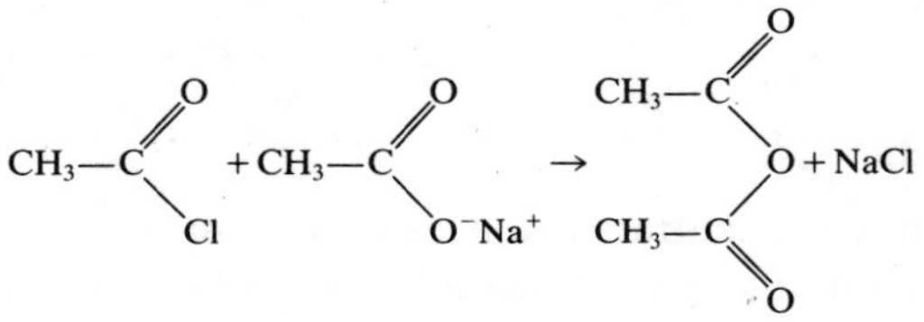

ethanoic anhydride

Acid anhydrides are high boiling point liquids which are not very soluble in water. They react in a similar way to acid chlorides, but they are considerably less reactive. Reactions with acid anhydrides can be slowed down further by addition of ethanoic acid.

The reactions include:

### (i) The reaction with water

Acid anhydrides are very strongly hydrolyzed by water to form the corresponding **carboxylic acid**.

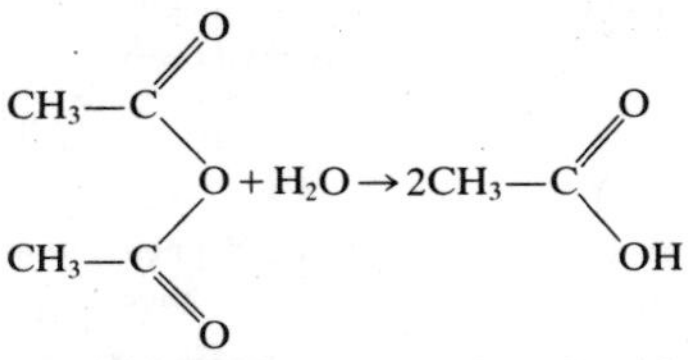

ethanoic anhydride       ethanoic acid

### (ii) Reactions with alcohols and phenols

When an alcohol or phenol is mixed with an acid anhydride, and the mixture heated, an **ester** is produced.

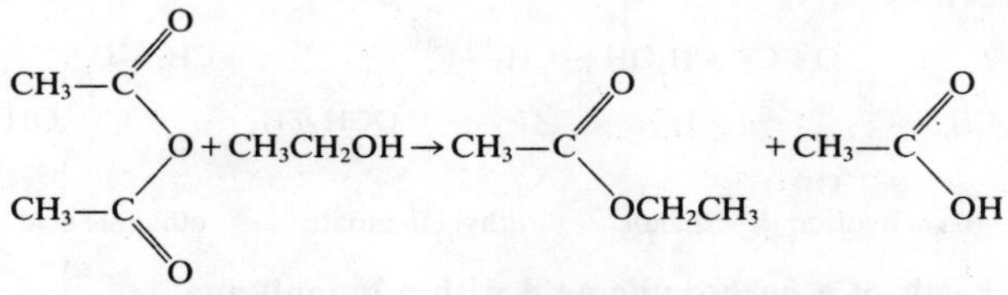

$$CH_3-C\!\!\bigvee_{O}^{O} + CH_3CH_2OH \rightarrow CH_3-C\!\!\bigvee_{OCH_2CH_3}^{O} + CH_3-C\!\!\bigvee_{OH}^{O}$$

ethanoic anhydride       ethyl ethanoate    ethanoic acid

### (iii) Reaction with ammonia

Refluxing a concentrated aqueous solution of ammonia with ethanoic anhydride produces an **amide**.

$$E.g. \quad CH_3-C\!\!\bigvee_{O}^{O} + 2NH_3 \rightarrow CH_3-C\!\!\bigvee_{NH_2}^{O} + CH_3-C\!\!\bigvee_{O^-NH_4^+}^{O}$$

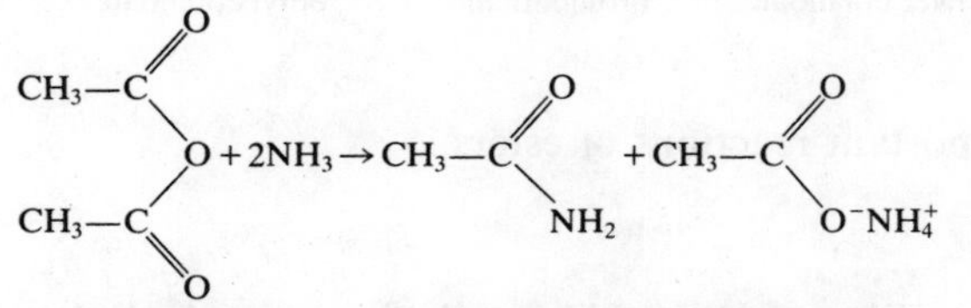

ethanoic anhydride     ethanamide     ammonium ethanoate

### (iv) Reaction with amines

Acid anhydrides react with primary or secondary amines to form **substituted amides**.
*E.g.*

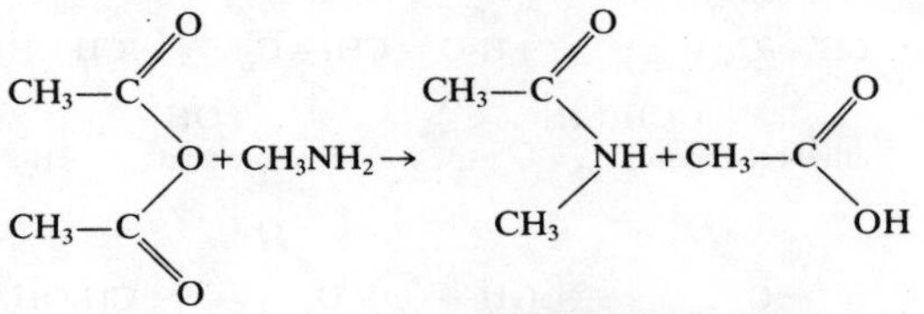

$$CH_3-C\!\!\bigvee_{O}^{O} + CH_3NH_2 \rightarrow \underset{CH_3}{C\!\!H_3-C}\!\!-NH + CH_3-C\!\!\bigvee_{OH}^{O}$$

*N*-methylethanamide

## 34.4 ESTERS

Esters are sweet-smelling liquids. Naturally occurring fats and oils contain organic esters.

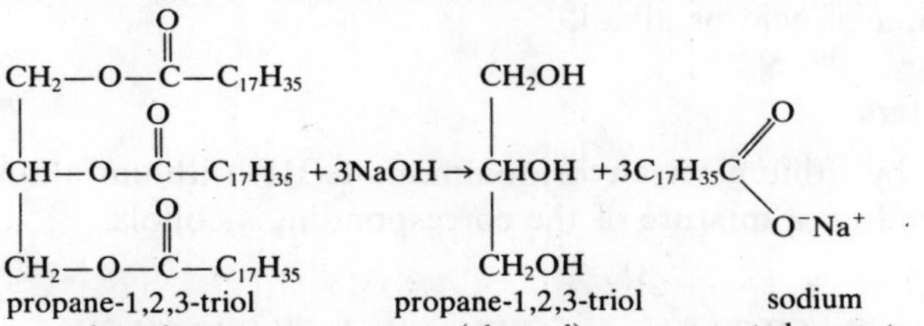

$$\begin{array}{ll} CH_2-O-\overset{O}{\overset{\|}{C}}-C_{17}H_{35} & CH_2OH \\ CH-O-\overset{O}{\overset{\|}{C}}-C_{17}H_{35} + 3NaOH \rightarrow CHOH + 3C_{17}H_{35}C\!\!\bigvee_{O^-Na^+}^{O} \\ CH_2-O-\overset{O}{\overset{\|}{C}}-C_{17}H_{35} & CH_2OH \end{array}$$

propane-1,2,3-triol    propane-1,2,3-triol    sodium
trioctadecanoate      (glycerol)       octadecanoate
(glyceryl tristearate)              (sodium stearate)

Heating these oils with sodium hydroxide hydrolyzes the esters to form triols and sodium salts of carboxylic acids. The process of hydrolysis is called **saponification** and the sodium salts produced are **soaps**. The most common component of soap is sodium stearate.

### Preparation of esters

### (i) Reaction of acid and alcohol in the presence of concentrated sulphuric acid.

$$E.g. \quad CH_3-C\!\!\bigvee_{OH}^{O} + CH_3CH_2OH \rightleftharpoons CH_3-C\!\!\bigvee_{OCH_2CH_3}^{O} + H_2O$$

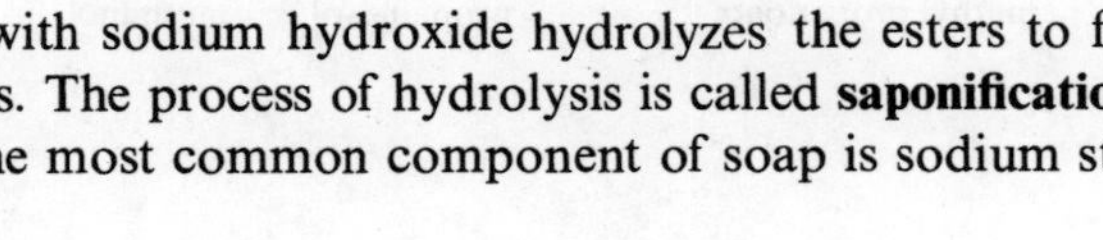

ethanoic acid     ethanol     ethyl ethanoate

The reactants are refluxed together. This method is suitable for preparing aliphatic esters.

### (ii) Treat an alcohol with acid chloride or acid anhydride

With an acid chloride, the reaction takes place without heating.

$$E.g. \quad CH_3CH_2CH_2OH + \bigcirc\!\!-C\!\!\bigvee_{Cl}^{O} \rightarrow \bigcirc\!\!-C\!\!\bigvee_{OCH_2CH_2CH_3}^{O} + HCl$$

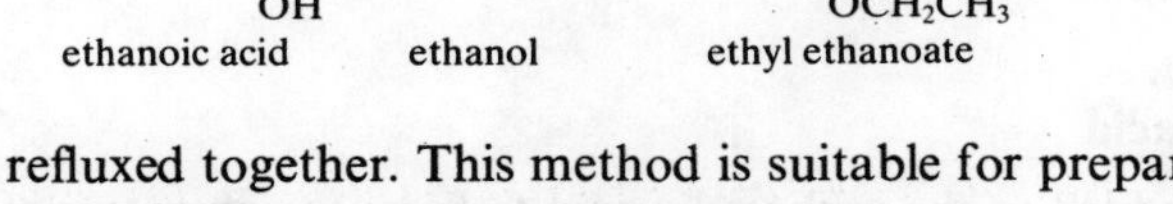

propan-1-ol   benzenecarbonyl chloride   propyl benzenecarboxylate

Refluxing an acid anhydride with an alcohol will produce an ester.

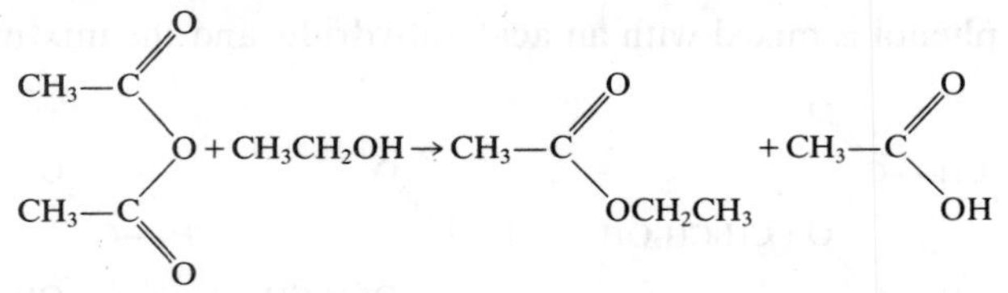

$$CH_3-C(O)O-C(O)CH_3 + CH_3CH_2OH \rightarrow CH_3-C(O)OCH_2CH_3 + CH_3-C(O)OH$$

ethanoic anhydride     ethanol     ethyl ethanoate     ethanoic acid

### (iii) Heating the silver salt of a carboxylic acid with a haloalkane

$$E.g. \quad CH_3-C(O)O^-Ag^+ + CH_3CH_2Br \rightarrow CH_3-C(O)OCH_2CH_3 + AgBr$$

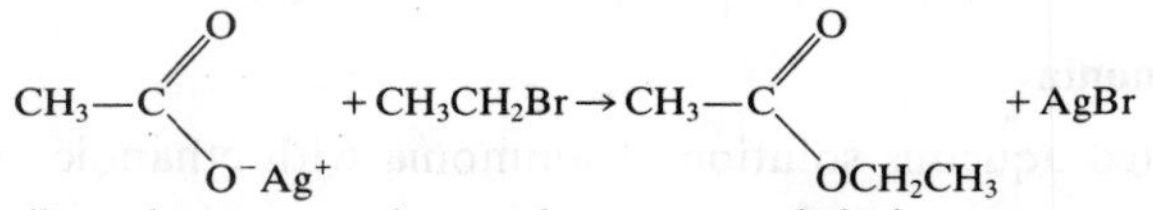

silver ethanoate     bromoethane     ethyl ethanoate

## Reaction of esters

There are only two important reactions of esters.

### (i) Hydrolysis of esters

Refluxing an ester with a dilute acid or a dilute alkali solution hydrolyzes the ester. If the acid is used, the products are an alcohol and a carboxylic acid. With sodium hydroxide solution, the products are an alcohol and the sodium salt of a carboxylic acid. The alkaline hydrolysis is called **saponification**.

$$E.g. \quad Acid\ hydrolysis \quad CH_3-C(O)OCH_2CH_3 + H_2O \rightleftharpoons CH_3-C(O)OH + CH_3CH_2OH$$

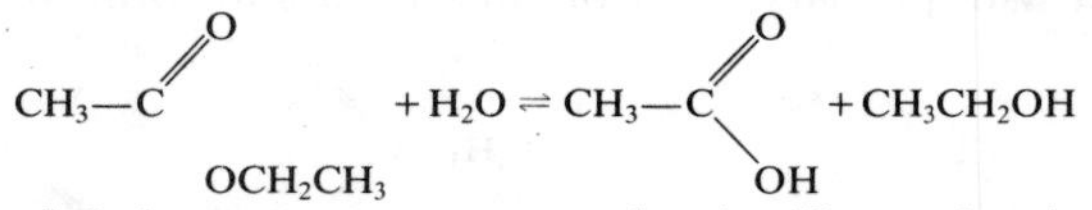

ethyl ethanoate     ethanoic acid     ethanol

*Alkaline hydrolysis*

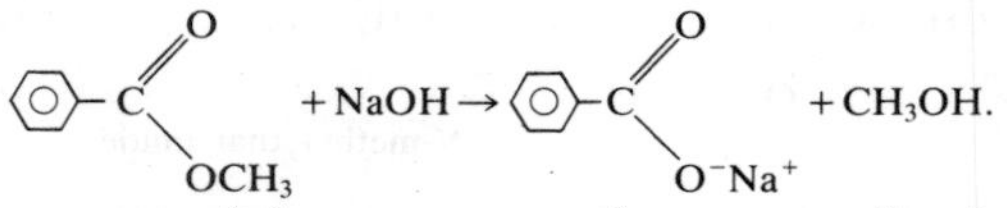

$$\bigcirc-C(O)OCH_3 + NaOH \rightarrow \bigcirc-C(O)O^-Na^+ + CH_3OH.$$

methyl
benzenecarboxylate     sodium
benzenecarboxylate     methanol

A **hydrolysis** reaction is the splitting up of a molecule with water. The rate of hydrolysis is increased by addition of acid or alkali.

### (ii) Reduction of esters

Esters are reduced by lithium tetrahydridoaluminate(III) (lithium aluminium hydride) in dry ethoxyethane to produce a mixture of the corresponding alcohols.

$$E.g. \quad CH_3CH_2C(O)OCH_3 + 4[H] \rightarrow CH_3CH_2CH_2OH + CH_3OH$$

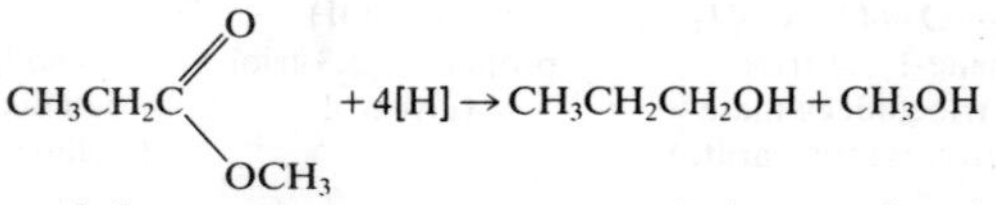

methyl propanoate     propan-1-ol     methanol

## *34.5   AMIDES

Amides are white, crystalline solids. They are the least reactive of the organic derivatives of carboxylic acids.

## Preparation of amides

As the name might suggest, amides are made from ammonia. There are two methods of producing amides.

### (i) From a carboxylic acid

Ammonia solution is added to the carboxylic acid to form the ammonium salt of the acid. This ammonium salt is heated with a small amount of free acid. The amide distils over. The acid prevents the dissociation of the ammonium salt.

$$E.g. \quad CH_3-C(O)OH + NH_3 \rightarrow CH_3-C(O)O^-NH_4^+ \qquad CH_3-C(O)O^-NH_4^+ \rightarrow CH_3-C(O)NH_2 + H_2O$$

ethanoic acid     ammonium ethanoate     ethanamide

**(ii) Reacting an acid chloride or acid anhydride with ammonia (34.2, 34.3)**

*E.g.*

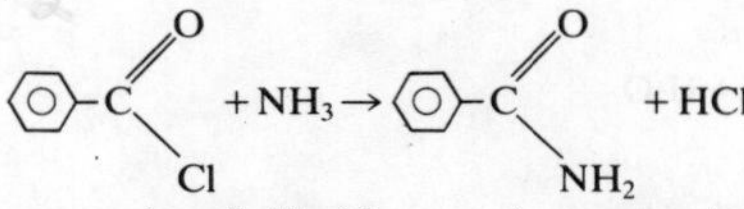

benzenecarbonyl chloride    benzenecarboxamide

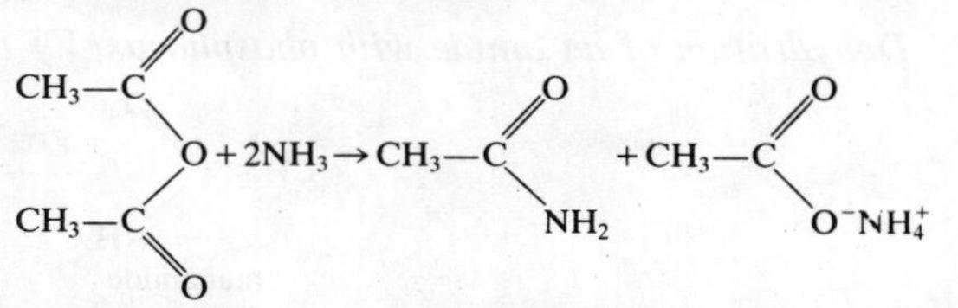

ethanoic anhydride    ethanamide    ammonium ethanoate

## Reactions of amides

Amides show slight basic properties. They form unstable salts with strong acids.

*E.g.* $\qquad CH_3CONH_2 + H_3O^+ \rightarrow CH_3CONH_3^+ + H_2O$

Other reactions of amides are:

### (i) Dehydration

When an amide is heated with phosphorus(V) oxide **dehydration** takes place to form the corresponding **nitrile**.

*E.g.*

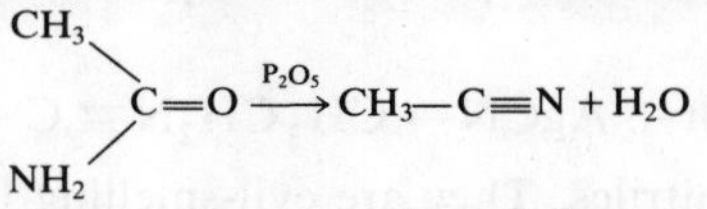

ethanenitrile

### (ii) Reaction with sodium hydroxide solution

Sodium hydroxide solution can be used to distinguish an amide from an ammonium salt. Amides evolve ammonia gas when a mixture of amide and sodium hydroxide solution is heated.

*E.g.*

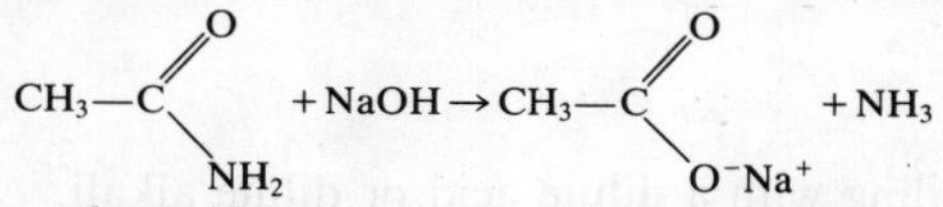

ethanamide    sodium ethanoate

Ammonium salts liberate ammonia on addition of sodium hydroxide without heating.

*E.g.*

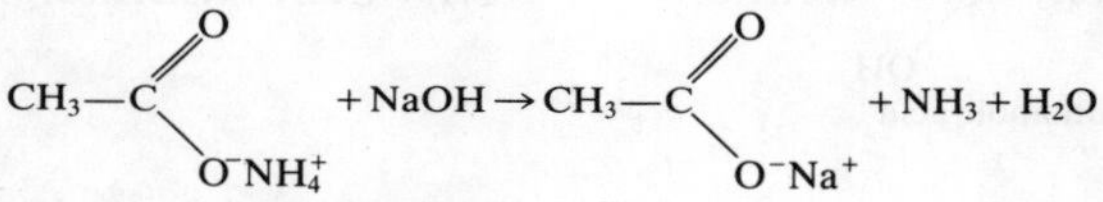

### (iii) Reaction with nitrous acid (nitric(III) acid)

Nitrous acid (nitric(III) acid) is prepared by adding sodium nitrite (sodium nitrate(III)) to dilute hydrochloric acid:

$$NaNO_2 + HCl \rightarrow NaCl + HONO$$

Nitrous acid (nitric(III) acid) reacts with amides without heating with the evolution of nitrogen gas to form the corresponding carboxylic acid.

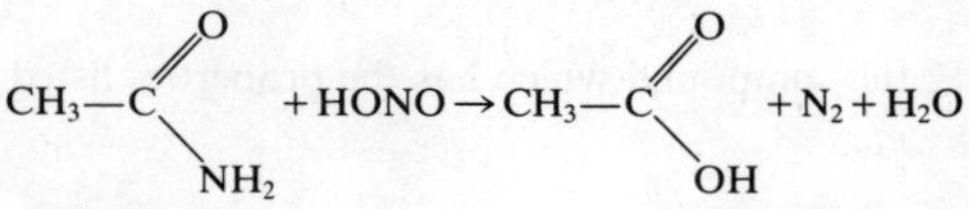

ethanamide    ethanoic acid

### (iv) Hofmann degradation

This reaction is important when descending an homologous series (Unit 37.4). The amide is treated with bromine and sodium hydroxide solution. The solution is heated and an amine is produced.

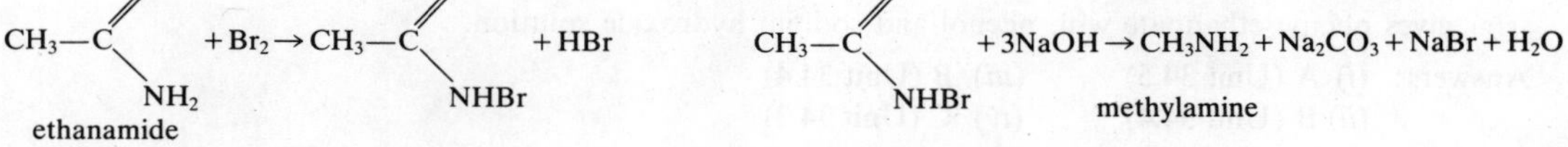

ethanamide    methylamine

## 34.6 NITRILES

Nitriles were frequently called **cyanides** since they contain the $-C{\equiv}N$ group.
The simplest nitriles are

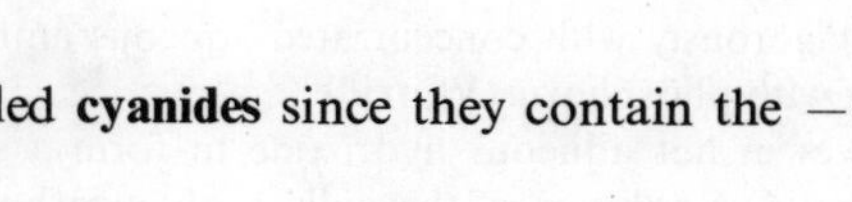

ethanenitrile    propanenitrile

**Preparation of nitriles**

*(i) Dehydration of an amide with phosphorus(V) oxide.*

$$CH_3-C\underset{NH_2}{\overset{O}{<}} \xrightarrow{P_4O_{10}} CH_3-C\equiv N + H_2O$$

ethanamide                ethanenitrile

*(ii) Action of cyanide on haloalkane (Unit 30.3)*

A haloalkane is refluxed with a solution of potassium cyanide dissolved in ethanol. A substitution reaction takes place producing a nitrile.

$$CH_3CH_2Br + KCN \rightarrow CH_3CH_2CN + KBr$$
bromoethane                propanenitrile

This reaction is important in ascending an homologous series (Unit 37.3). Potassium cyanide is predominately ionic and a nitrile is produced containing the $-C\equiv N$ group. If this reaction is repeated using silver cyanide, which is predominately covalent, the product is an isonitrile containing the $-N\rightleftharpoons C$ group.

$$CH_3CH_2Br + AgCN \rightarrow CH_3CH_2N \rightleftharpoons C + AgBr$$

**Isonitriles** are isomeric with nitriles. They are evil-smelling liquids and can be prepared by heating a primary amine, trichloromethane and potassium hydroxide dissolved in ethanol.

$$CH_3NH_2 + CHCl_3 + 3OH^- \rightarrow CH_3N \rightleftharpoons C + 3Cl^- + 3H_2O$$
isocyanomethane

**Reactions of nitriles**

*(i) Hydrolysis of nitriles*

This can be achieved by boiling with a dilute acid or dilute alkali.

$$CH_3-C\equiv N + 2H_2O + HCl \rightarrow CH_3-C\underset{OH}{\overset{O}{<}} + NH_4Cl$$

ethanenitrile                ethanoic acid

$$CH_3-C\equiv N + NaOH + H_2O \rightarrow CH_3-C\underset{O^-Na^+}{\overset{O}{<}} + NH_3$$

sodium ethanoate

*(ii) Reduction of nitriles*

Nitriles can be reduced to primary amines by sodium and ethanol.

$$CH_3C\equiv N + 4[H] \rightarrow CH_3CH_2NH_2$$
ethylamine

QUESTIONS

1 Choose, from the list A to E, the compound which has the properties listed in each of the questions (*i*) to (*iv*).
A ammonium ethanoate
B ethyl ethanoate
C ethanoic anhydride
D ethanamide
E ethanenitrile
  (*i*) gives ammonia on addition of sodium hydroxide solution without heating.
  (*ii*) gives ethanol when heated with sodium hydroxide solution.
  (*iii*) gives ethanol on reduction with lithium tetrahydridoaluminium(III) (lithium aluminium hydride).
  (*iv*) gives phenyl ethanoate with phenol and sodium hydroxide solution.
  Answers:  (*i*) **A** (Unit 34.5)    (*iii*) **B** (Unit 34.4)
            (*ii*) **B** (Unit 34.4)    (*iv*) **C** (Unit 34.3)

2 In each of the following cases give the name or structure of *one* compound which fits the information given. Explain your reasoning and write equations for the reactions involved.
(a) R is a liquid which reacts vigorously with concentrated aqueous ammonia forming a solid which is readily dehydrated by warming with phosphorus(V) oxide.
(b) S is a liquid which dissolves in hot aqueous hydroxide to form a solution which on cooling and acidifying precipitates a white solid. A solution of this solid evolves carbon dioxide with aqueous sodium carbonate.

(a) R is an acid chloride which reacts with ammonia to form an amide. Amides are dehydrated with phosphorus(V) oxide.

A suitable compound would be ethanoyl chloride

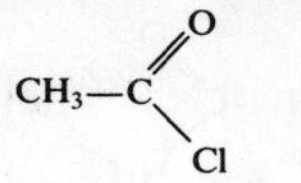

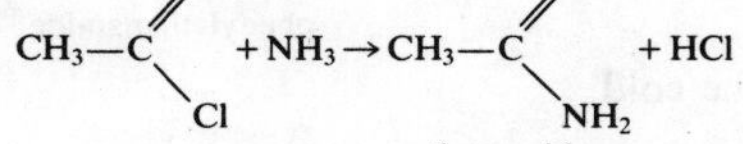

$$CH_3-C\begin{smallmatrix}O\\\\Cl\end{smallmatrix}$$

$$CH_3-C\begin{smallmatrix}O\\\\Cl\end{smallmatrix} + NH_3 \rightarrow CH_3-C\begin{smallmatrix}O\\\\NH_2\end{smallmatrix} + HCl$$

ethanamide

$$CH_3-C\begin{smallmatrix}O\\\\NH_2\end{smallmatrix} \xrightarrow{P_2O_5} CH_3-C\equiv N.$$

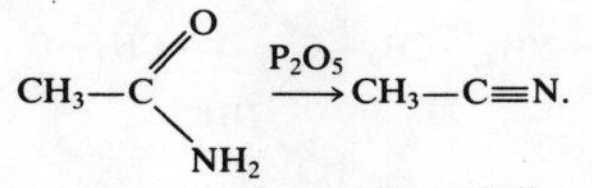

ethanenitrile

(*b*) The white solid which evolves carbon dioxide with aqueous sodium carbonate could be benzene-carboxylic acid (benzoic acid). The liquid S could be an ester such as ethyl benzoate or ethyl benzene-carboxylate

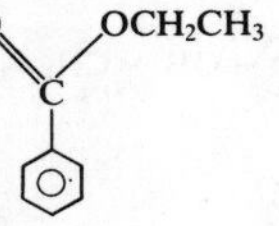

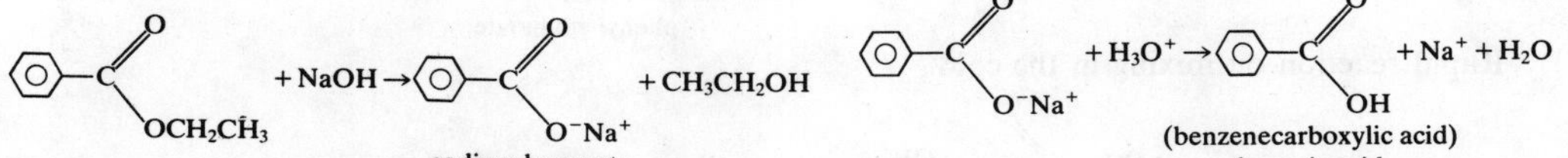

$$C_6H_5-C\begin{smallmatrix}O\\\\OCH_2CH_3\end{smallmatrix}$$

$$C_6H_5-C\begin{smallmatrix}O\\\\OCH_2CH_3\end{smallmatrix} + NaOH \rightarrow C_6H_5-C\begin{smallmatrix}O\\\\O^-Na^+\end{smallmatrix} + CH_3CH_2OH$$

sodium benzoate

$$C_6H_5-C\begin{smallmatrix}O\\\\O^-Na^+\end{smallmatrix} + H_3O^+ \rightarrow C_6H_5-C\begin{smallmatrix}O\\\\OH\end{smallmatrix} + Na^+ + H_2O$$

(benzenecarboxylic acid)
benzoic acid

**3** Explain how the following conversions could be achieved. In each case give the essential conditions and equations.
  (*i*) ethanoic acid to ethanoyl chloride
 (*ii*) ethanoic acid to ethanoic anhydride.

    Compare the reactions of ethanoyl chloride and ethanoic anhydride with water, phenylamine and phenol.

    Describe a simple test tube experiment which could be used to distinguish ethanoyl chloride from benzenecarbonyl chloride (benzoyl chloride).

(*i*) **Warm a mixture of ethanoic acid and sulphur dichloride oxide. Sulphur dioxide and hydrogen chloride are evolved as gases.**

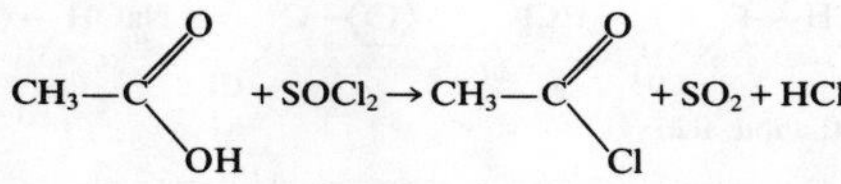

$$CH_3-C\begin{smallmatrix}O\\\\OH\end{smallmatrix} + SOCl_2 \rightarrow CH_3-C\begin{smallmatrix}O\\\\Cl\end{smallmatrix} + SO_2 + HCl$$

(*ii*) **Prepare ethanoyl chloride as in (*i*). Mix ethanoic acid and sodium hydroxide solutions to form sodium ethanoate.**

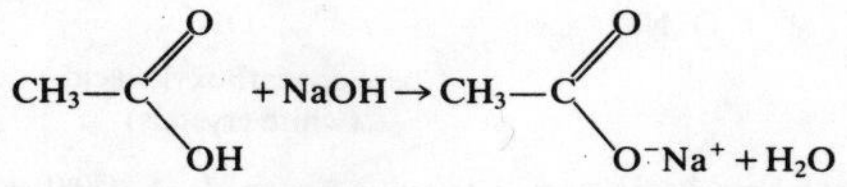

$$CH_3-C\begin{smallmatrix}O\\\\OH\end{smallmatrix} + NaOH \rightarrow CH_3-C\begin{smallmatrix}O\\\\O^-Na^+\end{smallmatrix} + H_2O$$

**Mix the anhydrous sodium ethanoate and ethanoyl chloride and distil the mixture.**

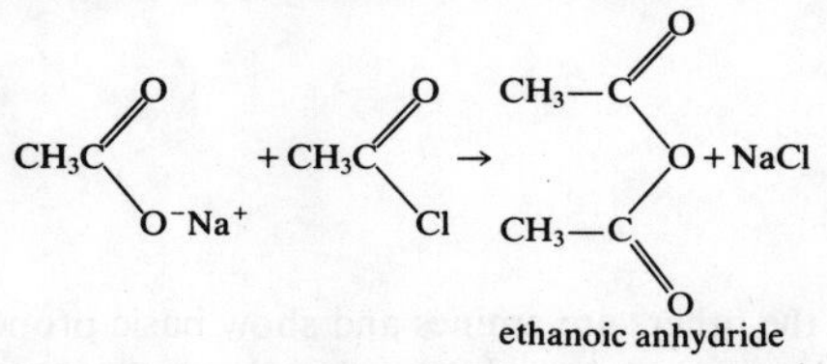

$$CH_3C\begin{smallmatrix}O\\\\O^-Na^+\end{smallmatrix} + CH_3C\begin{smallmatrix}O\\\\Cl\end{smallmatrix} \rightarrow \begin{smallmatrix}CH_3-C\\\\O\\\\CH_3-C\end{smallmatrix}O + NaCl$$

ethanoic anhydride

*Reaction with water*. Ethanoyl chloride reacts vigorously with cold water with the evolution of fumes of hydrogen chloride.

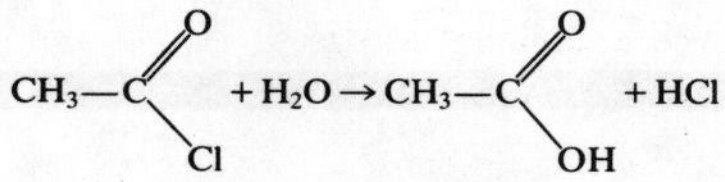

$$CH_3-C\begin{smallmatrix}O\\\\Cl\end{smallmatrix} + H_2O \rightarrow CH_3-C\begin{smallmatrix}O\\\\OH\end{smallmatrix} + HCl$$

ethanoic acid

    Ethanoic anhydride forms a separate layer when added to water. It reacts slowly with water, but the reaction can be speeded up by heating.

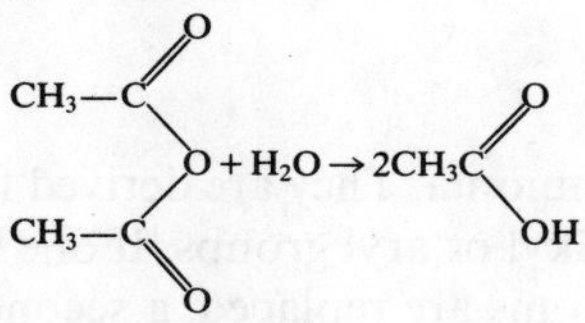

$$\begin{smallmatrix}CH_3-C\\\\O\\\\CH_3-C\end{smallmatrix}O + H_2O \rightarrow 2CH_3C\begin{smallmatrix}O\\\\OH\end{smallmatrix}$$

*Reaction with phenylamine.*

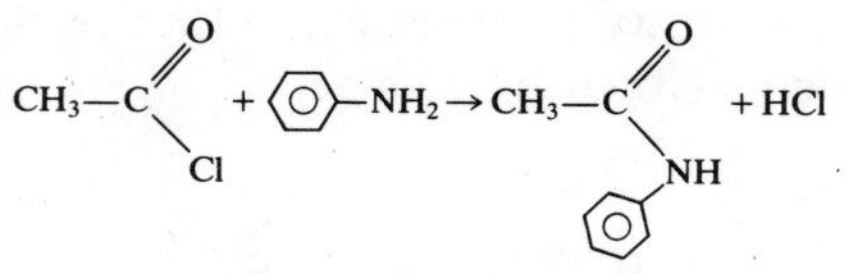

N-phenylethanamide

Rapid reaction on mixing in the cold.

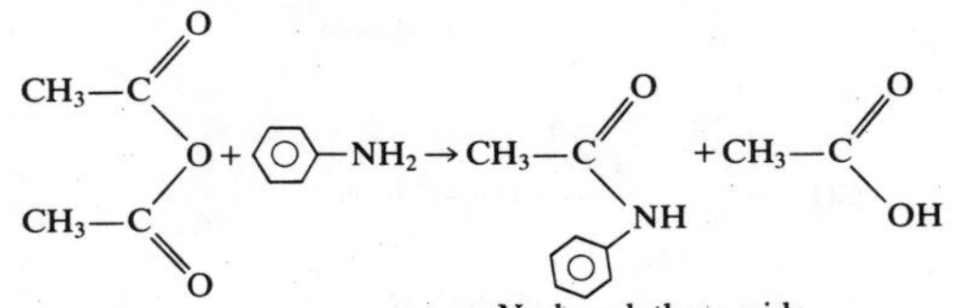

N-phenylethanamide

Reflux reagents together.
*Reaction with phenol.*

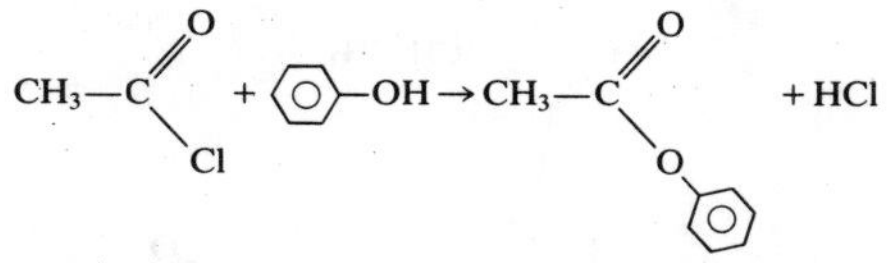

phenyl ethanoate

Rapid reaction on mixing in the cold.

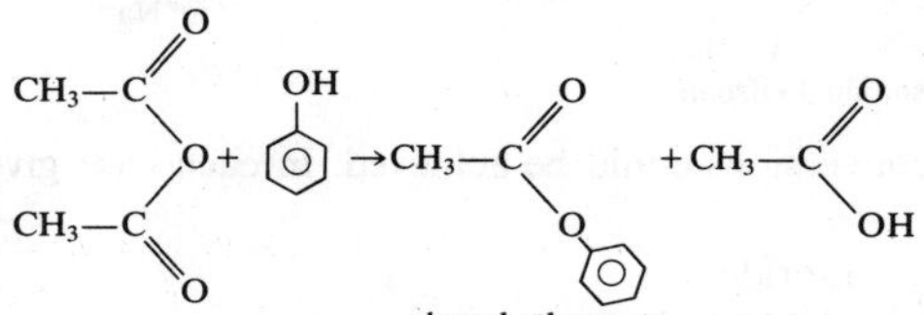

phenyl ethanoate

Reflux reagents together.

If ethanoyl chloride is added cautiously to water fumes of hydrogen chloride are produced. If benzenecarbonyl chloride is added to sodium hydroxide solution, a reaction takes place on warming to produce sodium benzenecarboxylate. On acidification with dilute hydrochloric acid, crystals of benzene-carboxylic acid are formed.

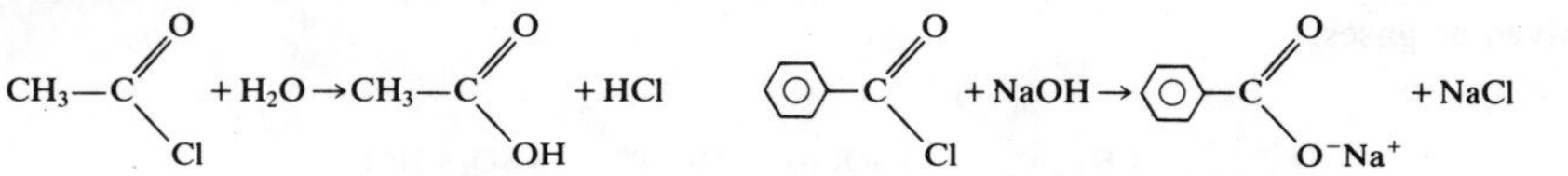

ethanoic acid                    sodium benzenecarboxylate

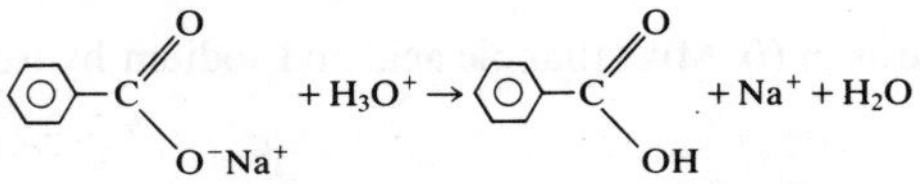

benzenecarboxylic acid
(white crystals)

4 Which one of the following compounds does **not** form a salt when a dilute aqueous solution of an acid is added?
A $C_6H_5NH_2$
B $C_6H_5CH_2NH_2$
C $C_2H_5CONH_2$
D $(C_2H_5)_2NH$
E $(C_2H_5)_3N$

The correct answer is C. All of the others are amines and show basic properties such as forming salts with acids. C is an amide and amides are amphoteric showing the ability to form salts with acids only with concentrated acids.

# 35  Amines

## 35.1  INTRODUCTION

Amines are very closely related to ammonia. They are derived from ammonia by the replacement of hydrogen atoms in ammonia by alkyl or aryl groups. If one hydrogen is replaced, the amine is a primary amine. If two hydrogen atoms are replaced, a secondary amine is the result and three replaced hydrogens produces a tertiary amine. Refer back to Unit 31 to note the difference between primary, secondary and tertiary alcohols, where the functional group remains the same, and amines where the functional group itself changes.

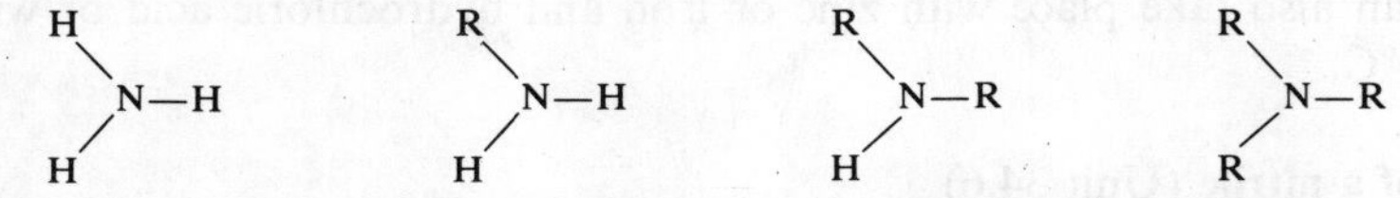

ammonia    primary amine    secondary amine    tertiary amine

Common amines include

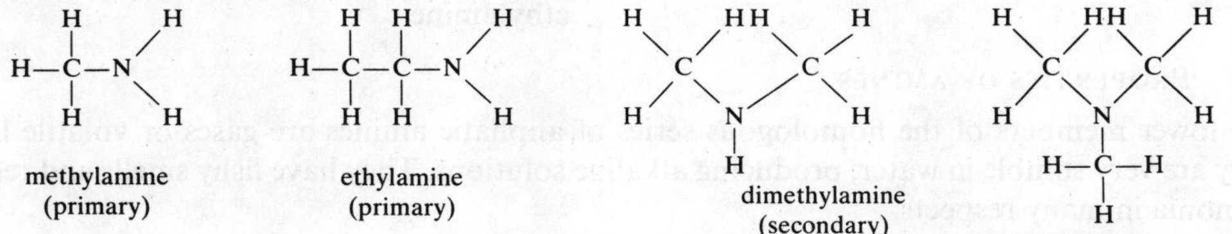

methylamine
(primary)

ethylamine
(primary)

dimethylamine
(secondary)

trimethylamine
(tertiary)

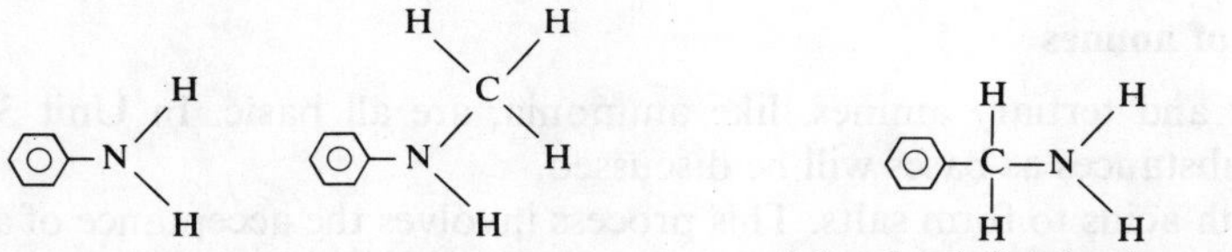

phenylamine
(primary)

N-methylphenylamine
(secondary)

(phenylmethyl)amine (benzylamine)
(primary)

When the amine group is attached to a benzene nucleus the resulting amine is called an **arylamine**. An **alkylamine** is an amine in which only alkyl groups or hydrogen atoms are attached to the nitrogen atom. (Phenylmethyl)amine behaves as an alkylamine because the $-NH_2$ group is not attached directly to the ring.

**Quaternary ammonium salts** are the organic equivalent of ammonium compounds.

*E.g.*         $NH_4^+Cl^-$         $(CH_3)_4N^+Cl^-$
         ammonium chloride    tetramethylammonium chloride

### 35.2  PREPARATION OF AMINES

#### (i)  Preparation from ammonia

This method of preparation emphasizes the close relationships between ammonia and the amines. It is used to produce alkylamines.

Haloalkanes are heated with ammonia under pressure. A series of reactions takes place and a mixture of products is usually obtained.

*E.g.* chloromethane and ammonia

$$CH_3Cl + NH_3 \rightarrow CH_3NH_2 + HCl$$
methylamine

$$CH_3NH_2 + CH_3Cl \rightarrow (CH_3)_2NH + HCl$$
dimethylamine

$$(CH_3)_2NH + CH_3Cl \rightarrow (CH_3)_3N + HCl$$
trimethylamine

$$(CH_3)_3N + CH_3Cl \rightarrow (CH_3)_4N^+Cl^-$$
tetramethylammonium chloride

These amines react with hydrogen chloride and the products isolated initially will be salts, *e.g.* $CH_3NH_3^+Cl^-$, $(CH_3)_2NH_2^+Cl$, $(CH_3)_3NH^+Cl^-$.

#### (ii)  Reduction of nitrocompounds

The reduction of nitrocompounds to form amines is most important for the production of aryl amines. The reduction is usually carried out with tin and concentrated hydrochloric acid.

*E.g.*      $\langle O \rangle-NO_2 + 6[H] \rightarrow \langle O \rangle-NH_2 + 2H_2O$

nitrobenzene         phenylamine

Following the reduction, the amine reacts to form $(C_6H_5NH_3^+)_2SnCl_6^{2-}$. The phenylamine is obtained by addition of excess sodium hydroxide solution to liberate the free phenylamine, followed by steam distillation.

Reduction can also take place with zinc or iron and hydrochloric acid or with nickel and hydrogen at 300°C.

### (iii) Reduction of a nitrile (Unit 34.6)

Reduction of a nitrile with sodium and ethanol produces a primary amine.

*E.g.*
$$CH_3C\equiv N + 4[H] \rightarrow CH_3CH_2NH_2$$
$$\text{ethylamine}$$

### 35.3 PROPERTIES OF AMINES

The lower members of the homologous series of aliphatic amines are gases or volatile liquids. They are very soluble in water, producing alkaline solutions. They have fishy smells and resemble ammonia in many respects.

Aromatic amines are much less volatile and are virtually insoluble in water.

### (i) Basic properties of amines

Primary, secondary and tertiary amines, like ammonia, are all basic. In Unit 39 the relative strengths of these substances as bases will be discussed.

Amines react with acids to form salts. This process involves the acceptance of a proton:

$$CH_3CH_2NH_2 + H_3O^+ \rightarrow CH_3CH_2NH_3^+ + H_2O$$
$$\text{ethylamine} \qquad\qquad \text{ethylammonium ion}$$

The free amine is liberated by the addition of alkali:

$$[CH_3NH_3^+]Cl^- + NaOH \rightarrow CH_3NH_2 + HCl$$
$$\text{methylammonium chloride} \quad \text{methylamine}$$

### (ii) Ethanoylation (acetylation) of amines (Unit 34.2, 34.3)

**Ethanoylation** of primary and secondary amines involves the replacement of a hydrogen atom attached to a nitrogen atom by the ethanoyl group ($CH_3CO-$). This process takes place with acid chlorides and anhydrides.

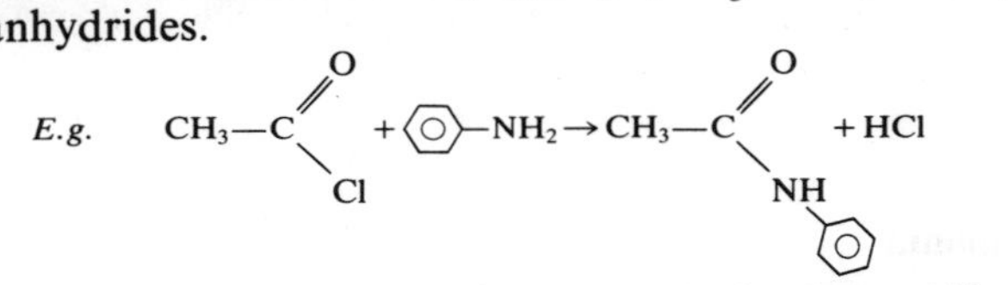

*N*-phenylethanamide

In the cold with careful addition of the acid chloride.

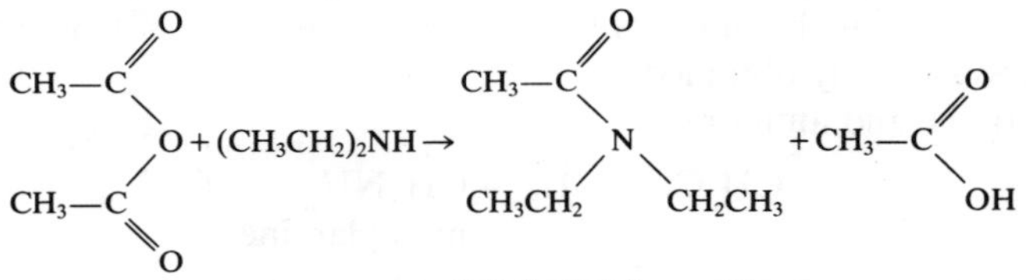

*N,N*-diethylethanamide

Refluxing amine with acid anhydride.

If a sample of an amine has to be identified, ethanoylation is most useful. The amine is probably a gas or a volatile liquid. Boiling points are difficult to do accurately and the actual value depends upon the atmospheric pressure.

The amine is ethanoylated by treatment with ethanoyl chloride or ethanoic anhydride. The solid derivative can be purified by recrystallization and an accurate melting point can be determined. By looking in tables, it is possible to find the ethanoyl derivative with the appropriate melting point.

Ethanoylation can also be used to protect amine groups during reaction. The amine can be restored by hydrolysis.

### (iii) Reaction with nitrous acid (nitric(III) acid)

Nitrous acid (nitric(III) acid) is an unstable acid which is prepared from sodium nitrite (sodium nitrite (III)) and dilute hydrochloric acid:

$$NaNO_2 + HCl \rightarrow NaCl + HONO$$

Primary arylamines react with nitrous acid (nitric(III) acid) at temperatures below 5°C to form unstable **diazonium compounds**. These compounds cannot be isolated but, like Grignard reagents, they provide a means of producing a range of other compounds.

$$\text{C}_6\text{H}_5\text{NH}_2 + \text{HONO} + \text{HCl} \rightarrow \text{C}_6\text{H}_5\overset{+}{\text{N}}\equiv\text{N Cl}^- + \text{H}_2\text{O}$$

benzenediazonium chloride

An alkylamine reacts with nitrous acid to form a primary alcohol. Diazonium compounds formed from alkylamines are not stable.

*E.g.* $\quad\quad \underset{\text{ethylamine}}{\text{CH}_3\text{CH}_2\text{NH}_2} + \text{HONO} \rightarrow \underset{\text{ethanol}}{\text{CH}_3\text{CH}_2\text{OH}} + \text{N}_2 + \text{H}_2\text{O}$

The difference in reaction with nitrous acid (nitric(III) acid) is used to differentiate between alkyl and aryl amines.

## 35.4 DIAZONIUM COMPOUNDS

Primary arylamines form diazonium compounds with nitrous acid (nitric(III) acid) below 5°C. The compounds are more stable than similar compounds formed by alkylamines because of the presence of the aromatic ring.

The reactions of diazonium compounds can be subdivided into reactions where nitrogen is replaced, and coupling reactions where nitrogen is retained.

### (i) Reactions where nitrogen is replaced

### (a) Replaced by hydroxyl group

Heating the diazonium compound above 10°C in aqueous solution produces a phenol.

*E.g.* $\quad \text{C}_6\text{H}_5\text{-}\overset{+}{\text{N}}\equiv\text{NCl}^- + \text{H}_2\text{O} \rightarrow \text{C}_6\text{H}_5\text{-OH} + \text{N}_2 + \text{HCl}$

benzene diazonium chloride $\quad\quad$ phenol

### (b) Replaced by a chlorine or bromine

This can be achieved by warming the diazonium compound with hydrogen chloride or hydrogen bromide and the corresponding copper(I) halide.

*E.g.* $\quad \text{C}_6\text{H}_5\text{-}\overset{+}{\text{N}}\equiv\text{NCl}^- \xrightarrow{\text{CuCl/HCl}} \text{C}_6\text{H}_5\text{-Cl} + \text{N}_2$

benzene diazonium chloride $\quad\quad$ chlorobenzene

Alternatively, the diazonium halide can be warmed with powdered copper.

$$\text{C}_6\text{H}_5\text{-}\overset{+}{\text{N}}\equiv\text{NBr}^- \xrightarrow{\text{Cu}} \text{C}_6\text{H}_5\text{-Br} + \text{N}_2$$

bromobenzene

### (c) Replaced with iodine

The diazonium compound is warmed with potassium iodide solution.

*E.g.* $\quad \text{C}_6\text{H}_5\text{-}\overset{+}{\text{N}}\equiv\text{NCl}^- + \text{KI} \rightarrow \text{C}_6\text{H}_5\text{-I} + \text{KCl} + \text{N}_2$

iodobenzene

### (d) Replaced with nitrile

The diazonium compound is heated with potassium cyanide solution in the presence of copper(I) cyanide as a catalyst.

$$\text{C}_6\text{H}_5\text{-}\overset{+}{\text{N}}\equiv\text{NCl}^- + \text{CN}^- \rightarrow \text{C}_6\text{H}_5\text{-C}\equiv\text{N} + \text{N}_2 + \text{Cl}^-$$

### (ii) Reactions where nitrogen is retained – coupling reaction

Diazonium compounds couple with phenols and aromatic amines to give coloured azo compounds.

### (a) With phenols

The diazonium compound is added to an alkaline solution of phenol. An orange precipitate of (4-hydroxyphenyl) azobenzene is formed.

$$\text{NaOH} + \text{C}_6\text{H}_5\text{-OH} + \text{C}_6\text{H}_5\text{-}\overset{+}{\text{N}}\equiv\text{NCl}^- \rightarrow \text{HO-C}_6\text{H}_4\text{-N}=\text{N-C}_6\text{H}_5 + \text{NaCl} + \text{H}_2\text{O}$$

(4-hydroxyphenyl)azobenzene

### (b) With primary amines

When a diazonium compound is added to phenylamine an orange precipitate of 4-(phenylazo) phenylamine is formed.

$$\text{\Large\bigcirc}-NH_2 + \text{\Large\bigcirc}-\overset{+}{N}\equiv NCl^- \rightarrow NH_2\text{\Large\bigcirc}-N=N-\text{\Large\bigcirc} + HCl$$

4-(phenylazo)phenylamine

## QUESTIONS

**1** Multiple completion question. (Summary of the possible responses on page 3.)
Which of the following amines is a primary aromatic amine?

(*i*) $CH_3C_6H_4NH_2$
(*ii*) $C_6H_5CH_2NH_2$
(*iii*) $C_6H_5NH(CH_3)$
(*iv*) $C_6H_5NH_2$

Correct answer — E, corresponding to answers (*i*) and (*iv*) correct. (*i*), (*ii*) and (*iv*) are primary amines and (*iii*) is a secondary amine. (*ii*) is not an aromatic amine as the $NH_2$ group is not attached to an aromatic ring.

**2** For each of the following pairs of compounds, describe one reaction which could be used to distinguish one from another. Describe the effect, if any, of the reagent (or reagents) on both compounds of the pair and explain what has happened.

(*i*) 1-aminobutane and phenylamine (aniline)
(*ii*) methylamine and ammonia

*(Associated Examining Board Syllabus II)*

(*i*) Treat both compounds with nitrous acid (nitric(III) acid) below 5°C. Nitrous acid (nitric(III) acid) is prepared from sodium nitrite (sodium nitrate(III)) and dilute hydrochloric acid.

$$CH_3CH_2CH_2CH_2NH_2 + HONO \rightarrow CH_3CH_2CH_2CH_2OH + N_2 + H_2O$$
butan-1-ol

These solutions are added separately to samples of alkaline solution of phenol. Benzene diazonium chloride forms an orange precipitate.

$$\text{\Large\bigcirc}-OH + \text{\Large\bigcirc}-\overset{+}{N}\equiv NCl^- \rightarrow HO\text{\Large\bigcirc}-N=N-\text{\Large\bigcirc} + NaCl + H_2O$$

(4-hydroxyphenyl)azobenzene

No precipitate is formed with butan-1-ol.

(*ii*) Ammonia does not burn in air but burns in oxygen.

$$4NH_3 + 3O_2 \rightarrow 2N_2 + 6H_2O$$

Methylamine burns in air.

$$4CH_3NH_2 + 9O_2 \rightarrow 4CO_2 + 10H_2O + 2N_2$$

The gas produced turns limewater milky.

**3** Identify A in the following passage and explain, with equations, the reactions taking place.

A is a gas which dissolves in water to form a basic solution. It burns in air to form carbon dioxide and water vapour. When A comes in contact with hydrogen chloride fumes, dense white fumes of a salt are observed.

When an aqueous solution of A is warmed with trichloromethane and a solution of potassium hydroxide in ethanol, an unpleasant smelling compound of relative molecular mass 41 is formed.

A is a gas which forms salts with acids. This suggests an amine. The reaction of a primary amine with trichloromethane and potassium hydroxide dissolved in ethanol produces an isocyanide containing the $-N\equiv C$ group (Unit 34.6).

Relative molecular mass of isocyanide = 41
Deduct 14 for the nitrogen and 12 for carbon.
Remainder = 15
This suggests a $CH_3-$ group.
Isocyanide is $CH_3 - N\equiv C$.
A is, therefore, methylamine $CH_3NH_2$

$$CH_3NH_2 + CHCl_3 + 3OH^- \rightarrow CH_3 - N\equiv C + 3Cl^- + 3H_2O$$

Methylamine burns in air:

$$4CH_3NH_2 + 9O_2 \rightarrow 4CO_2 + 10H_2O + 2N_2$$

dissolves in water:

$$CH_3NH_2 + H_2 \rightarrow CH_3NH_3^+OH^-$$

forms a salt with hydrogen chloride:

$$CH_3NH_2 + HCl \rightarrow CH_3NH_3^+Cl^-$$
methylammonium chloride

In this type of question attempt to explain all the points made in the question even if they seem trivial to you.

# 36 Large organic molecules

## 36.1 INTRODUCTION

In this unit organic macromolecules and polymers will be considered. Some of these are naturally occurring macromolecules, *e.g.* proteins, starch. Many of these large molecules, mostly polymers, are man-made or synthetic.

The small repeating units from which these polymers are made are called **monomers**. **Polymerization** is the process whereby these small units are joined together to form the **polymer**.

There are two types of polymerization – **addition polymerization** and **condensation polymerization**.

## 36.2 ADDITION (CHAIN) POLYMERIZATION

Ethene undergoes a number of addition reactions (Unit 29.6). A common factor in these reactions is the removal of the double bond. Addition polymerization involves a series of addition reactions. Examples are shown in Table 36.1.

**Table 36.1 Common polymers**

| Monomer | Structure of polymer | Name of polymer |
|---|---|---|
| 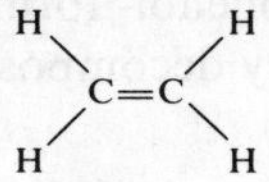 ethene | 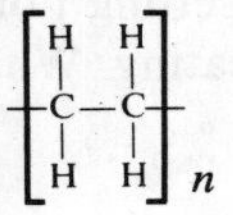 | poly(ethene) (Polythene) |
| 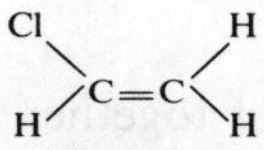 chloroethene (vinyl chloride) | 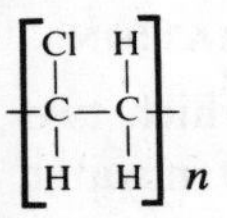 | poly(chloroethene) (Polyvinylchloride or PVC) |
| 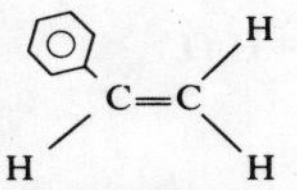 phenylethene (styrene) | 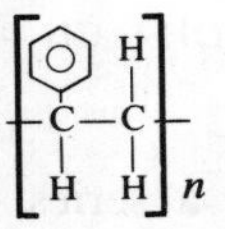 | poly(phenylethene) (polystyrene) |
| 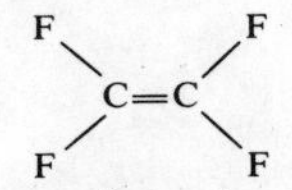 tetrafluoroethene | 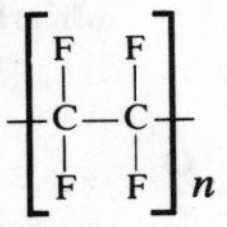 | poly(tetrafluoroethylene) (PTFE or Teflon) |

Poly(ethene) (trade name Polythene) was the first of these addition polymers to be made. At first it was prepared by heating ethene to 250°C and at a high pressure with a catalyst. Nowadays, polymerization of ethene can be carried out more easily by bubbling ethene through a hydrocarbon solvent containing complex catalysts discovered by Ziegler and Natta. These catalysts consist of $(C_2H_5)_3Al$ and $TiCl_4$. The polymerization process is usually a free radical process. A peroxide such as di(benzoyl)peroxide splits up into free radicals. If the free radical is represented by $R\cdot$, the following steps in the reaction occur:

$$R\cdot + CH_2 = CH_2 \rightarrow R - CH_2 - CH_2\cdot$$
$$R - CH_2 - CH_2\cdot + CH_2 = CH_2 \rightarrow R\,CH_2 - CH_2 - CH_2 - CH_2\cdot$$
$$R - CH_2 - CH_2 - CH_2 - CH_2\cdot + CH_2 = CH_2 \rightarrow R - CH_2 - CH_2 - CH_2 - CH_2 - CH_2 - CH_2\cdot$$
$$\text{etc}$$

Addition polymers of this type have a wide range of uses. Poly(ethene) can be manufactured in two forms.

(*i*) Low density poly(ethene). This is cheap to produce and is used for, *e.g.* wrapping food, plastic gloves for laboratory work etc. For these uses, great strength is not necessary.

(*ii*) High density poly(ethene). This is more expensive to produce but, in this form, the poly-

(ethene) is much stronger. The chains are more tightly packed (hence higher density). This form is used for making milk crates, bleach bottles etc.

(Phenylethene styrene) is very readily polymerized and has a wide range of uses. Poly(phenylethene) can be used in the ordinary form (*e.g.* flower pots, plastic model kits) or in the expanded form (ceiling tiles).

The importance of these addition polymers often depends upon their lack of chemical reactivity.

Where the alkene monomer is of the form $CH_2 = CHX$ (where X is $CH_3 -$, $C_6H_5 -$, $Cl-$ etc) when polymerization occurs, the X groups can be situated

(*i*) on the same side of the chain,

(*ii*) on the alternate sides of the chain,

(*iii*) randomly.

These three possibilities are called **isotactic, syndiotactic** and **atactic** polymers.

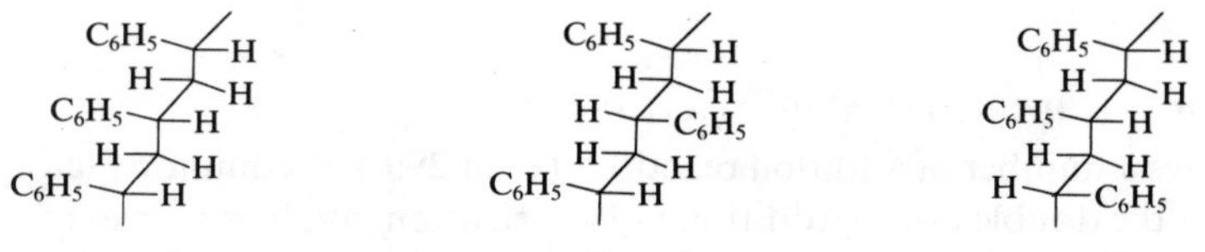

isotactic                    syndiotactic                    atactic

The type of polymer determines the properties. In isotactic polymers the chains can approach more closely. This increases the crystallinity of the polymer.

Polymers can also be classified as **thermosetting** and **thermoplastic**. All of the addition polymers are thermoplastic. This means that on heating they will soften and melt. As a result they can be moulded or extruded. Thermosetting polymers (*e.g.* phenol-'formaldehyde' resins – condensation polymers) do not melt on heating. When heated they decompose and, therefore, cannot be worked in the same way.

## 36.3    CONDENSATION (STEP) POLYMERIZATION

A condensation reaction is a reaction in which two molecules join together to form a larger molecule with the loss of a small molecule often water. An example of a condensation reaction is:

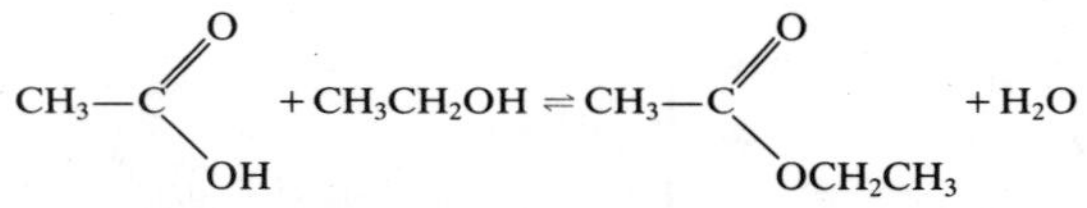

A condensation polymer is formed by a series of condensation reactions using starting materials containing two reactive groups.

*E.g.*    HO—●—OH          HOOC—□—COOH
              diol                   dicarboxylic acid

1st condensation reaction product

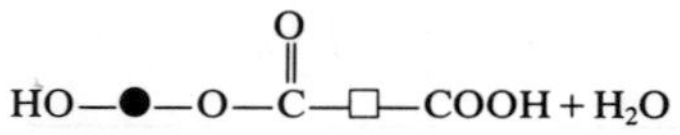

$$HO—●—O—\overset{\overset{O}{\|}}{C}—□—COOH + H_2O$$

reacts with another molecule of diol to produce

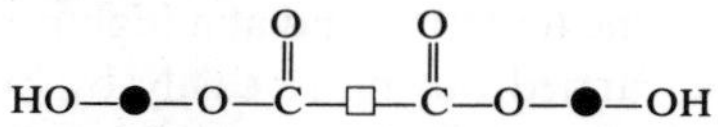

$$HO—●—O—\overset{\overset{O}{\|}}{C}—□—\overset{\overset{O}{\|}}{C}—O—●—OH$$

which reacts with another molecule of dicarboxylic acid.

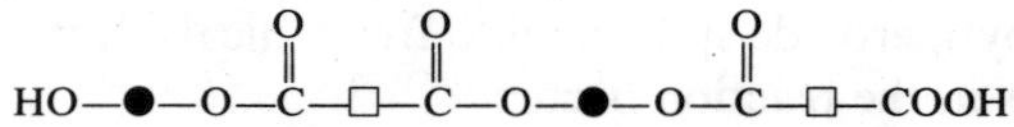

$$HO—●—O—\overset{\overset{O}{\|}}{C}—□—\overset{\overset{O}{\|}}{C}—O—●—O—\overset{\overset{O}{\|}}{C}—□—COOH$$

This continues until eventually a polymer is formed. In this case, after a series of esterification reactions, the polymer is called a **polyester**.

Terylene is a polyester prepared by heating ethane-1,2-diol with dimethylbenzene-1,4-dicarboxylate (dimethyl terephthalate). The small molecule lost in each step is methanol. The methyl ester is used in preference to the free benzene-1,4-dicarboxylic acid because it is easier to purify.

Terylene:

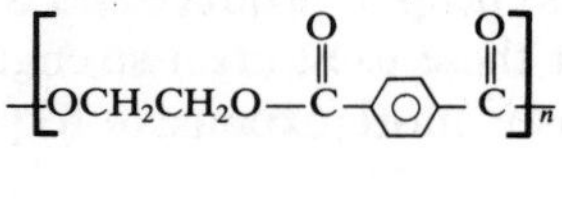

Nylon is a name given to synthetic polyamides formed by a series of reactions involving diamines and dicarboxylic acids. There are different types of nylon.

*Nylon 6, 6*

This is prepared by heating hexane-1,6-dioic acid (adipic acid) with hexane-1,6-diamine (hexamethylene diamine)

$$\text{Nylon 6, 6: } -[CO(CH_2)_4CO.NH(CH_2)_6NH]_n-$$

*Nylon 6*

This is prepared from phenol by a series of reactions.

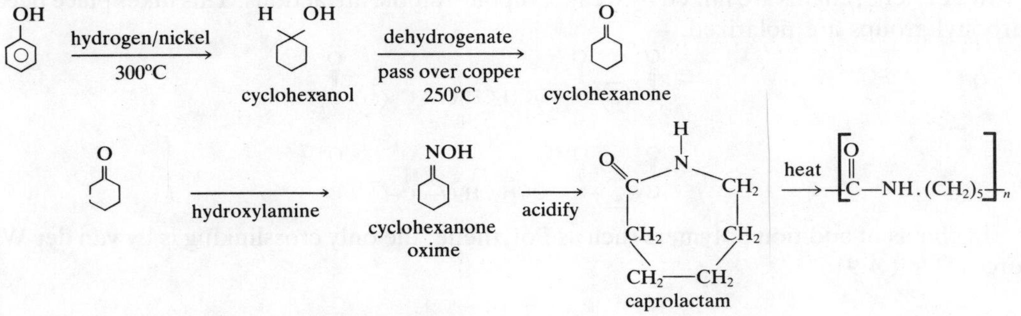


## 36.4 Amino acids and proteins

Amino acids are obtained by hydrolysis of proteins. The amino acids can be represented by

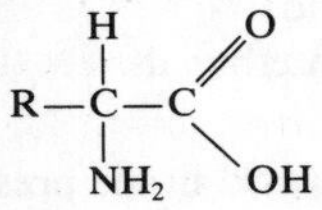


The simplest amino acids are

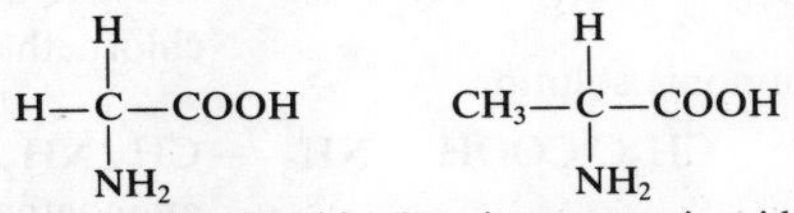


aminoethanoic acid (glycine) 2-aminopropanoic acid (alanine)

2-aminopropanoic acid and all amino acids except aminoethanoic acid contain an asymmetric carbon atom and exhibit **optical isomerism** (see Unit 7.12).

Since the amino acids contain amino and carboxyl groups, they show the properties of both groups. Amino acids have high melting points and are usually water-soluble.

Amino acids usually exist as **inner salts** or **zwitterions** (*i.e.* a species containing both positive and negative charges).

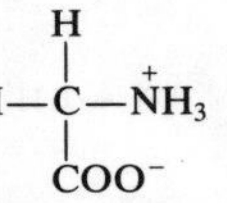


Amino acids are linked together by peptide linkages.

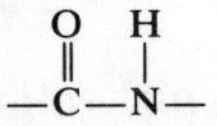


Amino acid polymers whose relative molecular masses are 10 000 or less are called **polypeptides**. Those with larger relative molecular masses are called **proteins**, *e.g.* egg albumen, 40 000.

The presence of protein can be shown by warming the suspected protein with dilute sodium hydroxide solution and copper(II) sulphate solution. A violet colouration confirms the presence of protein.

The particular amino acids present in a protein can be identified by hydrolysis with dilute acid followed by paper chromatography. Ninhydrin produces pink/purple spots on the chromatogram, showing where the amino acids are.

## 36.5 Crosslinking

Crosslinking between polymer chains alters the physical properties of the polymer. It increases the effective relative molecular mass of the polymer and affects its solubility. It also prevents the movement of a chain relative to nearby chains. Crosslinking of some kind is essential if the polymer is to act as a fibre.

Natural rubber is a natural addition polymer. It is unsuitable for use in making rubber tyres unless it is treated with sulphur. Treatment with 30% sulphur hardens the rubber by forming cross links between chains. This process is called **vulcanization**.

Polyamide chains are cross linked by hydrogen bonding (Unit 4.8).

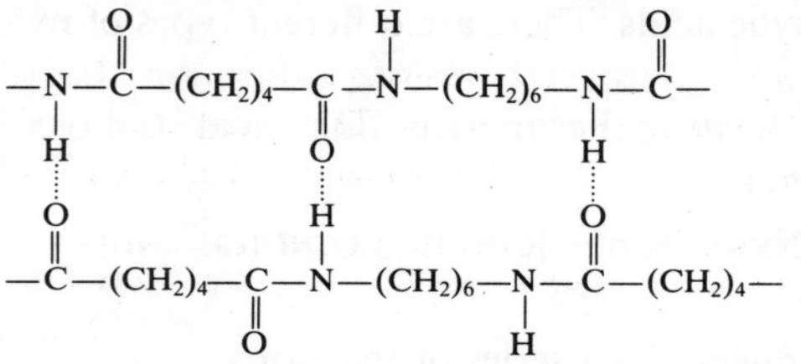

$$-N-C-(CH_2)_4-C-N-(CH_2)_6-N-C-$$

$$-C-(CH_2)_4-C-N-(CH_2)_6-N-C-(CH_2)_4-$$

In Terylene, chains are linked by weaker dipole–dipole attractions. This takes place because carbonyl groups are polarized.

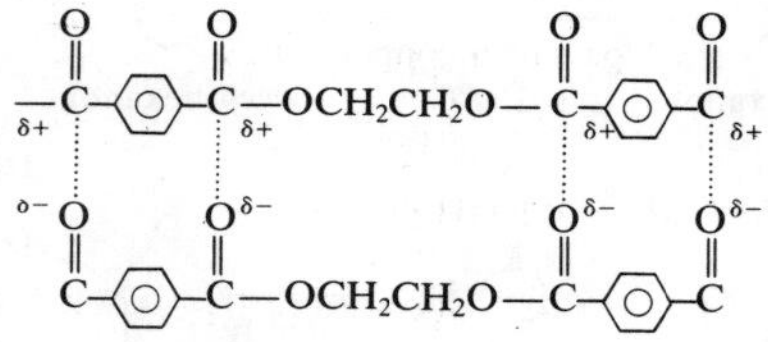

$$-C-\langle O \rangle-C-OCH_2CH_2O-C-\langle O \rangle-C-$$

$$C-\langle O \rangle-C-OCH_2CH_2O-C-\langle O \rangle-C$$

In chains of addition polymers such as Polythene, the only crosslinking is by van der Waals' forces (Unit 4.9)

## QUESTIONS

1  Explain briefly how the following amino acids could be prepared.

(*i*) Aminoethanoic acid from ethanoic acid.
(*ii*) 2-aminopropanoic acid (alanine) from ethanal.

(*i*)  This change requires two stages.
*Stage 1.* Bubble chlorine through ethanoic acid in the presence of ultraviolet light:

$$CH_3COOH + Cl_2 \rightarrow CH_2ClCOOH + HCl$$
chloroethanoic acid

*Stage 2.* Add concentrated ammonia solution:

$$CH_2ClCOOH + NH_3 \rightarrow CH_2(NH_2)COOH + HCl$$
aminoethanoic acid

(*ii*)  This requires four stages
*Stage 1.* Addition of hydrogen cyanide to ethanal:

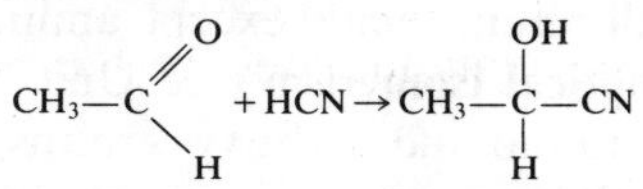

$$CH_3-C{\overset{O}{\underset{H}{}}} + HCN \rightarrow CH_3-\overset{OH}{\underset{H}{C}}-CN$$
2-hydroxypropanenitrile

*Stage 2.* React with phosphorus pentachloride:

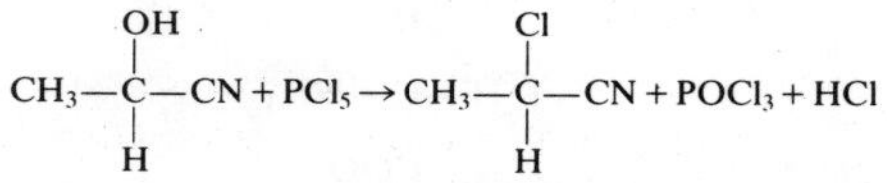

$$CH_3-\overset{OH}{\underset{H}{C}}-CN + PCl_5 \rightarrow CH_3-\overset{Cl}{\underset{H}{C}}-CN + POCl_3 + HCl$$
2-chloropropanenitrile

*Stage 3.* React with concentrated ammonia solution:

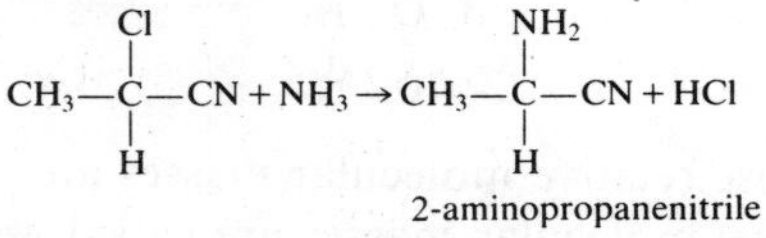

$$CH_3-\overset{Cl}{\underset{H}{C}}-CN + NH_3 \rightarrow CH_3-\overset{NH_2}{\underset{H}{C}}-CN + HCl$$
2-aminopropanenitrile

*State 4.* Hydrolyze with dilute acid:

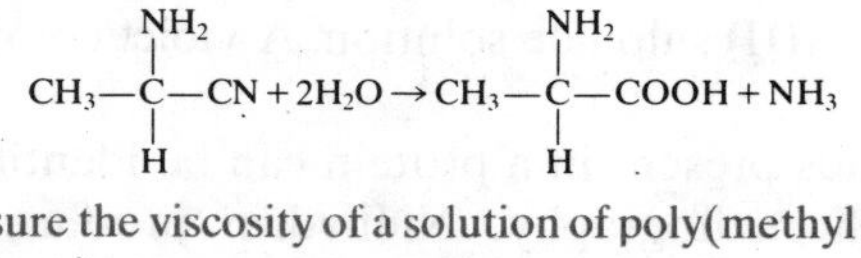

$$CH_3-\overset{NH_2}{\underset{H}{C}}-CN + 2H_2O \rightarrow CH_3-\overset{NH_2}{\underset{H}{C}}-COOH + NH_3$$

2  Describe how you could measure the viscosity of a solution of poly(methyl methacrylate) (Perspex) dissolved in trichloromethane (chloroform).

The 'intrinsic viscosity' (*n*) was measured as 10cm$^3$g$^{-1}$ for the polymer solution and is related to the average relative molecular mass (*M*) of the polymer by the relationship

$$n = KM^\alpha$$

The viscosity of a polymer solution is related to the relative molecular mass of the polymer. The longer the chains, the more viscous the solution will be.

A simple method of comparing the viscosities of two polymer solutions is to fill two metre long tubes with the polymer solutions. Drop a ball bearing into the top of each tube and time until the ball bearing reaches the bottom of the tube. The more viscous the solution the longer it will take for the ball bearing to fall.

The viscosity can be measured accurately using a viscometer. Liquids in a reservoir are allowed to pass through a capilliary tube into a lower reservoir. The viscosities can be measured by timing how long this process takes for each liquid. Further details would be found in an A-level Physics text.

Calculate $M$:

$$n = 10 \text{ cm}^3\text{g}^{-1}$$
$$n = KM^\alpha \qquad K = 0.0100 \text{cm}^3\text{g}^{-1}$$
$$\alpha = 0.5$$
(check all units are in agreement)

Substitute:

$$10 = 0.01 \times M^{0.5}$$
$$M^{0.5} = \frac{10}{0.01} = 1000$$

squaring both sides     $$M = 1000^2 = 1\,000\,000$$

NB This value is an average relative molecular mass. In any polymer the chains will be of different lengths.

# 37 Conversions

## 37.1 INTRODUCTION

It is important to develop an understanding of the inter-relationships between organic compounds in different homologous series. Many questions on A-level papers test your knowledge of these. Questions called conversions are particularly useful for this purpose. This unit should only be attempted after Units 29–36 have been studied.

This unit contains examples of common conversions and how this type of question should be attempted. It should develop your understanding of the basic framework of common organic reactions.

## 37.2 HOW TO ANSWER THESE QUESTIONS

The following points should be remembered:

(*i*) For each step in the answer give an equation, reaction conditions and names of the products.
(*ii*) The conversion should be completed in the smallest number of stages – usually three or four are sufficient. Lengthy routes will waste time and may not receive full marks.
(*iii*) If you cannot immediately see the route adopt the following routine.
    **A** List all the substances which can be made from the starting material in one step.
    **B** List all the substances which undergo reaction to give the final product in one step.
Then try to find reactions which will convert a substance in list **A** into a substance in list **B**.
(*iv*) Use steps which give a good yield of final product. For example chlorination of ethane to give 1,2-dichloroethane is not a good reaction as it is difficult to control the formation of other products.

The following sections detail common conversions.

## 37.3 METHANOL TO ETHANOL

This involves increasing the numbers of carbon atoms by one (called **ascending the homologous series**). Possible methods are summarized as follows:

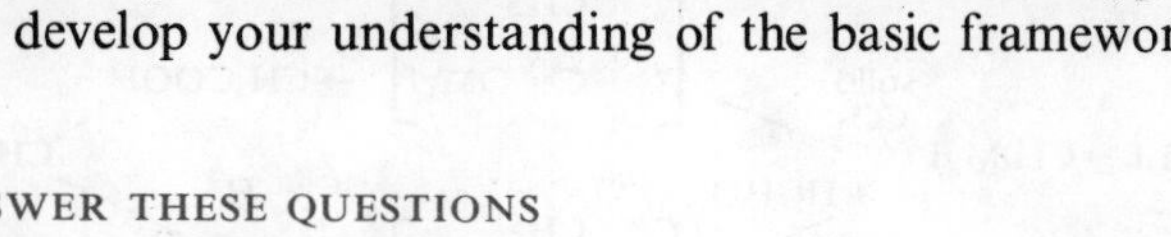

$$CH_3OH \xrightarrow{1} CH_3I \xrightarrow{2} CH_3CN \begin{array}{c} \xrightarrow{A3} CH_3CH_2NH_2 \xrightarrow{A4} \\ \searrow_{B3} \qquad \qquad \nearrow_{B4} \\ CH_3COOH \end{array} CH_3CH_2OH$$

*Step 1:* $CH_3OH \rightarrow CH_3I$

Methanol is refluxed with a mixture of red phosphorus and iodine (which produces phosphorus triiodide $PI_3$).

$$3CH_3OH + PI_3 \rightarrow 3CH_3I + H_3PO_3$$
iodomethane

*Step 2:* $CH_3I \rightarrow CH_3CN$

Iodomethane is refluxed with a solution of potassium cyanide dissolved in ethanol producing ethanenitrile (acetonitrile)

$$CH_3I + KCN \rightarrow CH_3CN + KI$$

There are two alternative routes from here.

**Alternative A**

*Step 3:* $CH_3CN \rightarrow CH_3CH_2NH_2$

Ethanenitrile is reduced to ethylamine using either sodium in ethanol or lithium tetra-hydridoaluminate(III) (lithium aluminium hydride) dissolved in dry ether as the reducing agent.

$$CH_3CN + 4[H] \rightarrow CH_3CH_2NH_2$$
$$\text{ethylamine}$$

*Step 4:* $CH_3CH_2NH_2 \rightarrow CH_3CH_2OH$

Ethylamine is dissolved in excess dilute hydrochloric acid and nitrous acid (nitric(III) acid) added.

$$CH_3CH_2NH_2 + HONO \rightarrow CH_3CH_2OH + H_2O + N_2$$

**Alternative B**

*Step 3:* $CH_3CN \rightarrow CH_3COOH$

Ethanenitrile is hydrolyzed by boiling with dilute hydrochloric acid.

$$CH_3CN + 2H_2O + HCl \rightarrow CH_3COOH + NH_4Cl$$
$$\text{ethanoic acid} \quad \text{ammonium chloride}$$

*Step 4:* $CH_3COOH \rightarrow CH_3CH_2OH$

Ethanoic acid is reduced to ethanol by lithium tetrahyridoaluminate(III) (lithium aluminium hydride)

$$CH_3COOH + 4[H] \rightarrow CH_3CH_2OH + H_2O$$

It is possible also to bring about this, and many other conversions, by means of Grignard reagents (Unit 30.3).

A Grignard reagent is prepared by adding dry magnesium turnings to a haloalkane dissolved in dry ethoxyethane. This produces an unstable compound which, when poured onto solid carbon dioxide and hydrolyzed with dilute acid, produces an acid which can be reduced to the alcohol. Alternatively, if the Grignard reagent is added to methanol and hydrolyzed the alcohol is formed directly

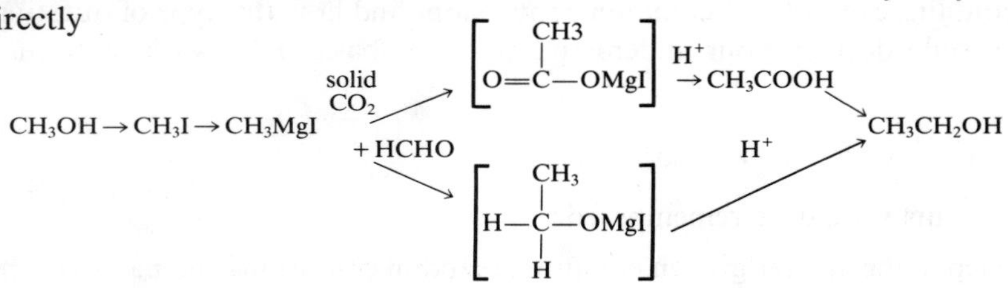

A similar sequence can be used to convert $CH_3COOH \rightarrow CH_3CH_2COOH$

## 37.4 ETHANOL TO METHANOL

This is the reverse of the conversion in 37.3 (**descending the homologous series**).

$$CH_3CH_2OH \xrightarrow{1} CH_3COOH \xrightarrow{2} CH_3CONH_2 \xrightarrow{3} CH_3NH_2 \xrightarrow{4} CH_3OH$$

*Step 1:* $CH_3CH_2OH \rightarrow CH_3COOH$

Ethanol is oxidized by heating with an excess of acidified potassium dichromate(VI) solution.

$$CH_3CH_2OH + 2[O] \rightarrow CH_3COOH + H_2O$$
$$\text{ethanoic acid}$$

*Step 2:* $CH_3COOH \rightarrow CH_3CONH_2$

Ethanoic acid is neutralized with excess ammonia solution and the ammonium ethanoate formed is heated to produce ethanamide (acetamide)

$$CH_3COOH + NH_3 \rightarrow CH_3COONH_4$$
$$CH_3COONH_4 \rightarrow CH_3CONH_2 + H_2O$$
$$\text{ethanamide}$$

*Step 3:* $CH_3CONH_2 \rightarrow CH_3NH_2$

This step removes a carbon atom. The reaction is called **Hofmann's degradation**. Ethanamide is treated with bromine and aqueous sodium hydroxide solution and the mixture is heated.

$$CH_3CONH_2 + Br_2 \rightarrow CH_3CONHBr + HBr$$
$$CH_3CONHBr + 3NaOH \rightarrow CH_3NH_2 + Na_2CO_3 + H_2O + NaBr$$
$$\text{methylamine}$$

*Step 4:* $CH_3NH_2 \rightarrow CH_3OH$

Methylamine is treated with nitrous acid (nitric(III) acid).

$$CH_3NH_2 + HONO \rightarrow CH_3OH + H_2O + N_2$$

(NB In practice an ether is the main product, however.)

### 37.5 ETHANAL TO PROPANONE

Two methods are possible.

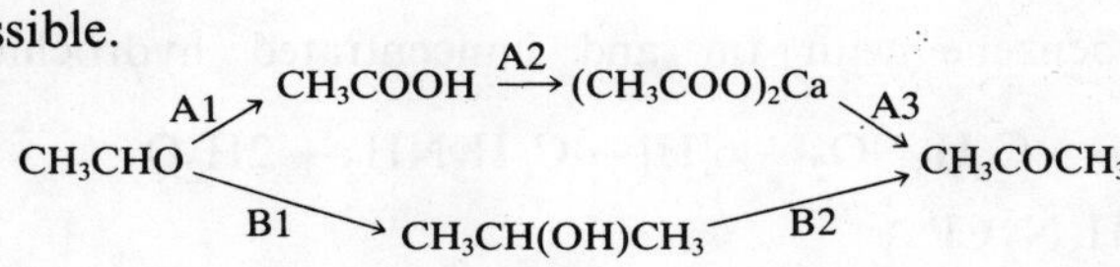

### Alternative A

*Step 1:* $CH_3CHO \rightarrow CH_3COOH$

Ethanal is oxidized by heating with potassium dichromate(VI) solution.

$$CH_3CHO + [O] \rightarrow CH_3COOH$$
$$\text{ethanoic acid}$$

*Step 2:* $CH_3COOH \rightarrow (CH_3COO)_2Ca$

Ethanoic acid is neutralized by adding solid calcium carbonate.

$$2CH_3COOH + CaCO_3 \rightarrow (CH_3COO)_2Ca + H_2O + CO_2$$
$$\text{calcium ethanoate}$$

*Step 3:* $(CH_3COO)_2Ca \rightarrow CH_3COCH_3$

Solid calcium ethanoate is heated (or dry distilled) and decomposes.

$$(CH_3COO)_2Ca \rightarrow CH_3COCH_3 + CaCO_3$$

### Alternative B

This again uses a Grignard reagent

*Step 1:* $CH_3CHO \rightarrow CH_3CH(OH)CH_3$

A Grignard reagent (*e.g.* $CH_3MgBr$) is prepared (Unit 30.3). Ethanal is added and an intermediate is formed which, on hydrolysis with a dilute acid, produces propan-2-ol (a secondary alcohol).

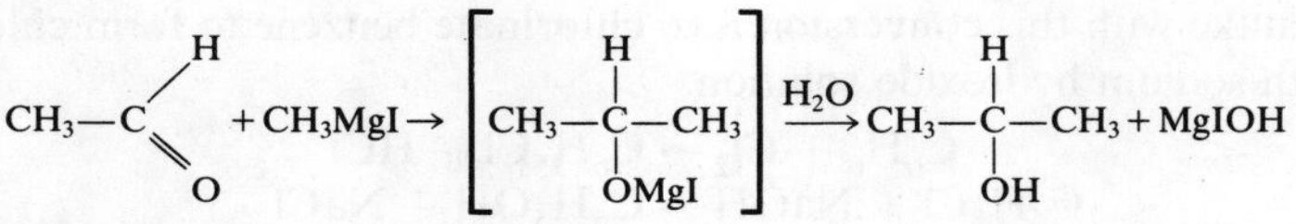

*Step 2:* $CH_3CH(OH)CH_3 \rightarrow CH_3COCH_3$

Propan-2-ol is oxidised to propanone by refluxing with acidified potassium dichromate(VI) solution.

$$CH_3CH(OH)CH_3 + [O] \rightarrow CH_3COCH_3 + H_2O$$
$$\text{propanone}$$

### 37.6 ETHANAL TO 2-HYDROXYPROPANOIC ACID

This conversion to 2-hydroxypropanoic acid (lactic acid) can be completed in two steps.

*Step 1:* $CH_3CHO \rightarrow CH_3CH(OH)CN$

This step involves the addition of hydrogen cyanide to ethanal.

$$HCN + CH_3CHO \rightarrow CH_3CH(OH)CN$$
$$\text{2-hydroxypropanenitrile}$$

*Step 2:* $CH_3CH(OH)CN \rightarrow CH_3CH(OH)COOH$

Hydrolysis of 2-hydroxypropanenitrile by refluxing with dilute hydrochloric acid produces the required product.

$$CH_3CH(OH)CN + 2H_2O + HCl \rightarrow CH_3CH(OH)COOH + NH_4Cl$$

### 37.7 BENZENE TO PHENOL

The first stage in any conversion starting with benzene is invariably one of the electrophilic substitution reactions (Unit 29.12). There are two possible routes.

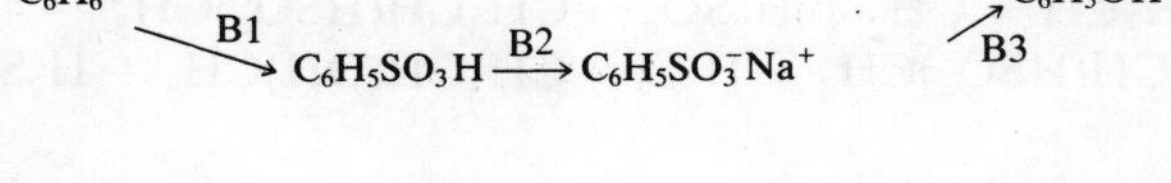

**Alternative A**

*Step 1:* $C_6H_6 \rightarrow C_6H_5NO_2$

Nitration of benzene with a mixture of concentrated nitric and sulphuric acids at a temperature below 55°C to prevent further substitution.

$$C_6H_6 + HNO_3 \rightarrow C_6H_5NO_2 + H_2O$$
$$\text{nitrobenzene}$$

*Step 2:* $C_6H_5NO_2 \rightarrow C_6H_5NH_2$

Reduction of nitrobenzene with tin and concentrated hydrochloric acid produces phenylamine.

$$C_6H_5NO_2 + 6[H] \rightarrow C_6H_5NH_2 + 2H_2O$$

*Step 3:* $C_6H_5NH_2 \rightarrow C_6H_5N_2^+Cl^-$

Phenylamine is converted to benzenediazonium chloride by adding nitrous acid (nitric(III) acid) to a solution of phenylamine in dilute hydrochloric acid with the temperature below 5°C. Benzenediazonium chloride is most useful in aromatic conversions as it is readily converted into a range of other compounds.

*Step 4:* $C_6H_5N_2^+Cl^- \rightarrow C_6H_5OH$

Phenol is produced when the solution of benzenediazonium chloride is warmed after the addition of more water.

$$C_6H_5N_2^+Cl^- + H_2O \rightarrow C_6H_5OH + N_2 + HCl$$

**Alternative B**

*Step 1:* $C_6H_6 \rightarrow C_6H_5SO_3H$

Benzene is sulphonated by heating with fuming sulphuric acid.

$$C_6H_6 + H_2SO_4 \rightarrow C_6H_5SO_3H + H_2O$$

*Step 2:* $C_6H_5SO_3H \rightarrow C_6H_5SO_3^-Na^+$

Benzenesulphonic acid is neutralized with sodium hydroxide to form sodium benzenesulphonate.

$$C_6H_5SO_3H + NaOH \rightarrow C_6H_5SO_3^-Na^+ + H_2O$$

*Step 3:* $C_6H_5SO_3^-Na^+ \rightarrow C_6H_5OH$

The solid sodium benzenesulphonate is then fused with solid sodium hydroxide

$$C_6H_5SO_3^-Na^+ + NaOH \rightarrow C_6H_5OH + Na_2SO_3$$

A common mistake with this conversion is to chlorinate benzene to form chlorobenzene and then hydrolyze with sodium hydroxide solution.

$$C_6H_6 + Cl_2 \rightarrow C_6H_5Cl + HCl$$
$$C_6H_5Cl + NaOH \rightarrow C_6H_5OH + NaCl$$

The second step however does not take place under ordinary laboratory conditions.

### 37.8    METHYLBENZENE TO BENZENECARBALDEHYDE

*Step 1:* $C_6H_5CH_3 \rightarrow C_6H_5CHCl_2$

Methylbenzene is mixed with chlorine. The mixture is heated in ultraviolet light and a free radical substitution reaction in the $CH_3$ group takes place. No halogen carrier should be present.

$$C_6H_5CH_3 + Cl_2 \rightarrow C_6H_5CH_2Cl + HCl$$
$$\text{(chloromethyl)benzene (benzyl chloride)}$$

$$C_6H_5CH_2Cl + Cl_2 \rightarrow C_6H_4CHCl_2 + HCl$$
$$\text{(dichloromethyl)benzene (benzal chloride)}$$

*Step 2:* $C_6H_5CHCl_2 \rightarrow C_6H_5CHO$

Hydrolysis of (dichloromethyl)benzene by refluxing with sodium hydroxide solution produces benzenecarbaldehyde (benzaldehyde)

$$C_6H_5CHCl_2 + 2NaOH \rightarrow C_6H_5CHO + 2NaCl + H_2O$$

### 37.9    PROPAN-1-OL TO PROPAN-2-OL

*Step 1:* $CH_3CH_2CH_2OH \rightarrow CH_3CH = CH_2$

Propan-1-ol is dehydrated by heating with excess concentrated sulphuric acid at 170°C to form propene.

$$CH_3CH_2CH_2OH \rightarrow CH_3CH = CH_2 + H_2O$$

*Step 2:* $CH_3CH = CH_2 \rightarrow CH_3CH(OH)CH_3$

Propene is passed into concentrated sulphuric acid and the product is then hydrolyzed with water.

$$CH_3CH = CH_2 + H_2SO_4 \rightarrow CH_3CH(HSO_4)CH_3$$
$$CH_3CH(HSO_4)CH_3 + H_2O \rightarrow CH_3CH(OH)CH_3 + H_2SO_4$$

QUESTIONS

**1** Complete the flow diagrams in Fig. 37.1 and 37.2 by inserting the correct structural formulae and names.

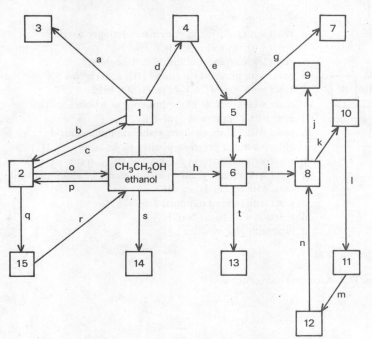

Label this diagram as follows:

a  react with trioxygen (ozone) and hydrolyse
b  react with HBr
c  reflux with soln. of KOH in ethanol
d  reaction with bromine
e  heat with soln. of KOH in ethanol
f  pass into warm dil $H_2SO_4$ in presence of $HgSO_4$
g  pass through heated copper tube
h  warm with acidified potassium dichromate (VI)
i  further oxidation
j  heat with ethanol in the presence of conc sulphuric acid
k  react with ammonia solution
l  heat
m heat with $P_2O_5$
n  boil with dil HCl
o  reflux with solution of KOH in water
p  reflux with mixture of red P and bromine
q  heat with ammonia
r  react with nitrous (nitric (III) acid)
s  excess ethanol heated with conc $H_2SO_4$ at 140°C
t  react with $PBr_5$

**Fig. 37.1** Aliphatic conversions

The compounds in Fig. 37.1 are

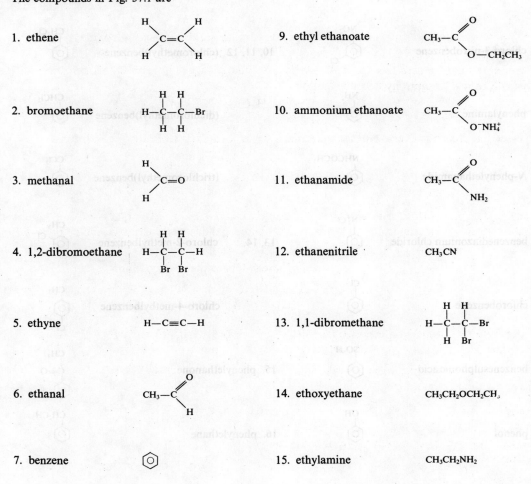

1. ethene

2. bromoethane

3. methanal

4. 1,2-dibromoethane

5. ethyne            $H—C\equiv C—H$

6. ethanal

7. benzene

8. ethanoic acid

9. ethyl ethanoate

10. ammonium ethanoate

11. ethanamide

12. ethanenitrile      $CH_3CN$

13. 1,1-dibromethane

14. ethoxyethane     $CH_3CH_2OCH_2CH_3$

15. ethylamine       $CH_3CH_2NH_2$

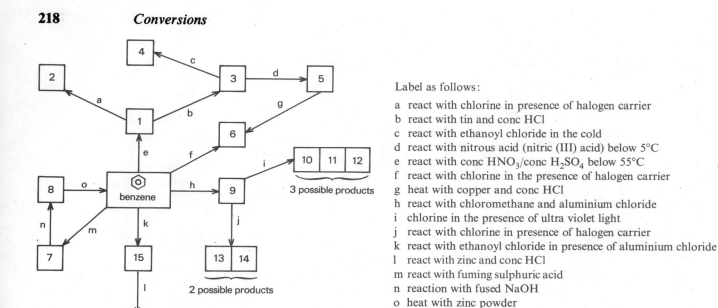

**Label as follows:**

a   react with chlorine in presence of halogen carrier
b   react with tin and conc HCl
c   react with ethanoyl chloride in the cold
d   react with nitrous acid (nitric (III) acid) below 5°C
e   react with conc $HNO_3$/conc $H_2SO_4$ below 55°C
f   react with chlorine in the presence of halogen carrier
g   heat with copper and conc HCl
h   react with chloromethane and aluminium chloride
i   chlorine in the presence of ultra violet light
j   react with chlorine in presence of halogen carrier
k   react with ethanoyl chloride in presence of aluminium chloride
l   react with zinc and conc HCl
m   react with fuming sulphuric acid
n   reaction with fused NaOH
o   heat with zinc powder

**Fig. 37.2** Aromatic conversions

The compounds in Fig. 37.2 are

1.  nitrobenzene

2.  chloro-3-nitrobenzene

3.  phenylamine

4.  *N*-phenylethanamide

5.  benzenediazonium chloride

6.  chlorobenzene

7.  benzenesulphonic acid

8.  phenol

9.  methylbenzene

10, 11, 12   (chloromethyl)benzene

(dichloromethyl)benzene

(trichloromethyl)benzene

13, 14.   chloro-2-methylbenzene

chloro-4-methylbenzene

15.  phenylethanone

16.  phenylethane

Having completed this exercise it is worthwhile spending some time looking at these two frameworks.
Draw out the frameworks again, missing out the reaction conditions but putting in the names and formulae.
Then list the conditions for each step and write a balanced equation.

**2** Describe how the following conversions could be carried out. In each case you should give reagents, conditions of the reactions and the structures of the products.

(a) $CH_2 = CH_2$ to $CH_2CO_2H$
$\phantom{CH_2 = CH_2 to} CH_2CO_2H$

(b) $CH_3CHBrCH_3$ to $CH_3CHBrCH_2Br$
(c) $C_2H_5CN$ to $C_2H_5OH$
(d) $C_6H_6$ to $C_6H_5NHCOCH_3$

(a) This involves the addition of two carbon atoms (Unit 37.3).
Addition of bromine to ethene forms 1,2-dibromoethane.

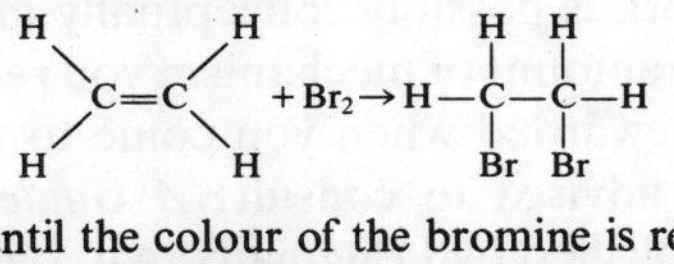

$$C=C + Br_2 \rightarrow H-C-C-H$$

Pass ethene into liquid bromine until the colour of the bromine is removed. Cool the liquid bromine and cover with a small volume of water to prevent evaporation.
React with potassium cyanide dissolved in ethanol with gentle refluxing.

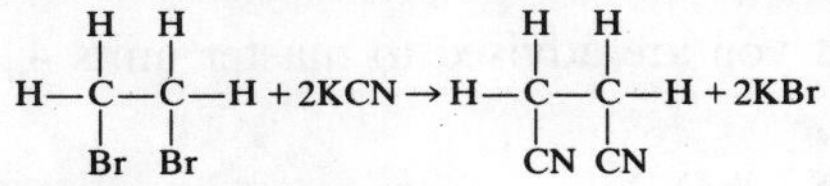

$$H-C-C-H + 2KCN \rightarrow H-C-C-H + 2KBr$$

Hydrolyse by boiling with dilute hydrochloric acid.

$$H-C-C- + 4H_2O + 2HCl \rightarrow HOOC.CH_2CH_2.COOH + 2NH_4Cl.$$

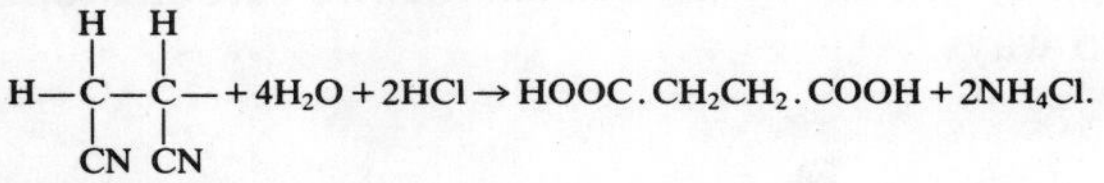

(b) Elimination of hydrogen bromide.
Reflux 2-bromopropane with a solution of potassium hydroxide in ethanol.

$$CH_3CHBrCH_3 + KOH \rightarrow CH_3CH = CH_2 + KBr + H_2O$$
$$\text{propene}$$

Addition of bromine to propene.
Pass propene into liquid bromine until the colour is removed.

$$CH_3CH = CH_2 + Br_2 \rightarrow CH_3CHBrCH_2Br$$

(c) This conversion involves descending a homologous series (Unit 37.4). The nitrile is hydrolyzed to the corresponding acid by refluxing with dilute hydrochloric acid.

$$C_2H_5CN + 2H_2O + HCl \rightarrow C_2H_5COOH + NH_4Cl$$
$$\text{propanoic acid}$$

The propanoic acid is treated with ammonia solution to form ammonium propanoate and this is heated to dryness to produce propanamide.

$$C_2H_5COOH + NH_3 \rightarrow C_2H_5COONH_4$$
$$C_2H_5COONH_4 \rightarrow C_2H_5CONH_2 + H_2O$$

Hofmann's degradation of propanamide by heating the mixture of propanamide, bromine and potassium hydroxide solution produces ethylamine

$$C_2H_5CONH_2 + Br_2 \rightarrow C_2H_5CONHBr + HBr$$
$$C_2H_5CONHBr + 3NaOH \rightarrow C_2H_5NH_2 + Na_2CO_3 + H_2O + NaBr$$

Finally the ethylamine is treated with nitrous acid (nitric(III) acid) formed from sodium nitrite (sodium nitrate(III)) and dilute hydrochloric acid.

$$C_2H_5NH_2 + HONO \rightarrow C_2H_5OH + H_2O + N_2$$

(d) This involves the first two steps of the conversion of benzene to phenol (37.7). The product phenylamine is finally treated with either
(i) ethanoyl chloride in the cold with careful mixing, or
(ii) ethanoic anhydride dissolved in ethanoic acid. In this case the reaction is slower and refluxing is necessary

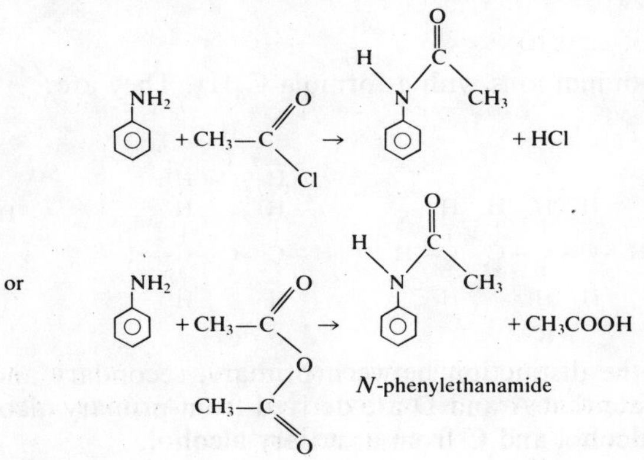

*N*-phenylethanamide

# 38 Organic reaction mechanisms

## 38.1 INTRODUCTION

It is useful to assist your understanding of organic chemistry to consider the way in which particular chemical reactions take place. This a fairly recent requirement of certain A-level syllabuses (see page 12). This work is possibly conceptually more difficult than the rest of the organic chemistry at A level. The amount of mechanism you require is limited and the time you spend on this Unit will be well rewarded when you come to take the examination.

For further reading you are advised to consult: *A Guidebook to Mechanism in Organic Chemistry* by Dr Peter Sykes (published by Longman). Although this book is written to a level above A level, it is very clearly written and you would benefit from Chapters 1, 3–8 and 10 in particular.

Before attempting this unit you are advised to master units 4, 5, 7, 29–37.

## 38.2 BREAKING A COVALENT BOND

Consider the breaking of the covalent bond between the two carbon atoms in an ethane molecule – it can be broken in two ways.

### (i) Homolytic fission

The bond is broken so that, of the two electrons in the bond, one electron goes to each carbon atom.

$$E.g. \quad H-\overset{\overset{\displaystyle H}{|}}{\underset{\underset{\displaystyle H}{|}}{C}}-\overset{\overset{\displaystyle H}{|}}{\underset{\underset{\displaystyle H}{|}}{C}}-H \rightarrow H-\overset{\overset{\displaystyle H}{|}}{\underset{\underset{\displaystyle H}{|}}{C}}\cdot + \cdot\overset{\overset{\displaystyle H}{|}}{\underset{\underset{\displaystyle H}{|}}{C}}-H$$

The resulting species are methyl **free radicals** and they each contain a single unpaired electron. Free radicals are extremely short-lived and readily undergo further reaction. This type of bond breaking usually occurs in the gas phase and in non-polar solvents especially in strong light.

### (ii) Heterolytic fission

The bond is broken so that one carbon atom receives both electrons while the other receives none.

$$E.g. \quad H-\overset{\overset{\displaystyle H}{|}}{\underset{\underset{\displaystyle H}{|}}{C}}-\overset{\overset{\displaystyle H}{|}}{\underset{\underset{\displaystyle H}{|}}{C}}-H \rightarrow H-\overset{\overset{\displaystyle H}{|}}{\underset{\underset{\displaystyle H}{|}}{C}}:^- + {}^+\overset{\overset{\displaystyle H}{|}}{\underset{\underset{\displaystyle H}{|}}{C}}-H$$

carbanion    carbocations (carbonium ions)

The positively charged ion is called a carbocation (**carbonium ion**) and the positively charged carbon atom is liable to attack by a negatively charged species which is called a **nucleophile**. (Remember nucleophile and negative both begin with the letter n.)

The negatively charged ion is called the **carbanion** and this is liable to attack by a positively charged species called an **electrophile**. This kind of species is likely to exist in solutions in polar solvents.

Although free radicals, carbonium ions and carbanions may only exist for the shortest possible period of time, they have a particular importance in controlling how a reaction takes place.

## 38.3 STABILITY OF CARBONIUM IONS

There are four possible carbonium ions with a formula $C_4H_9^+$. They are:

$$\underset{A}{H-\overset{\overset{\displaystyle H}{|}}{\underset{\underset{\displaystyle H}{|}}{C}}-\overset{\overset{\displaystyle H}{|}}{\underset{\underset{\displaystyle H}{|}}{C}}-\overset{\overset{\displaystyle H}{|}}{\underset{\underset{\displaystyle H}{|}}{C}}-\overset{\displaystyle H}{\underset{\displaystyle H}{C}}{}^+}$$

$$\underset{B}{H-\overset{\overset{\displaystyle H}{|}}{\underset{\underset{\displaystyle H}{|}}{C}}-\overset{\overset{\displaystyle H}{|}}{\underset{\underset{\displaystyle H}{|}}{C}}-\overset{\overset{\displaystyle H}{|}}{\underset{\underset{\displaystyle +}{|}}{C}}-\overset{\overset{\displaystyle H}{|}}{\underset{\underset{\displaystyle H}{|}}{C}}-H}$$

$$\underset{C}{H-\overset{\overset{\displaystyle H}{|}}{\underset{\underset{\displaystyle H}{|}}{C}}-\overset{\overset{\overset{\displaystyle H}{|}}{\underset{\displaystyle H-C-H}{}}}{\underset{\underset{\displaystyle +}{|}}{C}}-\overset{\overset{\displaystyle H}{|}}{\underset{\underset{\displaystyle H}{|}}{C}}-H}$$

$$\underset{D}{H-\overset{\overset{\displaystyle H}{|}}{\underset{\underset{\displaystyle H-C-H}{|}}{C}}-\overset{\overset{\displaystyle H}{|}}{\underset{\underset{\displaystyle +}{}}{C}}-\overset{\displaystyle H}{\underset{\displaystyle H}{C}}-H}$$

If you have understood the distinction between primary, secondary and tertiary alcohols (Unit 31.2), you will appreciate that A and D are derived from primary alcohols (by removing OH⁻), B from a secondary alcohol and C from a tertiary alcohol.

A and D are called primary carbonium ions, B is a secondary carbonium ion and C is a tertiary carbonium ion.

In general tertiary carbonium ions are more stable than secondary carbonium ions which are, in turn, more stable than primary carbonium ions. It is possible for rearrangement of carbonium ions to take place. Carbonium ions can eliminate an $H^+$ ion and produce an alkene.

*E.g.*

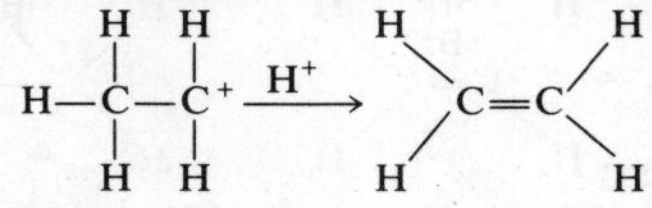

$$H-\underset{\underset{H}{|}}{\overset{\overset{H}{|}}{C}}-\overset{\overset{H}{|}}{\underset{\underset{H}{|}}{C}}{}^{+} \xrightarrow{H^+} \quad \overset{H}{\underset{H}{}}C=C\overset{H}{\underset{H}{}}$$

## 38.4 FREE RADICAL CHAIN REACTIONS

A most commonly quoted free radical chain reaction is the reaction between methane and chlorine (Unit 29.3). This reaction takes place when a mixture of the gases is subjected to ultra-violet light.

The light energy breaks a few chlorine–chlorine bonds to produce chlorine free radicals (chlorine atoms)

$$Cl_2 \rightarrow Cl \cdot + Cl \cdot \quad \textit{Initiation}$$

The following steps then take place

$$\left.\begin{array}{l} Cl \cdot + CH_4 \rightarrow CH_3 \cdot + HCl \\ CH_3 \cdot + Cl_2 \rightarrow CH_3Cl + Cl \cdot \end{array}\right\} \textit{Propagation}$$

The following reactions also take place but they do not promote further reaction

$$\left.\begin{array}{l} CH_3 \cdot + Cl \cdot \rightarrow CH_3Cl \\ Cl \cdot + Cl \cdot \rightarrow Cl_2 \\ CH_3 \cdot + CH_3 \cdot \rightarrow CH_3CH_3 \end{array}\right\} \textit{Termination}$$

The mechanism explains why, in practice, small amounts of ethane are detected in the products.

The reactions between hydrogen and chlorine and methylbenzene and chlorine have similar mechanisms.

## 38.5 ADDITION REACTIONS

In this section a comparison will be made between addition to carbon–carbon double bonds in alkenes (Unit 29.6) and carbon–oxygen double bonds in aldehydes and ketones (Unit 32.4).

### (i) Addition to alkenes (Unit 29.6) – electrophilic addition

The carbon–carbon double bond, as in ethene, can be represented by

The double bond between the two carbon atoms consists of a sigma ($\sigma$) bond and a pi ($\pi$) bond formed by the overlap of *p* orbitals on the two carbon atoms. There is a concentration of negative charge between the two carbon atoms making it susceptible to attack by electrophiles (positive species).

Consider the addition of hydrogen bromide to ethene:

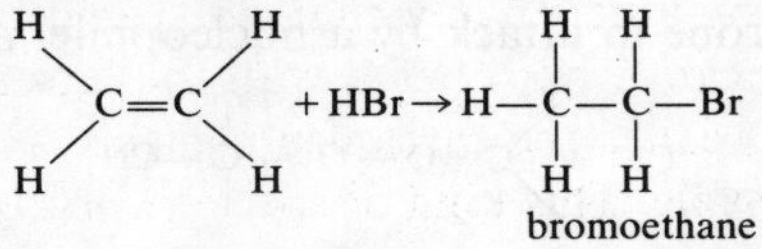

$$\overset{H}{\underset{H}{}}C=C\overset{H}{\underset{H}{}} + HBr \rightarrow H-\underset{\underset{H}{|}}{\overset{\overset{H}{|}}{C}}-\underset{\underset{H}{|}}{\overset{\overset{H}{|}}{C}}-Br$$

bromoethane

There is a dipole within the hydrogen bromide molecule giving the hydrogen a slight positive charge (shown as $\delta+$) and the bromine a slight negative charge ($\delta-$).

The first step involves the formation of a weak complex between the positive end of the HBr molecule and the electrons of the carbon–carbon double bond.

$$\overset{H}{\underset{H}{}}C=C\overset{H}{\underset{H}{}} \quad \begin{array}{c} Br^{\delta-} \\ | \\ H^{\delta+} \end{array}$$

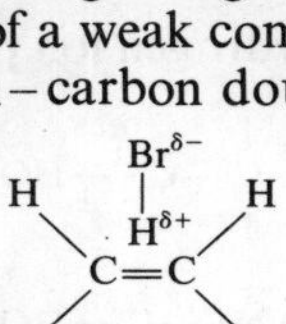

This weak complex may then convert into a more stable carbonium ion.

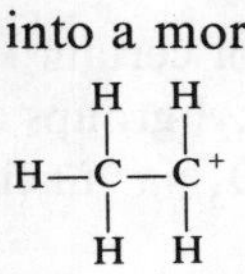

$$H-\underset{\underset{H}{|}}{\overset{\overset{H}{|}}{C}}-\underset{\underset{H}{|}}{\overset{\overset{H}{|}}{C}}{}^{+}$$

This then reacts rapidly with a bromide ion to form the product.

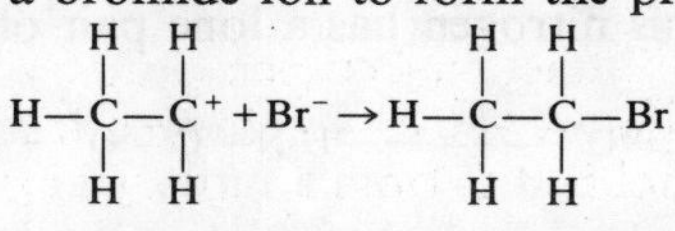

$$H-\underset{\underset{H}{|}}{\overset{\overset{H}{|}}{C}}-\underset{\underset{H}{|}}{\overset{\overset{H}{|}}{C}}{}^{+} + Br^{-} \rightarrow H-\underset{\underset{H}{|}}{\overset{\overset{H}{|}}{C}}-\underset{\underset{H}{|}}{\overset{\overset{H}{|}}{C}}-Br$$

The reaction between ethene and bromine has a similar mechanism. The bromine molecule Br-Br does not contain a permanent dipole. An induced dipole is formed when it approaches an ethene molecule. This is due to repulsion between electrons in the carbon–carbon double bond and the electrons in the covalent bond between the bromine atoms.

The reaction between propene and hydrogen bromide can, in theory, lead to two products:

1-bromopropane          2-bromopropane

This can occur because the groups attached to the two carbon atoms, joined by the double bond, are different. The alkene is not symmetrical.

In practice, addition produces 2-bromopropane only according to **Markownikoff's rule**. Markownikoff's rule was based upon observations of a large number of addition reactions. The rule states that during addition the more negative part of the molecule adding to the alkene (Br in this case) adds to the carbon atom attached to the lesser number of hydrogen atoms.

We can now explain Markownikoff's rule in terms of the stability of carbonium ions. Depending upon which carbon atom the $H^+$ of the hydrogen bromide attaches to, there are two possible carbonium ions which can be formed.

secondary          primary

The secondary carbonium ion is more stable than the primary carbonium ion. It is, therefore, formed in preference to the primary carbonium ion and leads to the formation of 2-bromopropane.

Addition of hydrogen bromide to propene to produce 1-bromopropane is called **anti-Markownikoff addition**. It can be achieved by carrying out the reaction in the presence of peroxides. Free radicals are produced rather than carbonium ions.

### (ii) Addition to aldehydes and ketones – nucleophilic addition

The carbon–oxygen double bond has a permanent shift of electrons because oxygen is more electronegative than carbon. As a result there is a slight positive charge on the carbon and a slight negative charge on the oxygen.

The carbon atom is then prone to attack by a nucleophile, *e.g.* $CN^-$ or $SO_3Na^-$

$$CN^- + \, \rangle C{=}O \rightarrow O^- - \overset{|}{\underset{|}{C}} - CN$$

$$SO_3Na^- + \, \rangle C{=}O \rightarrow O^- - \overset{|}{\underset{|}{C}} - SO_3Na$$

The ion produced then reacts with a hydrogen ion present in the solution to form the product.

$$H^+ + O^- - \overset{|}{\underset{|}{C}} - CN \rightarrow HO - \overset{|}{\underset{|}{C}} - CN$$

$$O^- - \overset{|}{\underset{|}{C}} - SO_3Na + H^+ \rightarrow HO - \overset{|}{\underset{|}{C}} - SO_3Na$$

The slow reaction or lack of reaction of certain ketones with sodium hydrogensulphite can be explained by steric hindrance. If big alkyl groups are attached to the carbonyl group there is insufficient space available for the large $SO_3Na^-$ nucleophile to penetrate and attack the carbon atom.

Condensation reactions involve attack by a molecule containing a nitrogen atom, *e.g.* hydrazine, phenylhydrazine. This nitrogen has a lone pair of electrons.

$$-H_2N: + \overset{\delta+}{\underset{|}{C}}{=}O^{\delta-} \longrightarrow -H_2\overset{+}{N} - \overset{|}{\underset{|}{C}} - O^- \rightarrow -HN - \overset{|}{\underset{|}{C}} - OH$$

Finally there is a loss of water.

$$-HN-\overset{|}{\underset{|}{C}}-OH \rightarrow -N=\overset{|}{C}$$

## 38.6 SUBSTITUTION REACTIONS

A substitution reaction is a reaction where an atom or group of atoms replaces an atom or group of atoms already in the molecule.

$$AB + C \rightarrow AC + B$$

The reaction between methane and chlorine (38.4) is a free radical substitution reaction. There are examples of nucleophilic and electrophilic substitution reactions.

### (i) Nucleophilic substitution reactions

The hydrolysis of haloalkanes by refluxing with aqueous alkali (Unit 30.3) is a nucleophilic substitution reaction.

$$RX + KOH \rightarrow ROH + KX$$
$$CH_3CH_2Br + OH^- \rightarrow CH_3CH_2OH + Br^-$$

The carbon atom attached to the halogen is attacked by the nucleophile $OH^-$.

The rate of hydrolysis of iodoalkanes is greater than the rate of hydrolysis of bromoalkanes which is, in turn greater than the rate of hydrolysis of chloroalkanes. The rate of hydrolysis is related to the strength of the $C-X$ bond. The stronger is the bond between the carbon atom and the halogen, the slower will be the reaction.

Nucleophilic substitution reactions occur more readily with tertiary haloalkanes than secondary, and more readily with secondary than with primary. This can be explained when it is realised that in many case there is an intermediate carbonium ion formed.

The reaction of a haloalkane with aqueous hydroxide ions can occur in two ways.

$S_N1$ mechanism. This is a two-stage process that first involves the loss of an $X^-$ ion and the formation of a carbonium ion. The carbonium ion reacts rapidly with an $OH^-$ ion.

$$CH_3-\overset{\overset{\displaystyle CH_3}{|}}{\underset{\underset{\displaystyle CH_3}{|}}{C}}-X \xrightarrow[-X^-]{slow} CH_3-\overset{\overset{\displaystyle CH_3}{|}}{\underset{\underset{\displaystyle CH_3}{|}}{C^+}} + OH^- \xrightarrow{fast} CH_3-\overset{\overset{\displaystyle CH_3}{|}}{\underset{\underset{\displaystyle CH_3}{|}}{C}}-OH$$

The first stage is the rate-determining step and is independent of the concentration of $OH^-$ ions.

Rate $\propto$ [haloalkane]

The total order of the reaction is one.

$S_N2$ mechanism. This is a one-stage process involving a simultaneous loss of $X^-$ and a gain of $OH^-$ ion. The rate of reaction is not independent of the concentration of $OH^-$ ions.

Rate $\propto$ [haloalkane] $[OH^-]$

The total order of the reaction is two.

Elimination reactions are also possible when haloalkanes are refluxed with sodium hydroxide dissolved in ethanol (Unit 30.3). There is always competition between substitution and elimination reactions. When a haloalkane is heated with *aqueous* alkali, both substitution and elimination are possible but the substitution is so much more favourable that elimination does not, in practice, occur. In the reaction of a haloalkane with a solution of alkali in ethanol, the substitution does not occur, for the reason given shortly, and so elimination takes place.

When the alkali is dissolved in ethanol, the hydroxide ions are solvated with ethanol molecules. The resulting species is too large to approach the positive centre in the carbonium ion and, as a result, substitution is impossible. Instead, the hydroxide ion acts as a base and abstracts a proton.

$$E.g. \quad CH_3-\overset{\overset{\displaystyle CH_3}{|}}{\underset{\underset{\displaystyle CH_3}{|}}{C}} + \xrightarrow{-H^+} CH_2=\overset{\overset{\displaystyle CH_3}{|}}{\underset{\underset{\displaystyle CH_3}{}}{C}}$$

### (ii) Electrophilic substitution reactions (Unit 29.12)

Common examples of electrophilic substitution are found with benzene and similar aromatic compounds. The common examples are nitration, sulphonation, halogenation and Friedel-Crafts reactions.

Benzene has a ring of negative charge above and below the ring (Unit 29.11). The following sequence takes place, using $X^+$ to represent an electrophile.

In the first stage a weak complex is formed between the negative ring on the benzene and the electrophile. A more stable complex is then formed. This complex breaks down with the loss of $H^+$ to form the product. By doing this the stability of the benzene nucleus is restored.

There are four common electrophilic substitution reactions of benzene.

*A Nitration*

The nitrating mixture is a mixture of concentrated nitric and sulphuric acids. They react to produce the nitryl ion (nitronium ion) $NO_2^+$

$$HNO_3 + 2H_2SO_4 \rightleftharpoons NO_2^+ + H_3O^+ + 2HSO_4^-$$

It is possible to isolate salts such as $NO_2^+ClO_4^-$ nitryl chlorate(VII) containing the nitryl ion.

The steps in this reaction are:

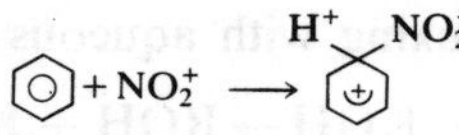

*B Sulphonation*

The effective electrophile in concentrated sulphuric acid is sulphur(VI) oxide (sulphur trioxide).

$$2H_2SO_4 \rightleftharpoons SO_3 + H_3O^+ + HSO_4^-$$

There is a partial positive charge on the sulphur atom and this bonds to the ring

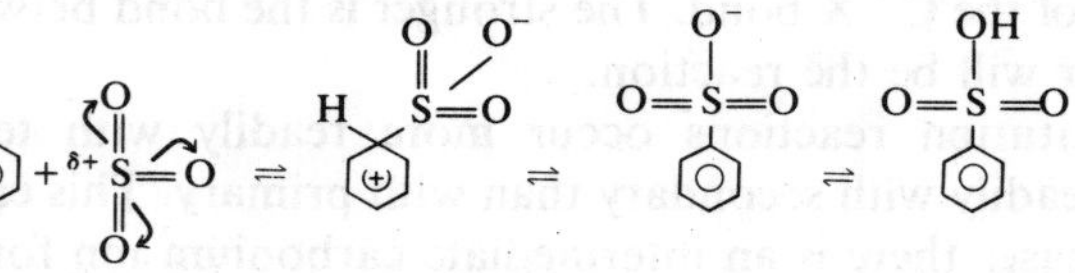

*C Halogenation*

Substitution of chlorine or bromine into a benzene ring requires a halogen carrier. The halogen carrier is a catalyst and is usually a Lewis acid (Unit 16.2). Suitable substances include aluminium bromide, $AlBr_3$, and iron(III) bromide, $FeBr_3$.

The halogen carrier causes a polarization in the halogen molecule.

*E.g.*

$$\overset{\delta+}{Br}\cdots\cdots Br\cdots\cdots \overset{\delta-}{AlBr_3}$$

The positive end of this complex attacks the benzene ring and forms a complex with the electrons of the benzene ring.

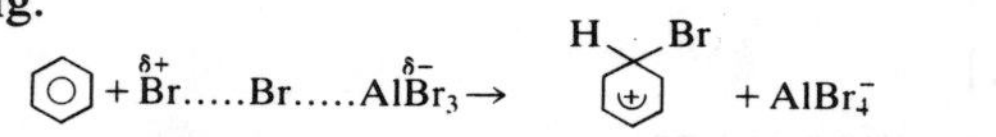

This complex then breaks down to give the product.

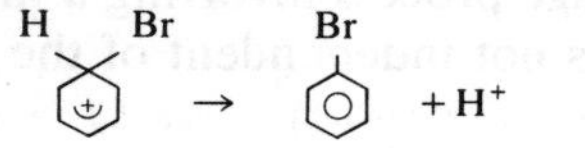

*D Friedel-Crafts reaction*

These reactions resemble halogenation because they require a halogen carrier.

*E.g.*

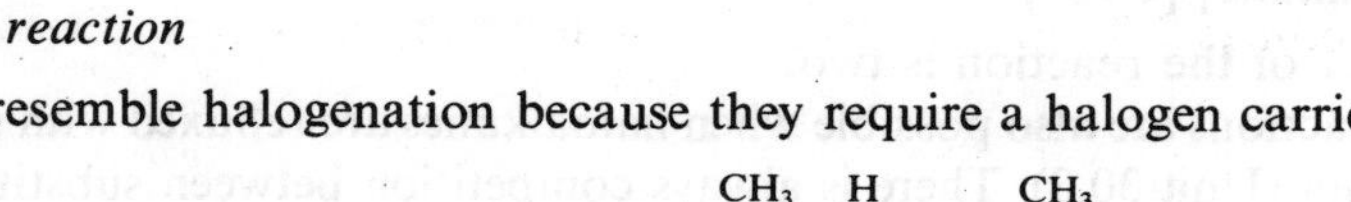

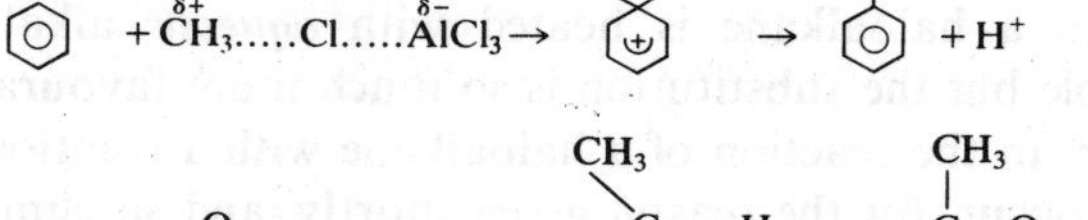

## 38.7 Introducing a second substituent into a benzene ring

The rate at which a second substituent can be introduced into a benzene ring and the position

**Table 38.1 Common substituents in order of electron donating or withdrawing power**

| | | X | |
|---|---|---|---|
| | **Electron donating** | – OH | |
| | *decreasing* ↓ | – $NH_2$ | |
| X  | *power* | – $OCH_3$ | Decreasing |
| | **Electron withdrawing** | – $CH_3$ | rate of |
| | | – Cl | reaction ↓ |
| | | – CHO | |
| | *increasing power* ↓ | – $SO_3H$ | |
| | | – $NO_2$ | |

of the second substituent, relative to the first, is determined by the substituent already present in the ring.

### (i) Rate of substitution

Since the substitution reactions in benzene are electrophilic, if the substituent already present donates electrons to the ring, the rate of reaction will be increased.

Table 38.1 lists the common substituents in order of power of electron donating or withdrawing from a benzene ring.

Phenol undergoes electrophilic substitution reactions far more easily than benzene. When bromine water is added to a solution of phenol, a white precipitate of 2,4,6-tribromophenol is formed immediately on mixing in the cold.

$$\text{C}_6\text{H}_5\text{OH} + 3\text{Br}_2 \rightarrow \text{Br-C}_6\text{H}_2\text{OH-Br}_2 + 3\text{HBr}$$

### (ii) Position of substitution

The position of the second substituent is determined by the nature of the substituent already present. There are three possible isomers:

| 1,2- | 1,3- | 1,4- |
| (or *ortho*) | (or *meta*) | (or *para*) |

In practice, when carrying out electrophilic substitution with a monosubstituted benzene, either the *meta* (1,3-) product is obtained or some mixture of the *ortho* (1,2-) and *para* (1,4-) is formed. Table 38.2 lists substituents which are *meta*-directing and those that are *ortho-para* directing.

**Table 38.2 Directive effects of some common substituents**

| *Meta* directing | *Ortho-para* directing |
|---|---|
| $-NO_2$ | $-CH_3$ |
| $-COOH$ | $-OCH_3$ |
| $-CHO$ | $-Cl$ |
| $-CN$ | $-NH_2$ |
| $-SO_3H$ | $-OH$ |

NB The *meta*-directing substituents all contain at least one double or triple bond, *e.g.* nitration of nitrobenzene produces 1,3-dinitrobenzene as the nitro group is *meta* directing.

$$\text{C}_6\text{H}_5\text{NO}_2 + \text{HNO}_3 \rightarrow \text{C}_6\text{H}_4(\text{NO}_2)_2 + \text{H}_2\text{O}$$

Nitration of phenol produces a mixture of 2-nitrophenol and 4-nitrophenol because the $-OH$ group is *ortho-para* directing.

$$\text{C}_6\text{H}_5\text{OH} + \text{HNO}_3 \rightarrow \text{NO}_2\text{-C}_6\text{H}_4\text{OH} \text{ and } \text{NO}_3\text{-C}_6\text{H}_4\text{OH} + \text{H}_2\text{O}$$

### 38.8 FORMATION OF CARBONIUM IONS FROM ALCOHOLS

When ethanol is treated with concentrated sulphuric acid, a carbonium ion is produced

This carbonium ion can behave in two ways.

(*i*) At a temperature of 140°C in the presence of excess ethanol, ethoxyethane is produced.

(*ii*) At a higher temperature of 170°C, the bond between carbon and hydrogen is broken forming ethene

$$\text{H—}\underset{\underset{\text{H}}{|}}{\overset{\overset{\text{H}}{|}}{\text{C}}}\text{—}\underset{\underset{\text{H}}{|}}{\overset{\overset{\text{H}}{|}}{\text{C}}}{}^{+}\xrightarrow{\text{—H}^{+}}\underset{\text{H}}{\overset{\text{H}}{}}\text{C=C}\underset{\text{H}}{\overset{\text{H}}{}}$$

## QUESTIONS

**1** Multiple completion question – see instructions on page 3.
  In which of the following cases is the initial attacking species an electrophile?
   (*i*) Propene and hydrogen bromide.
   (*ii*) Propane and chlorine in the presence of ultraviolet light.
   (*iii*) Propanone and hydroxylamine.
   (*iv*) Benzene and a mixture of concentrated nitric and sulphuric acids.
   (*i*) This is an example of an electrophilic addition reaction.
   (*ii*) This is a free radical substitution reaction.
   (*iii*) $CH_3COCH_3$ is a ketone and these generally undergo nucleophilic addition reactions initially.
   (*iv*) Benzene is giving here an electrophilic substitution reaction.
   The answer is therefore E corresponding to (*i*) and (*iv*) correct.

**2** (*a*) State what you understand by the terms addition reaction, substitution reaction, nucleophile, electrophile and free radical in organic chemistry.
(*b*) By using the relevant terms from (*a*) and by drawing structures, describe the mechanisms involved in the reactions between:
  (*i*) ethanal and hydrogen cyanide,
  (*ii*) benzene and chlorine at room temperature in the presence of iron filings,
  (*iii*) ethene and bromine,
  (*iv*) iodoethane and aqueous sodium hydroxide,
  (*v*) methane and chlorine in ultraviolet light.

*(Associated Examining Board)*

(*a*) An addition reaction is a reaction where two substances join together to form a single product. During an addition reaction a double bond is converted to a single bond, a triple bond to a single bond or a triple bond to a double bond.

$$E.g.\quad \underset{\text{H}}{\overset{\text{H}}{}}\text{C=C}\underset{\text{H}}{\overset{\text{H}}{}} + H_2 \rightarrow \text{H—}\underset{\underset{\text{H}}{|}}{\overset{\overset{\text{H}}{|}}{\text{C}}}\text{—}\underset{\underset{\text{H}}{|}}{\overset{\overset{\text{H}}{|}}{\text{C}}}\text{—H}$$
$$\quad\quad\quad\text{ethene}\quad\quad\quad\quad\text{ethane}$$

A substitution reaction is a reaction where an atom or group of atoms replaces an atom or a group of atoms already in the molecule:

$$\mathbf{AB + C \rightarrow AC + B}$$
*E.g.*
$$\mathbf{CH_3I + OH^- \rightarrow CH_3OH + I^-}$$
$$\text{iodomethane}\quad\quad\text{methanol}$$

A nucleophilic reagent or nucleophile is a negatively charged or electron-rich atom or group of atoms. A nucleophile will attack a positive centre in a molecule. Nucleophiles include $OH^-$ or $NH_3$.

An electrophilic reagent or electrophile is a positively charged or electron deficient atom or group of atoms. An electrophile will attack a negative centre in the molecule.
Electrophiles include $NO_2^+$ and $SO_3$.

A free radical is an atom or group of atoms containing an odd unpaired electron. They are formed by the homolytic fission of a covalent bond.
*E.g.*
$$\mathbf{CH_3CH_3 \rightarrow 2CH_3\cdot}$$
$$\text{ethane}\quad\text{methyl free radicals}$$

**NB** It is useful to include an example after each definition. It clarifies your answer for the examiner.
(*b*) (*i*)

$$\underset{\text{H}}{\overset{\text{CH}_3}{}}\text{C=O} + HCN \rightarrow \text{H—}\underset{\underset{\text{CN}}{|}}{\overset{\overset{\text{CH}_3}{|}}{\text{C}}}\text{—OH}$$
$$\text{ethanal}\quad\quad\quad\quad\quad\text{2-hydroxypropanenitrile}$$

As only one product is formed, this is an addition reaction.

$$\underset{\text{H}}{\overset{\text{CH}_3}{}}\text{C}\overset{\curvearrowleft}{=}\text{O} + CN^- \rightarrow \text{CN—}\underset{\underset{\text{H}}{|}}{\overset{\overset{\text{CH}_3}{|}}{\text{C}}}\text{—O}^- \longrightarrow \text{CN—}\underset{\underset{\text{H}}{|}}{\overset{\overset{\text{CH}_3}{|}}{\text{C}}}\text{—OH}$$

The $CN^-$ ion is a nucleophile.
(*ii*) Iron filings act as a halogen carrier. The reaction is an electrophilic substitution reaction.

$$2Fe + 3Br_2 \rightarrow 2FeBr_3$$
$$\text{iron(III) bromide}$$

$$\bigcirc + Br_2 \rightarrow \overset{\delta+}{\underset{}{\bigcirc}}\text{Br}....\text{Br}....\overset{\delta-}{\text{FeBr}_3} \rightarrow \overset{\text{Br}\ \ \text{H}}{\underset{}{\bigcirc}} \rightarrow \overset{\text{Br}}{\bigcirc}$$

The electrophile is a complex formed by a bromine molecule and iron(III) bromide.

(*iii*) Electrophilic addition

$$\underset{H}{\overset{H}{\diagdown}}C=C\underset{H}{\overset{H}{\diagup}} + Br_2 \rightarrow Br-\underset{H}{\overset{H}{C}}-\underset{H}{\overset{H}{C}}-Br$$

Charges induced in a bromine molecule forms a carbonium ion.

$$Br-\underset{H}{\overset{H}{C}}-\underset{H}{\overset{H}{C}}{}^{+} \qquad \underset{H}{\overset{H}{\diagdown}}C\overset{Br}{\overset{+}{\underset{\diagup}{}}}C\underset{H}{\overset{H}{\diagup}}$$

This undergoes nucleophilic attack by a $Br^-$ ion.

(*iv*)

$$H-\underset{H}{\overset{H}{C}}-\underset{H}{\overset{H}{C}}-I + OH^- \rightarrow H-\underset{H}{\overset{H}{C}}-\underset{H}{\overset{H}{C}}-OH + I^-$$
$$\text{iodoethane} \qquad\qquad\qquad \text{ethanol}$$

This is a nucleophilic substitution reaction. The $OH^-$ is the nucleophile.

(*v*) Methane and chlorine in the presence of ultraviolet light. This reaction is a free radical substitution reaction.

Formation of chlorine free radicals

$$Cl_2 \rightarrow 2Cl\cdot$$

Then $Cl\cdot + H-\underset{H}{\overset{H}{C}}-H \rightarrow H-\underset{H}{\overset{H}{C}}-Cl + H.$    Substitution reaction.

**3** Methylbenzene, $\langle O \rangle$–CH$_3$, can be nitrated using a mixture of nitric acid and sulphuric acid. In an attempt to find out what nitration products result, an experiment was conducted as follows: "10 drops of concentrated sulphuric acid were carefully added to 10 drops of concentrated nitric acid in a test-tube, while shaking the mixture and cooling the test-tube under a stream of cold water. The mixture of acids was then added to 5 drops of methylbenzene in another test-tube, again shaking this mixture under a stream of cold water. The mixture was poured into a beaker containing 10 cm$^3$ of cold water. The contents of the beaker were transferred to a separating funnel and the organic layer separated off,"

(*a*) When methylbenzene is nitrated:
 (*i*) what *type of reaction* (addition, elimination, etc.) is said to take place?
 (*ii*) what *class of reagent* (electrophile, nucleophile, etc.) attacks methylbenzene?
 (*iii*) what nitrogen-containing ion is believed to be involved in the actual nitration stage of the reaction?
 (*iv*) why is the reaction mixture cooled?

(*b*) Suggest briefly how the organic layer resulting from the nitration might be:
 (*i*) treated to remove residual acids
 (*ii*) dried.

(*c*) The methyl group in methylbenzene is said to direct incoming nitro groups to the 2- and 4-positions of the benzene ring. Using this information, draw structural formulae to show THREE of the possible products of nitration of methylbenzene.

(*Nuffield*)

(*a*)  (*i*) Substitution
   (*ii*) electrophile
   (*iii*) $NO_2{}^+$
   (*iv*) Reaction is very exothermic. Therefore cooled to prevent reactants/products from vapourizing, or to prevent acid spitting out.
(*b*)  (*i*) Shake with aqueous sodium carbonate solution
   (*ii*) Add anhydrous sodium sulphate.
(*c*) Any three of

# 39 The strengths of organic acids & bases

## 39.1 INTRODUCTION

In this Unit the relative strengths of organic acids and bases will be compared. The factors which affect the strength of acids and bases will be considered.

Before attempting this Unit, Units 15 and 16 should be studied. In Unit 16 the introduction of $pK_a$ and $pK_b$ will be encountered. You should remember that $pK_a$ is a measure of the strength (*i.e.* degree of ionization) of an acid. The smaller is the numerical value of $pK_a$ the more ionized (and stronger) the acid is.

Similarly, $pK_b$ is a measure of the strength of a base. The smaller the numerical value of $pK_b$, the stronger the base is.

In this Unit $pK_a$ and $pK_b$ values will be given to compare the strengths of acids or bases.

## 39.2 ORIGIN OF ACIDITY IN ORGANIC COMPOUNDS

When a compound H–X acts as an acid, the H–X bond breaks heterolytically to form $H^+$ ions. The extent of the acidity depends upon the number of H–X bonds broken in a sample.

The acidity of an organic compound H–X depends upon three factors:
(*i*) the strength of the H–X bond,
(*ii*) the electronegativity of X,
(*iii*) any stabilization of the $X^-$ ion compared with HX – this will be shown by the existence of a number of possible resonance structures.

### Comparison of methanol and phenol

Phenol is very slightly acidic. The difference is due to the stabilization of the $C_6H_5O^-$ ion produced by the loss of the $H^+$ ion.

The $CH_3O^-$ ion is not stabilized.

The $pK_a$ values of methanol and phenol are 16 and 9.95 respectively.

### Comparison of methanol and ethanoic acid

The carbonyl group in ethanoic acid has an electron withdrawing effect from the O—H bond.

This weakens the bond. Also the ethanoate ion is stabilized by the existence of two resonance structures.

The $pK_a$ values of methanol and ethanoic acid are 16 and 4.76 respectively.

## 39.3 RELATIVE STRENGTHS OF CARBOXYLIC ACIDS

The $pK_a$ values of some carboxylic acids are given in Table 39.1

**Table 39.1 The p$K_a$ values of some carboxylic acids**

| Acid | | p$K_a$ | Acid | | p$K_a$ |
|---|---|---|---|---|---|
| 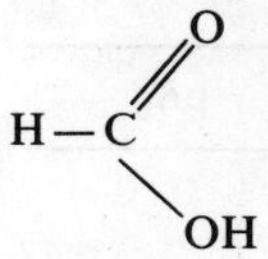 H—C(=O)OH | methanoic acid | 3.77 | 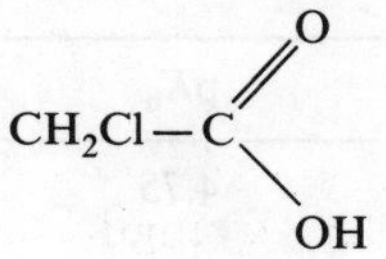 CH$_2$Cl—C(=O)OH | chloroethanoic acid | 2.86 |
| 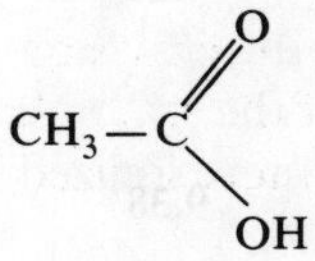 CH$_3$—C(=O)OH | ethanoic acid | 4.76 | 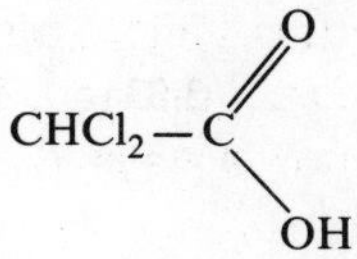 CHCl$_2$—C(=O)OH | dichloroethanoic acid | 1.29 |
| 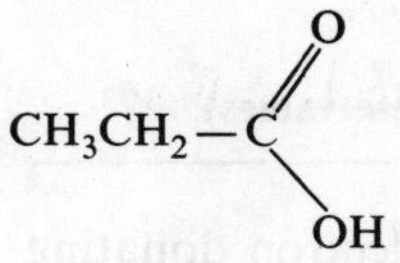 CH$_3$CH$_2$—C(=O)OH | propanoic acid | 4.88 | 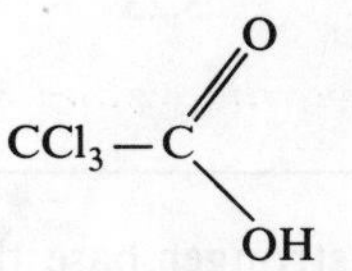 CCl$_3$—C(=O)OH | trichloroethanoic acid | 0.65 |

**(Do not attempt to remember these values)**

Replacing the non-hydroxylic hydrogen atom of methanoic acid by an alkyl group decreases the strength of the acid. There is a shift of electrons from the alkyl group towards the carboxyl group which reduces the tendency for the O–H bond to break heterolytically.

*E.g.*     CH$_3 \rightarrow$ C(=O)(O—H)

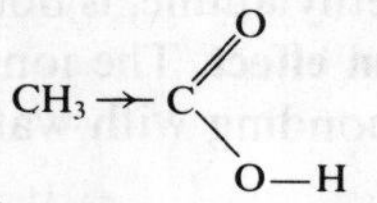

When a hydrogen atom in ethanoic acid is replaced by a chlorine atom to produce chloro-ethanoic acid, the acid strength increases. This is due to the electron withdrawing from the carboxyl group caused by the greater electronegativity of chlorine compared to hydrogen. This assists in the ionization of the OH group.

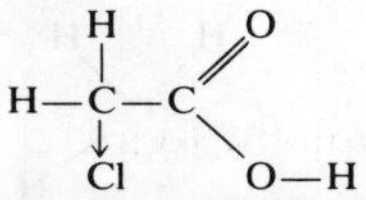

H—C(H)(Cl↓)—C(=O)(O—H)

Further substitution of chlorine atoms to produce dichloroethanoic and trichloroethanoic acids produce stronger acids.

Benzenecarboxylic acid is a weaker acid than methanoic acid. This is explained by the over-all electron donating effect of the phenyl group compared to hydrogen.

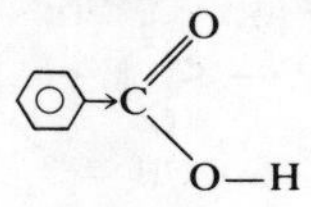

C$_6$H$_5$—C(=O)(O—H)

## 39.4 COMPARISON OF THE p$K_b$ VALUES OF AMINES AND AMIDES

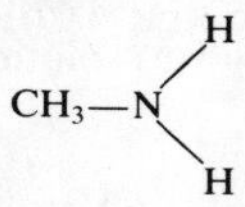

CH$_3$—N(H)(H)
methylamine
p$K_b$ 3.36

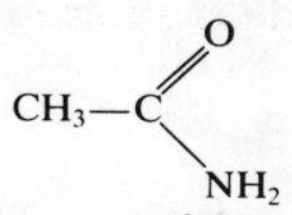

CH$_3$—C(=O)NH$_2$
ethanamide
p$K_b$ 14.5

The strength of a nitrogenous base is related to the readiness of the compound to accept protons or the availability of the unshared pair of electrons on the nitrogen atom.

The carbonyl group adjacent to the nitrogen atom in ethanamide has a tendency to withdraw electrons and reduce the availability of the pair of electrons.

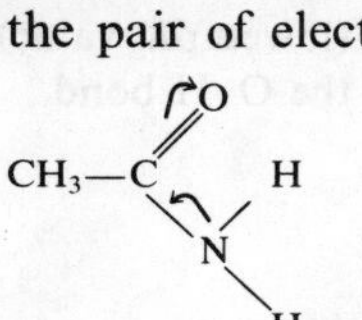

CH$_3$—C(=O)...N(H)(H)

As a result, ethanamide is only very weakly basic in water. Methylamine is much more strongly basic than ethanamide.

**39.5**    RELATIVE STRENGTHS OF ORGANIC BASES

The $pK_b$ values of some common bases are given in Table 39.2

**Table 39.2 The $pK_b$ values of some common bases**

| Base | | $pK_b$ | Base | | $pK_b$ |
|------|---|--------|------|---|--------|
| $NH_3$ | ammonia | 4.75 | | trimethylamine | 4.20 |
| $CH_3NH_2$ | methylamine | 3.36 | | | |
| $CH_3CH_2NH_2$ | ethylamine | 3.33 | | | |
| | dimethylamine | 3.23 | 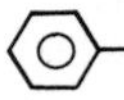 | phenylamine | 9.38 |

**(Do not attempt to remember these values)**

Methylamine is a stronger base than ammonia. This is explained by the electron donating effect of the methyl group increasing the availability of the spare pair of electrons on the nitrogen atom.

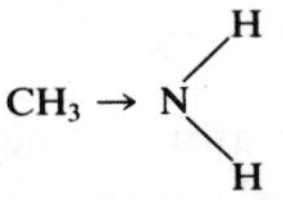

$$CH_3 \rightarrow N \overset{\displaystyle H}{\underset{\displaystyle H}{}}$$

The strength of the base is further increased by the substitution of a second methyl group. However, the trisubstituted amine, trimethylamine, is not a stronger base than methylamine and dimethylamine. This is due to a **solvation effect**. The ions produced by methylamine and dimethylamine are stabilized by hydrogen bonding with water molecules.

*E.g.*   $(CH_3)_2\overset{+}{N}$

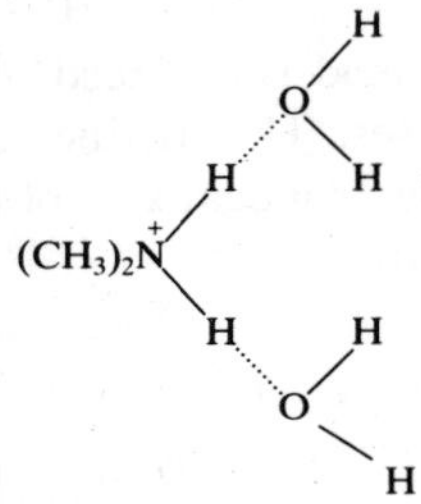

This solvation is not possible with trimethylamine.

Phenylamine is a much weaker base than ammonia because of the electron withdrawing effect of the phenyl group. The lone pair can be stabilized by delocalization in the ring.

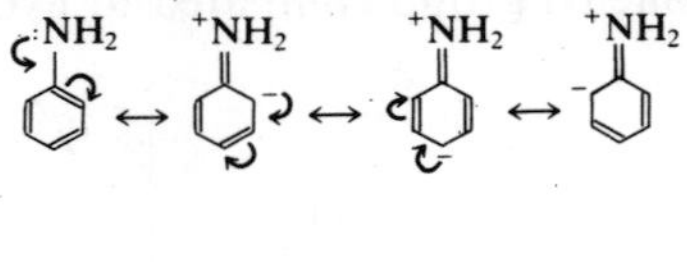

QUESTIONS

1 Which one of the following acids is most highly ionised in aqueous solution?

    A  HCOOH
    B  $CH_3COOH$
    C  $CH_2ClCOOH$
    D  $CHCl_2COOH$
    E  $CCl_3COOH$

**The correct answer is E.**

2 The $pK_a$ values for fluoroethanoic acid, chloroethanoic acid and bromoethanoic acid are 2.66, 2.90 and 3.16 respectively. Explain these results.

The fluoroethanoic acid is the strongest acid (lowest $pK_a$ value). Fluorine is the most electronegative element and withdraws electrons more, weakening the O–H bond.